Innovative Strategies to Mitigate the Impact of Mining

Innovative Strategies to Mitigate the Impact of Mining

Topic Editors

Chongchong Qi
Qiusong Chen
Danial Jahed Armaghani

MDPI

Basel • Beijing • Wuhan • Barcelona • Belgrade • Novi Sad • Cluj • Manchester

Topic Editors
Chongchong Qi
School of Environmental Science and Engineering
Tianjin University
Tianjin
China

Qiusong Chen
School of Resources and Safety Engineering
Beijing University of Science and Technology
Beijing
China

Danial Jahed Armaghani
School of Civil and Environmental Engineering
University of Technology Sydney
Ultimo
Australia

Editorial Office
MDPI AG
Grosspeteranlage 5
4052 Basel, Switzerland

This is a reprint of the Topic, published open access by the journals *Applied Sciences* (ISSN 2076-3417), *Minerals* (ISSN 2075-163X), *Mining* (ISSN 2673-6489) and *Toxics* (ISSN 2305-6304), freely accessible at: https://www.mdpi.com/topics/6EQC0N737S.

For citation purposes, cite each article independently as indicated on the article page online and as indicated below:

Lastname, A.A.; Lastname, B.B. Article Title. *Journal Name* **Year**, *Volume Number*, Page Range.

ISBN 978-3-7258-8049-2 (Hbk)
ISBN 978-3-7258-8050-8 (PDF)
https://doi.org/10.3390/books978-3-7258-8050-8

Contents

mining

Article
A Mine Closure Risk Rating System for South Africa

Megan J. Cole

Future Water Institute, University of Cape Town, Cape Town 7700, South Africa; megan.cole@uct.ac.za

Abstract: Mine closure is a growing concern in mining countries around the world due to the associated environmental and social impacts. This is particularly true in developing countries like South Africa where poverty, social deprivation and unemployment are widespread and environmental governance is not strong. South Africa has 230 operating mines located in diverse natural and social settings. Over 6 million people live in urban and rural mining host communities who will be significantly affected by mine closure. The national, provincial and local governments need guidance in identifying high-risk areas and relevant policy and programmatic interventions. This paper describes the development of a quantitative mine closure risk rating system that assesses the likelihood of mine closure, the risk of social impact and the risk of environmental impact of mine closure for every operating mine in the country. The paper visualises the high likelihood of closure and environmental impacts for numerous coal and gold mines, and the significant social risks in the deprived rural platinum and chrome mining areas. The rating system was tested with 10 mines and 19 experts, and the resulting maps are communicated in an online South African Mine Closure Risk and Opportunity Atlas. The risk ratings could be used in mine closure planning and management by mining companies, consultancies, governments and affected communities. While this risk rating system has been designed for South Africa, the methodology and framework could be applied to any mining country in the world.

Keywords: mine closure; mining; social risk; environmental risk; likelihood of closure; indicators; Africa; just transition; evidence-based decision-making

1. Introduction

1.1. Mining in South Africa

South Africa has a long history of mining which has played a significant role in the location and development of settlements, infrastructure and the economy [1]. The mining sector contributes 7.53% of GDP, employs 475,561 people (4.8% of formal sector) and paid company tax of R73.6bn (USD4bn), royalties of R14.2bn (USD0.76bn) and employee earnings of R174.9bn (USD9.4bn) in 2022 [2]. South Africa is a globally leading producer of platinum and palladium (74% of 38% of global production respectively) [3], chromium (36%), manganese (36%) [4], zirconium (23%) [5], vanadium (9%) [6], diamonds (8%) [7], fluorspar (5%) [8], gold (3.5%) [9] and coal (3.2%) [10] and also produces iron ore, nickel, copper, lead, zinc and phosphate rock [11,12]. These commodities are mined at 230 underground and open pit mines operated by 104 private mining companies, nearly a third of which are coal mines while a quarter are platinum or chrome mines. These mines are supported or hosted by over 360 diverse urban and rural communities, across South Africa, which are home to over 6 million people. These communities are situated in three metropolitan municipalities and 635 local municipalities that have a total population of 27.1 million [13].

Since large-scale commercial mining began in South Africa in 1852, mine closure has occurred on a regional scale in the Okiep copperbelt (1852–1970s) and West Coast diamonds (1926–2010) in the western Northern Cape province; three asbestos fields in Mpumalanga, Limpopo and Northern Cape (1930–1980s); the Natal coalfields (1880–1980s); the Wits goldfields (1886–1990s) in Gauteng, North West and Free State provinces; and

Rooiberg tin (1907–1994) in Limpopo province (see Figure 1) [1]. The country's mineral wealth has not benefited communities equally and has had negative public health, safety and environmental impacts [14–16]. The derelict and ownerless asbestos and gold mines have been the focus of government rehabilitation efforts, but more needs to be done and the widely held view is that industry and government are not effectively managing mine closure [17–19] or providing sufficient socio-economic benefits to mining regions [20–22].

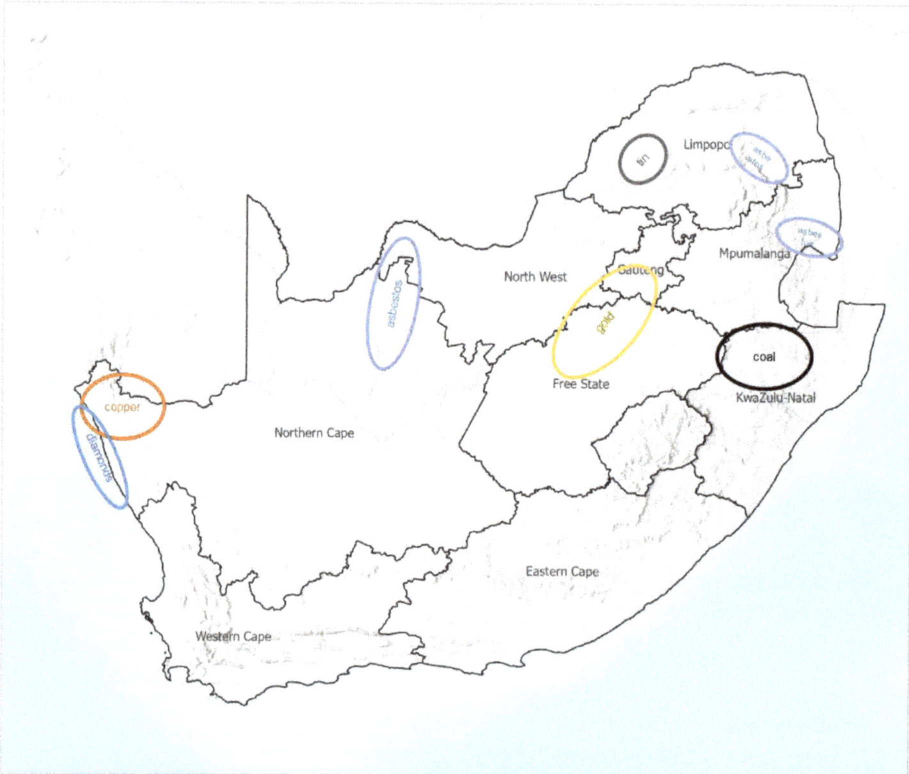

Figure 1. Map of South Africa's provinces and areas of historical mine closure.

Numerous coal mines are expected to close in the near future as the coal fields in Mpumalanga province are depleted and climate change limits coal investment; there is significant domestic and international pressure for a 'just transition' to clean energy that ensures that the mine workers and host communities are not left behind [23]. Similarly, the shift to electric vehicles may see a decline in platinum demand and result in mine closures in the high-cost underground mines in North West and Limpopo provinces. In other regions within South Africa, the increased global demand for so called 'energy transition minerals' is seeing increased exploration and development of manganese, copper, iron ore, base metals, vanadium and rare earth minerals, which are necessary for green energy and e-mobility technologies [24].

1.2. Mine Closure

Traditionally, mine closure has been seen as the end of a process—the final stage in the mining life cycle. There is a shift underway to seeing mine closure as a transition that acknowledges the multi-faceted role that mines play in local and regional economies, as well as the complex impacts, contributions and interrelationships that occur with society and governments [25]. The importance of mining companies building and sustaining relationships with local government and communities affected by mine closures is growing [26], and early stakeholder participation and continuous mine closure planning increases the success of social investment programmes [27].

Mining operations are affected by numerous environmental, social and governance (ESG) risks at multiple levels [28], and there are several factors that can influence when a mine closes. The major causes of mine closure are financial and economic, namely, price drops, high costs, resource exhaustion or depletion, declining grades and loss of markets; while other causes are adverse geological and/or geotechnical factors, voluntary administration, flooding, political decisions, and health and safety concerns [29]. In South Africa, gold and platinum mines have been negatively affected by price drops and escalating costs of deep underground mines, while gold, copper and coal mines have closed due to resource exhaustion [30,31], tin mines closed when the price collapsed [32,33] and asbestos mines were closed when the serious health implications were understood [34].

Globally, there are major concerns around potential impacts of future mine closures [35], and current best practice is that mine closure planning starts at the onset of mining activity and continues throughout mining operations [36]. Closure planning must consider environmental and socio-economic impacts and include economic diversification and succession planning, which have historically received little attention [15].

1.3. Mine Closure Risk

Managing risk is central to enabling a transitions-based approach to mine closure [25]. A key goal of mine closure for regulatory authorities and industry is the removing or reducing liability for residual risks and generating confidence for future environmental, social and economic management (ibid). Risk assessment and its application to mine closure have therefore increasingly become more important in mine planning [36,37]. Mine closure guidelines generally agree that effective data collection ensures that risks are managed and monitored [38–40], although the emphasis continues to be on environmental risk performance indicators, with socio-economic risks often understated [40–42].

The environmental and socio-economic risks associated with mine closure are significant and are affected by several inter-related factors. Mine closure can have significant environmental impacts on biodiversity, land and water resources involving water pollution, air pollution and land degradation. The most well-known risks in South Africa are acid mine drainage from gold and coal mines and tailings dam failures (most recently, Jagersfontein diamond tailings dam). Global assessments show that mining often operates near biodiversity-protected areas [43], water bodies [44,45] and sensitive land [46]. Mine closure can have significant social and economic impacts, including direct and indirect job losses; economic downturn in the communities and even the wider region; basic services disruption if the mining company was supporting or supplying services; and illegal mining which creates an unsafe environment for local residents [36,47,48]. Illegal mining in South Africa is focused on closed gold and diamond mines near populated areas.

The exposure to these hazards depends on how many people are affected, i.e., the population of mining host communities, and it is therefore very important how they are defined. The vulnerability or resilience of a community affects how well it can cope with the economic shock of mine closure, and local infrastructure and business activity determine post-closure economic prospects. Analysis of job prospects for coal mine workers in South Africa leaving mining show the majority are not employable or have low employability prospects [49]. Similarly, the financial and human resource capacity of local government authority affects its ability to cope with mine closure, and the resulting loss of revenue and analysis of audit reports in South Africa show most municipalities will struggle to cope [23]. Given these complexities, guidance is needed for national, provincial and local governments in identifying high-risk areas to mitigate these risks on a case-by-case basis.

This paper describes the development and proposition of a new Mine Closure Risk Rating System for South Africa, which has three components—likelihood of mine closure, social risk and environmental risk of closure. Section 2 provides the methodology for the development of the rating system which includes diverse influencing factors and their measurable indicators, weightings and categorisation, case studies and expert input. Section 3 provides the quantified results for each risk rating for all operating mines in the country, which are mapped in ArcGIS Pro. Section 4 discusses the measurement of mine closure risk, data gaps and future research, communication, stakeholder engagement and evidence-based decision-making, while Section 5 concludes the paper.

2. Methodology

The methodology used to calculate mine closure risk ratings for all operating mines in South Africa is captured in Figure 2, which gives an overview of the risk rating system developed in this study. Each element is described below.

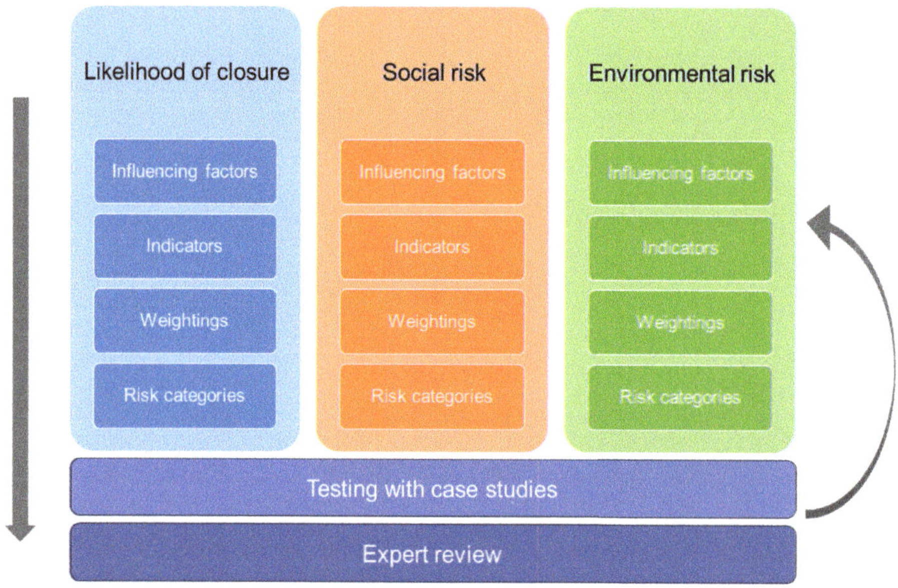

Figure 2. Mine closure risk rating framework.

2.1. Influencing Factors, Indicator Selection and Data Collection

Risk is the possibility or likelihood of loss or harm and incorporates exposure to a hazard. In this study three components of risk are assessed for all operating mines in South Africa: the likelihood of mine closure, the likelihood and impact of social loss/harm from mine closure to mining host communities, and the likelihood and impact of environmental loss/harm from mine closure. Each component has a set of influencing factors which determine overall risk and were identified through literature review, case studies, a stakeholder workshop and expert interviews. Each influencing factor was measured using one or two indicators, where the criteria for indicator selection were: 'Is the indicator the best available direct measure of the influencing factor?' and 'Are there sufficient reliable data that are measured for the whole country?'

The primary datasets for the rating systems are the 221 operating mines, the 360 mining communities, and 61 municipalities which have been built up over the past five years by the author [23,28,50,51] and expanded based on more recent mining company reports and community profiles in the 2011 Census. Secondary data were collected from numerous open-source national datasets identified below.

2.1.1. Likelihood of Closure

Seven factors influencing the likelihood of mine closure in South Africa were identified and are described below. Two factors, 'social licence to operate' and 'political dynamics', could not be quantified. Data were collected largely from company websites, company Social and Labour Plans, SLPs, (a South African regulatory requirement), and company annual reports on mineral resources and reserves (MRR).

Life of Mine: Each mine has a stated design life, the Life of Mine (LOM), which is the expected years of mine production based on the latest assessment of a mine's mineral reserves. It is essential information for mine planning and investor reporting and should be shared with mining host communities. It includes a detailed assessment of economic, financial, technical and environmental factors and is normally reviewed annually, so is subject to change during the design life of the mine. Mine extensions (often from open pit to underground) can increase the life of mine considerably. It is the most important factor in determining likelihood of closure. Not all mines publish their life of mine figures, and where there were gaps, the highest risk rating was given, assuming the mine is operated by a junior miner on a small scale with a short life of mine.

Mineral Resources and Reserves: South Africa's SAMREC Code provides minimum standards, guidelines and recommendations for the public reporting of exploration results, mineral resources and reserves [52]. The framework differentiates between an 'exploration result' with little information and low confidence in the basic geology; a 'Mineral Resource' estimate where a geological model is developed, with reasonable prospects for economic extraction; and a 'Mineral Reserve', which is an economically feasible project based on technical, economic, marketing, legal, environmental, social and governmental factors (Modifying Factors). Different commodities use different units in their MRR reporting due to the nature of the orebodies (e.g., million tonnes for coal but ounces for gold); therefore, five categories ranging from 'very large' to 'very limited' were developed in order to compare and rate the mines. Where no MRR was reported, the highest risk rating was given as it is likely that it is a smaller mine with limited reserves.

Commodity markets: Each commodity has different domestic and international markets that influence the commodity price and future demand, and thus the economic value of each mine's MRR. Changes in the commodity markets have a significant impact on when a mine closes, and whether that is permanent or temporary closure. For example, when the tin price crashed in the 1970s, tin mines closed down rapidly across the world [32], and the huge variations in the gold price have seen mines close and reopen as viability changes [47]. With the threat of climate change and the need to reduce greenhouse gas emissions, coal mines are likely to close sooner than expected, whereas copper mines are reopening as demand for energy transition metals rapidly increases [53], and new manganese mines have opened up rapidly in South Africa [54]. As there is so much uncertainty in the future commodity markets, each commodity was given a category: 'high growth', 'growth', 'stable', 'decline' or 'rapid decline', based on reliable market forecasts [55–57].

Operating costs and mining methods: The operating costs of mining operations vary significantly due to grade, mining method, depth of operation, recovery, labour efficiency, the use of subcontractors and input prices [58]. Conventional deep underground operations (like South African gold and platinum mines) are high-cost operations as they require refrigeration for ventilation and rock support, while mechanised operations are lower cost operations due to economies of scale. Mineral processing or treatment costs vary depending on the type of ore and the complexity of the process [59]. Industry cost curves can be plotted at a company or site level to facilitate cost comparisons [60]. Due to insufficient cost curves to generate a rating, mining method was used as a proxy for operating cost. The five main kinds of mining methods used in South Africa were rated, namely, open pit mining (usually low cost, low grade deposits), also called open cast mining; strip mining (most common method in coal mining); dredging (aquatic environments like ocean or river floors, often used for diamonds and heavy mineral sands); conventional underground mining used for deeper deposits (usually higher grade, higher cost deposits for coal, gold and platinum); and large-scale mechanised underground mining, which includes long-wall mining, sub-level caving and block caving (used for coal, diamonds and copper).

Company type: The type and size of a company partly determines its ability to accommodate changes in price, demand, operating costs and other factors that affect a mine's financial viability. The larger companies are much more likely to weather a downturn than the smaller mining companies. Internationally, mining companies are categorised as majors, mid-tier producers and juniors, with majors usually have a market capitalisation of over USD 1 billion. The term "juniors" generally refers to exploration or prospecting companies only involved in the early stages of mining development, but in South Africa the term has a wider meaning and includes exploration companies and mid-tier producers. The term "emerging miners" is used in South Africa to refer to smaller companies involved in the early phases of mining exploration or in the early developmental stage i.e., smaller producing companies and contractors [61]. Three categories were used for the rating—majors, juniors and emerging miners—and are based on the Minerals Council South Africa report on juniors and emerging miners, which categorises all operating mines listed by the Department of Mineral Resources and Energy (ibid).

Social licence to operate: The relationship between the mining company and the local community can be a risk to mine operations. Protests about environmental pollution, damage, noise, jobs and local procurement are common in South Africa and are known to bring operations to a halt and result in significant loss of revenue, jeopardising the viability of the mine. Ideally the number and type of protests per mine would be measured across the country; however, these data are not currently available in a comprehensive form and could not be included.

Political dynamics: National and local politics can promote or hinder investment in mining operations, with corruption, political interference, political instability and regulatory uncertainty all deterring investment in mine expansions, thus leading to premature mine closure. The regulatory uncertainty and national politics apply equally to all mines in the country, so are not included as an indicator. It is difficult to measure political dynamics at the local level; thus, an indicator has not been included at this stage but should be investigated further.

2.1.2. Social Risk of Mine Closure

Mine closure can have significant social impacts on host communities and municipalities. The vulnerability or resilience of a community affects how well it can cope with the economic shock of mine closure—whether it is sudden or a long time in coming. Similarly, the capacity of the local government authority affects its ability to cope with mine closure and the resulting loss of revenue. Eight factors influencing the social risk of mine closure in South Africa were identified, and six were quantified with indicators. Data were collected from Statistics South Africa (StatsSA) household surveys, company websites and SLPs, local municipality publications and Auditor-General reports.

Mining host community population: The mining host community for each mine was identified based on its location and proximity to the mine, whether it is home to mine employees or is a beneficiary of the mine's SLPs, and an assessment of the community and area in Google Earth. Setting a specific radius was not used as it can exclude communities that are close neighbours to other communities included in the definition. For data collection, 'main places' and 'sub places' defined by StatsSA for the 2011 census were used. Main places are small towns, townships, villages and suburbs in large towns and cities. They are subdivided into 'sub places', which can be specific areas like mines and compounds in urban and non-urban areas.

Social vulnerability: Fourteen indicators were used (see Table A1) to determine a score for social well-being in each mining host community, expanding on previous work by the author [50,51]. The indicator selection was based on the Sustainable Development Goals (SDGs), the South African Index of Multiple Deprivation (SAIMD) [62] and on data availability at the local level, i.e., in the national census. They cover income, household goods, health, education, gender equality, water access, sanitation, electricity access, clean cooking fuel, employment, housing, refuse removal and internet access. Census data at main place and sub place level in StatsSA's SuperCross Census 2011 Community Profile Database were obtained from DataFirst, which provides access to household survey data in Africa (DataFirst, 2015). The individual values were equally weighted and combined into an overall well-being score for each community and assigned to each mine.

Direct mining jobs: The more mine workers there are in a community, the bigger the social impact will be when a mine closes. Although most of the major mining companies report the current number of employees in their annual reports and on their websites, the junior and emerging miners do not. Instead, the Census 2011 data on employment in mining and quarrying per main place were used. This is not ideal as it usually gives higher numbers than the individual mine accounts for, and thus it has been given a low weighting. This dataset should be updated with the Census 2022 community database. An alternative indicator is the percentage of jobs in mining and quarrying to reflect the varying impact on the area. Future research should seek to obtain job numbers for individual mines.

Indirect mining jobs: Operating mines form part of a much bigger value chain, relying on upstream and downstream industries. They have a multiplying effect on employment which can be estimate at the national and regional level. Data do not exist yet at the mine site level and therefore are not included in the rating.

Dependency ratio: The dependency ratio relates the number of children (aged 0–14) and the number of older persons (aged 65 and over) to the working age population (persons aged 15–64). The ratio highlights the potential dependency burden on workers—a high dependency ratio indicates that the economically active population and the overall economy face a greater burden to support and provide the social services needed by children and by older persons who are often economically dependent [63]. This ratio was calculated for each community using the Census 2011 dataset and should be updated with the 2022 census data at the main place level.

Skills and education levels: The skills and education levels of mine workers and community members will affect their ability to find employment or create a business after the mine has closed down. SLPs provide detailed information about skills levels of all mine employees; however, the majority of SLPs are not publicly available. There is no available national skills dataset that could be used for the mining communities; thus, research is needed before this can be include in the rating.

Local economy: Mines can have a significant impact on a local economy through supply chain procurement and mine employee expenditure. Gross Value Added (GVA) measures the contribution of a corporate subsidiary, company, or municipality to an economy, producer, sector, or region, and is often reported on an annual basis in local municipality Integrated Development Plans (IDPs). As not all IDPs report this figure, an alternative dataset was used. stepSA provides local economic data in a national spatial dataset using CSIR's mesozones (a complete grid of 25,000 spatial units ~50 km^2) and economic data by sector [64]. This gives an indicator of economic production per sector expressed in Rands per mesozone. The mine dataset was overlaid on the mesozones in ArcGIS Pro.

Local municipality audit findings: A local government's ability to cope with mine closure and to mitigate the risks can be assessed through its current financial performance. Each year, the Auditor-General of South Africa (AGSA) assesses the financial management and performance management of all the country's municipalities and rates them according to set definitions [65]. These ratings were assigned to each mine based on the local municipality that they are located in.

Crime and Safety: Illegal mining and the associated crime can be a consequence of mine closure. Some commodities like diamonds and gold are more prone to illegal mining as the valuable minerals and metals are easy to extract. The approach taken by the national and local police force is critical in whether illegal mining takes place but is hard to measure. Illegal mining often opens up mine shafts and pits to the public, which poses a safety hazard. Similarly, poorly rehabilitated mines can also pose safety and health risk to mining communities, but this is hard to predict. There is no available national dataset, which is an area for future research.

2.1.3. Environmental Risk of Mine Closure

Mine closure can have significant environmental impacts on biodiversity, land and water resources involving water pollution, air pollution and land degradation. Determining a risk rating for each mine is complicated due to the mismatch of scales and the uncertainty over extent of impact. Nine influencing factors were identified, and seven measurable indicators were defined.

Duration of mining: The longer a mine has been operating, the greater the likelihood of environmental impact. The start date of each mine, and any periods of temporary closure, were determined from in-depth research into mines and host communities based on the author's previous research [1,51] to calculate the total duration of mining for each mine in years.

Threat status of terrestrial ecosystems: The National Biodiversity Assessment of 2018 delineated areas of different ecosystem threat status across the country for terrestrial, aquatic and marine ecosystems [66]. The author's mine location dataset was overlaid on the threat status for terrestrial ecosystems shapefile provided by the South African National Biodiversity Institute (SANBI) BGIS portal in ArcGIS Pro (desktop GIS software developed by Esri) to provide a status for each mine.

Distance to protected areas: In South Africa, protected areas are defined and maintained by the South African Department of Forestry, Fisheries, and the Environment (DFFE) and SANBI. The author's mine location dataset was overlaid on the latest 2023 protected areas shapefiles [67], and the distance from each mine to the nearest protected area was measured in ArcGIS Pro.

Distance to Strategic Water Source Areas: Mining often negatively impacts on local and regional water resources. In South Africa, surface and groundwater Strategic Water Source Areas (SWSA) have been identified [68]. The authors' mine location dataset was overlaid on the SWSA shapefiles provided by SANBI's BGIS, and the distance from each mine to the nearest SWSA was measured in ArcGIS Pro.

Mine Water Threat: The South African Mine Water Atlas provides risk ratings for each quaternary river catchment that lies within the boundary of assessed mineral provinces [69]. It is presented as key thematic maps including "mineral risk" indicating the assessed risk of acid production and/or leaching of constituents, "groundwater vulnerability" and "surface water threat" reflecting the vulnerability of those water sources to the type of mining activities (depth of mining is considered to have major varying impacts on aquifer systems) and "mine water threat" which combines all the factors into a risk rating [69]. These threats are rated from 'low or insignificant' to 'moderate low' to 'moderate' to 'high' and to 'very high' (ibid). The groundwater vulnerability ratings are separated into surface mining (<100 m below ground level) and underground mining (>100 m below ground level). The mine water threat risk ratings for groundwater (surface mining), groundwater (underground mining) and surface water were determined for each mine in ArcGIS Pro by overlaying the mines dataset on the relevant shapefiles and then combined to give a cumulative mine water threat rating.

Agricultural production and land capability: Agriculture can be significantly hindered by mining and mine closure through water and air pollution and land degradation. The current agricultural production gives an indication of the potential negative environmental impact of mine closure. Insufficient data could be sourced to measure agricultural production at the local level, so the national land capability classification was used instead. It uses soil, climate and terrain to differentiate between arable land suitable for cultivation (classes I, II and III), marginal land suitable for light cultivation (classes IV), grazing land (classes V, VI and VII) and wilderness (class VIII) [70]. Mine locations were overlaid on the national land capability map in ArcGIS Pro [71] to determine the land capability class for each mine.

Waste stability: Many large-scale mines produce significant amounts of waste rock and billions of tonnes of mine tailings (a slurry of rock, water and chemicals) stored in dams or tailings storage facilities (TSFs), which have the potential to fail if not designed or managed properly [72]. Factors affecting the stability of TSFs include exposure to earthquakes, tropical cyclones, high winds, heavy rainfall and steep terrain, and these five factors can be measured with global datasets and combined to form a TSF failure risk rating [73]. In addition, post-closure design life of the TSF, key design elements of the TSF such as spillways and management of the TSF are important factors to consider. A waste stability rating was not included in this version of the rating system due to an incomplete TSF and waste rock dump dataset, and this is an area of future research.

Waste volume: The total volume of rock waste dumps and TSFs at each mine affects the potential environmental impact in terms of water pollution, dust pollution and risk of failure [72]. This is a difficult parameter to measure although the Global Tailings Portal now provides information for the major mining companies [74]. Separate research projects are underway to develop this further for South Africa (e.g., [54]), and an indicator was not included in this rating.

Capacity and approach of mining company: The capacity and management approach of the mining company responsible for closure will determine the effectiveness of rehabilitation and water and waste management during the closure process. This is difficult to measure, and an indicator was not included at this stage but should be investigated. The presence of a risk assessment in the mine closure plan could be assessed. The financial provisions for each mine, a regulatory requirement in South Africa, should give an indication of the ability of the mine to undertake effective rehabilitation, but the data are hard to find and require further research. The capacity of government to monitor and enforce environmental regulations was not included as a factor as it is a national government function and therefore the same for all mines.

2.2. Calculations, Weightings and Visualisation

The risk rating for each component was calculated for all operating mines based on the indicators and data described in the previous section. A rating scale from 1 (very low risk) to 5 (very high risk) was developed by the author for each indicator, and a weighting was applied based on the literature, statistical analysis of all mines in Microsoft Excel, and expert input. The five rating categories were assigned based on analysis of all the mines and sub-division into five quintiles of roughly equal numbers. The weighted indicators for each risk rating were combined to create an overall score, shown in the equation below, which was categorised in the same way. The results were plotted on maps in ArcGIS Pro and colour-coded with warm to cool colours to enable communication and analysis.

$$Risk\ rating = F1\ xw1 + F2xw2 + F3xw3 + F4xw4 + F5xw5 \quad (1)$$

where

F is the influencing factor indicator;
w is the weighting for the influencing factor indicator.

2.3. Case Studies

The risk ratings were tested against four case studies of post-closure land use that included 10 operating mines with very diverse geographical contexts, mining methods, life of mine, and commodities [75]. The Impact Catalyst in Limpopo province project explores integrated game farming, agriculture and agro-processing initiatives to align with the platinum mine rehabilitation plan. The Community of Practice project explored the use of industrial crops for gold mine rehabilitation and downstream industries using multi-products from fibre and phytoremediation. The regional Bokamoso Ba Rona project in the West Rand goldfield proposes a multi-stakeholder collaborative approach for post-closure land use to create large-scale regenerative agriculture projects. The Green Engine in the Witbank coalfield in Mpumalanga province integrates post-mining land use, wate management, agriculture water treatment and industrial development. This testing using case studies improved the rating systems and ensured that the results reflected reality.

2.4. Expert Review

A draft risk rating was presented to the research project reference group—comprising eight academic, government and industry experts with significant experience in the field of mine closure—for feedback. The risk rating system was revised based on their input and presented in an online stakeholder workshop in March 2023. Over 70 mine closure experts in the mining industry, consulting, academia, civil society, and government were invited, and 32 confirmed their attendance. Although only 15 people attended on the day, they represented a balanced spread across academia, consulting, civil society and the mining industry. Participants were asked to answer questions about each draft risk rating using Zoom's Advanced Poll functionality. At the end of the workshop, the participants were asked whether they would be willing to participate in expert interviews and if they could suggest further resources that could contribute to the development of the risk rating system. All of the participants completed all of the polls, and many of them gave additional feedback at the end of the workshop. One major advantage of holding an online workshop was that it ensured that all stakeholders voices were heard. It also meant that the workshop was recorded, and all the input was received in a digital format that is easy to analyse. The results from the workshop were collated and analysed and used to improve the risk rating system. Follow-up semi-structured interviews were conducted with four experts in local government, multilateral agencies, and an industry association to further refine and improve the rating system. The expert feedback on influencing factors and indicators from all the engagements was used to revise and improve the risk rating system.

3. Results

3.1. Operating Mines and Mining Communities

The risk ratings were based on an evolving database of operating mines (not quarries or alluvial diggings) and mining host communities in South Africa, developed over the past five years by the author [23,28,50,51]. The operating mines dataset includes mine name, province, local municipality, location, commodity, owner, mining duration, life of mine, host communities and population, well-being score, and information required to generate the risk ratings described in this paper. The list of operating mines, their data and the ratings generated in this study are given in Supplementary Material Table S1. The operating mines and host communities are summarised in Tables 1 and 2 and mapped in Figure 3.

Table 1. Summary of all operating mines and their host communities in South Africa.

Commodity	Operating Mines	Mining Companies	Host Communities *	Population in 2011
Gold	33	8	61	1,538,739
Coal	78	32	89	2,638,804
Platinum	46	14	127	1,081,691
Chromium	21	11	50	680,638
Diamonds	9	6	21	498,645
Iron ore	7	8	6	92,339
Manganese	20	18	8	37,765
Vanadium	2	2	3	70,055
Copper	1	1	18	70,312
Nickel	1	2	1	582
Lead-Zinc	2	2	2	2262

Table 1. *Cont.*

Commodity	Operating Mines	Mining Companies	Host Communities *	Population in 2011
Titanium	4	3	3	63,193
Fluorspar	3	3	2	16,092
Phosphate rock	2	2	9	138,377
Rare earth minerals	1	1	1	650
TOTAL	230	104	367	6,243,010

* Note some communities have more than one commodity.

Table 2. Summary of all mining host local municipalities and metropolitan municipalities.

Province	Number of Municipalities	Municipal Population in 2022	Number of Operating Mines
Free State	5	859,736	13
Gauteng	9 *	14,966,638	19
KwaZulu-Natal	8	1,910,224	10
Limpopo	10	2,008,369	38
Mpumalanga	14	3,360,406	70
North West	8	2,481,414	46
Northern Cape	10	750,156	30
Western Cape	2	223,678	4
TOTAL	66	26,534,859	230

* Includes 3 metropolitan municipalities.

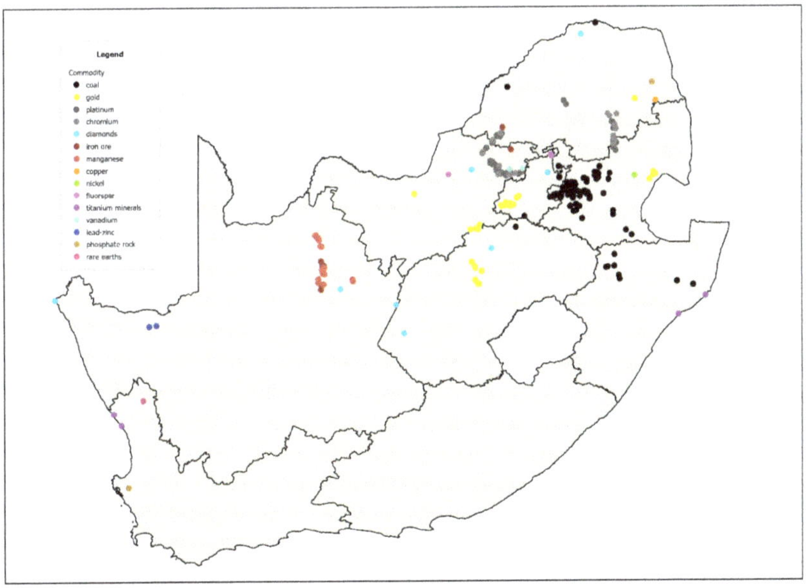

Figure 3. All operating mines in South Africa.

3.2. *Likelihood of Closure*

The likelihood of closure risk rating is shown in Table 3, and results for all operating mines are displayed on the map in Figure 4. The map shows clearly the high risk of closure in the Mpumalanga coal fields, the majority of gold mines and a few chrome mines on the Western Limb of the Bushveld Complex in North West province. The minimum rating was 1.2, the maximum rating was 4.8 and the average was 3.3. Of the 32 mines with a 'very high' rating, 29 are coal mines and three are gold mines. The mines with 'high' ratings are predominantly coal, gold and chromium mines, along with 10 relatively new manganese mines operated by emerging junior miners with limited published information, which may skew the results as the highest rating was assumed for data gaps.

Table 3. Risk rating system for likelihood of mine closure.

Influencing Factor/Indicator	Risk Rating					Weighting
	1—Very Low	2—Low	3—Medium	4—High	5—Very High	
Life of Mine	>40	20–40	11–20	5–10	<5	0.50
Ore reserves and Resources	Very large	Large	Moderate	Limited	Very limited	0.20
Commodity market	High growth	Growth	Stable	Decline	Rapid decline	0.10
Mining method (operating cost)	Dredging	Strip	Open pit	Mechanised underground	Conventional underground	0.10
Company type	Major		Junior		Emerging miner	0.10

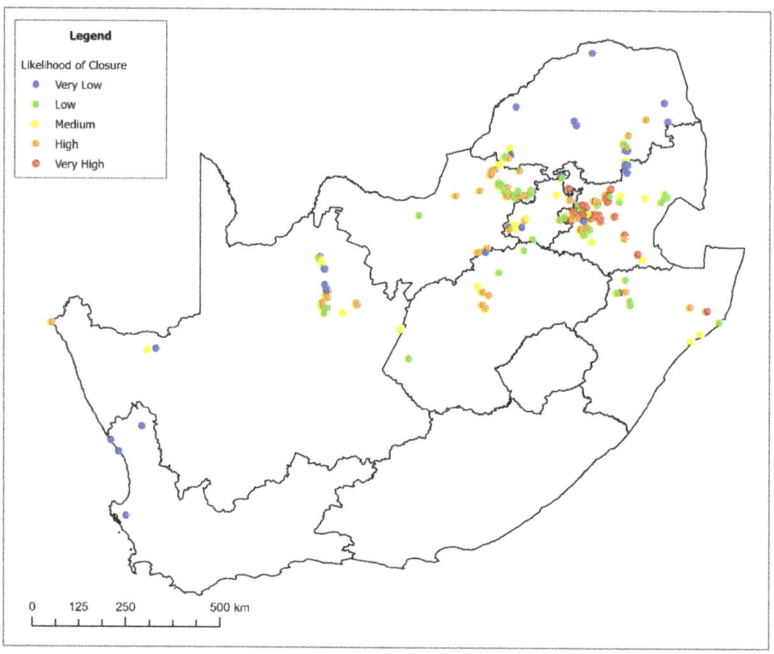

Figure 4. Likelihood of mine closure rating for all operating mines in South Africa.

3.3. *Social Risk of Closure*

The socialrisk rating system is shown in Table 4 and the results calculated for all operating mines are displayed on the map in Figure 5. The map shows highest risk for mines surrounded by rural villages and lowest for mines near to cities. The mines with the 'very high' ratings are almost all platinum and chromium mines in North West and Limpopo provinces, as they are surrounded by deprived rural villages and have very large workforces. Notable exceptions are the copper and phosphate mines in Phalaborwa, which are two of the most remote mines, and two coal mines in eastern KwaZulu-Natal, which all have deprived rural communities. The manganese mines in Northern Cape have a 'high rating' due to their deprived rural communities and remote location. Although most gold mines have large workforces, their communities are better off with lower dependency ratios, and they are close to cities, so they have a 'Medium' rating. Similarly, the remaining coal mines are categorised as 'Low' or 'Very low' because they are hosted by small towns near cities, with higher levels of social wellbeing and lower dependency, and their workforce is relatively small compared to gold and platinum mines. These ratings are a first attempt at a national comparison of social risk and will require further testing and analysis in the future.

Table 4. Risk ratings for socio-economic impacts of mine closure.

Influencing Factor	Indicator	Risk Rating					Weighting
		1—Very Low Risk	2—Low Risk	3—Medium Risk	4—High Risk	5—Very High Risk	
Population at risk	Community population	<5000	5000–30,000	30,000–60,000	60,000–100,000	>100,000	0.2
Social vulnerability	Social well-being score	>8	7–8	6–7	5–6	<5	0.1
Direct mining jobs	Number of jobs	<500	500–1000	1000–3000	3000–5000	>5000	0.3
Local economy dependence	GVA in local municipality	<10%	10–25%	25–50%	50–75%	>75%	0.2
Remoteness	Distance to nearest city	0–25 km	25–50 km	50–100 km	100–200 km	>200 km	0.05
Government capacity	Local municipality audit finding	Clean	Unqualified	Qualified	Adverse	Disclaimed or Not finalised	0.1

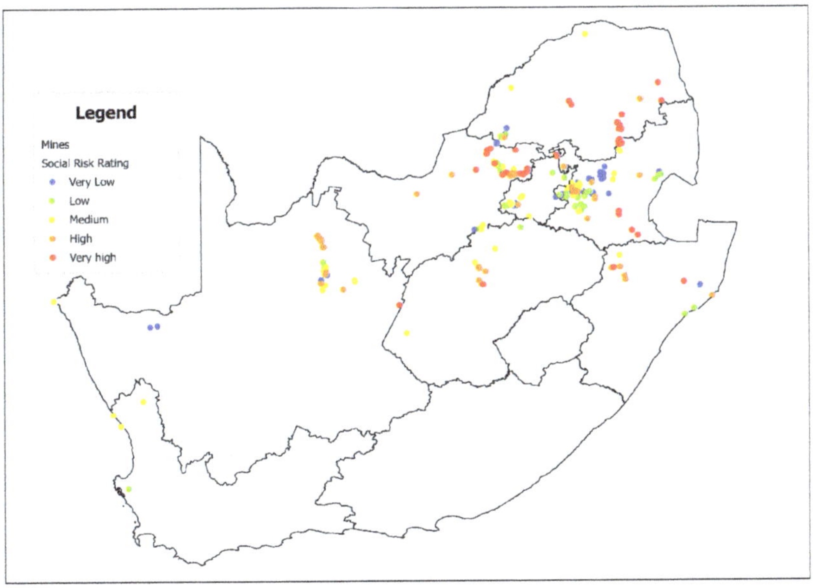

Figure 5. Social risk of mine closure rating for all operating mines in South Africa.

3.4. Environmental Risks

The environmental risk rating system is shown in Table 5 and the results calculated for all operating minesare displayed on the map in Figure 6. The map shows very high and high risk for gold and coal mines because of the risks of acid mine drainage to water resources and degradation of arable land. There are several platinum mines on the Western Limb in the North West province that also receive a high rating largely due to the threat to arable land, terrestrial ecosystems and protected areas. The lowest-risk areas are in the Northern Cape, where there is no arable land and the least threatened ecosystems, and Limpopo, where the mine water threat is low and there are no endangered ecosystems.

Table 5. Risk ratings for environmental impacts of mine closure.

Influencing Factor	Indicator	Risk Rating					Weighting
		1–Very Low Risk	2—Low Risk	3—Medium Risk	4—High Risk	5—Very High Risk	
Mining duration	Total years of production	0–25 years	26–50 years	51–75 years	76–100 years	>100 years	0.20
Threat to biodiversity	Terrestrial ecosystem threat status	Least concern	Vulnerable		Endangered	Critically endangered	0.05
	Distance to nearest protected area	>40 km	20–40 km	10–20 km	5–10 km	0–5 km	0.05
Threat to water resources	Distance to nearest Strategic Water Source Area	>40 km	20–40 km	10–20 km	1–10 km	0–1 km	0.10
	Mine water threat—total	Low	Moderate Low	Moderate	High	Very high	0.10
Threat to arable land	Land capability	VIII Wilderness	VI–VIIGrazing land	IV Marginal land	III Moderate potential	I–II High-potential arable land	0.10

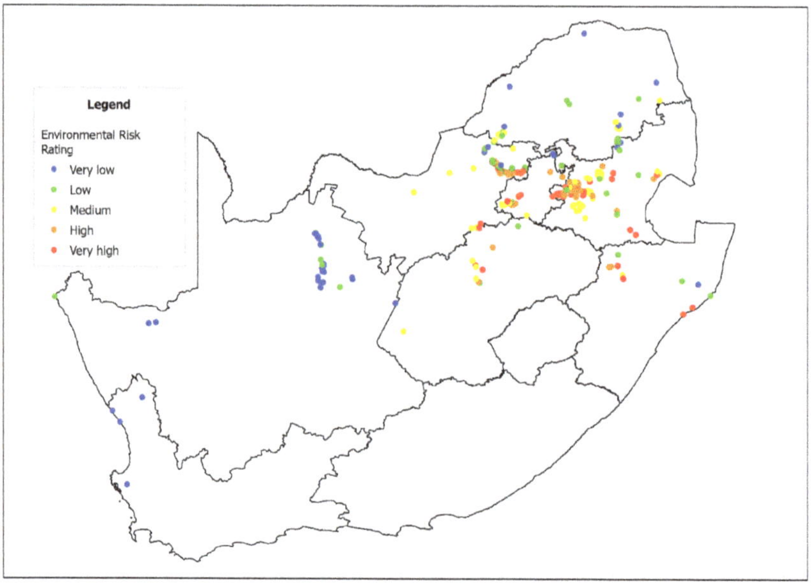

Figure 6. Environmental Risk Rating Map for all operating mines in South Africa.

3.5. Case Studies

In order to test the draft risk rating system, the three ratings were calculated for all 10 mines in case studies. These mines have very diverse mine plans, social contexts and environmental settings, ensuring robust testing of the ranges and categories. The results given in Table 6 show 'very high' likelihood of closure for Khwezela coal mine and 'high' likelihood for Kusasalethu gold mine, which is due to close next year but has ore reserves that could extend the mine plan to 2037 if economic and political conditions improve. They show 'medium' likelihood for four other conventional underground gold mines, 'low' likelihood for the three new mines (two being highly mechanised gold mines that previously closed due to low gold prices and high operating costs), and 'very low' likelihood for Mogalakwena, which has a life of mine of over 70 years and low operating costs. The social impact risk ranges from low risk in some gold mines, with relatively better off communities located near the Gauteng metropolitan areas, to very high risk in Mogalakwena platinum mining areas in Limpopo, where the mines are surrounded by many very deprived communities with high dependency ratios. The environmental risk ratings range from very high for the gold mines located in Strategic Water Source Areas or areas with more threatened ecosystems or better arable land to low in the Mogalakwena platinum mining area, where there are low threats to water resources and ecosystems and the mines have been operating for a much shorter time.

Table 6. Risk ratings for all mines in the case studies.

Mine	Commodity	Mine Type	Mining Region	Likelihood of Closure Rating	Social Risk Rating	Environmental Risk Rating
Mogalakwena	Platinum	Open pit	Northern Limb, Limpopo	Very low	Very high	Low
Ivanplats		Mechanised underground		Low	High	Low
Mponeng	Gold	Conventional underground	Far West Rand, Gauteng	Medium	Low	Medium
Kloof				Medium	Medium	High
Driefontein				Medium	Medium	Very high
Kusasalethu				High	Low	Medium
South Deep		Mechanised underground		Low	High	High
Blyvoor				Low	Low	Medium
Doornkop		Conventional underground	West Rand, Gauteng	Medium	Medium	Very high
Khwezela	Thermal coal	Open pit	Witbank, Mpumalanga	Very high	Medium	High

4. Discussion

4.1. Measuring Risks Related to Mine Closure

This paper has described the development of a comprehensive mine closure risk rating system for South Africa and the quantification of three mine closure risk ratings for all operating mines in the country.

The likelihood of closure map identifies mines and areas where mine closure is highly likely and needs immediate attention. Out of the 221 operating mines that were rated, 48 are expected to close within 10 years, based on their current life of mine. This map is beneficial for government officials at local, provincial and national levels for policy, planning and budgeting, as well as for mining host communities that will be affected by closure. The map also shows that there are significant ranges in likelihood of closure within commodities and between commodities, and assumptions should not be made about imminent mine closure purely based on narratives around certain commodities. For example, gold mines that have been converted from conventional to mechanised mining now have long life of mines unlike their neighbouring conventional mines that are facing closure in the short term. This is important for the contested 'just transition' narrative in South Africa where it is assumed that coal mines will be forced to close, whereas it is much more likely that they will operate until the end of their current life of mine [23].

The social risk map reflects the potential negative impact that mine closure will have on neighbouring communities. There are at least 360 mining host communities in South Africa, but they have very diverse characteristics in terms of living standards and location. For example, those communities located in urban hubs with diverse economies will be less affected than the rural communities in remote areas. More importantly, historically disadvantaged communities (rural villages and townships) are still more deprived today and are much more vulnerable to mine closure and its resulting job losses and economic downturn. Given South Africa's extreme inequalities, it is essential that, when site-specific studies are done on social risk of closure, the socio-economic baseline of each individual community is taken into account.

The environmental risk map reflects the potential negative impact that mine closure will have on the natural environment and enables the prioritisation of mitigation and intervention measures by mining companies and government. The focus of this risk rating system is to compare risks across all mine sites, and therefore it requires comprehensive, national standardised datasets. It does not account for localised environmental concerns although the general themes are captured in the risk rating. Ideally, once the high-risk mines are identified, detailed environmental assessments should be undertaken for each site. In South Africa, many of these reports commissioned by mining companies for mining licences are publicly available and can be used for further investigations.

4.2. Data Gaps and Future Research

It is important to note that this risk rating system is intended as a first version of a proposed risk framework for South Africa, and it is expected that it will be reviewed and improved over the years to come. While 17 indicators could be measured and quantified with existing or developed national datasets, there were gaps and opportunities for adding further aspects to the risk ratings. For likelihood of closure, ideally the operating cost of every mine should be captured instead of the mining method proxy used here, and more definition on mining companies would improve the accuracy of the company stability indicator. For social risk, the social data on communities came from the 2011 census, the only local-level national dataset. This should be updated when the 2022 census database is published. For environmental risk, threat to agricultural production, waste volumes and tailings dam failure risk all require comprehensive national datasets to be used in the rating system. Potentially a separate National Waste Atlas could be developed to focus on the risks and opportunities related to mine waste facilities [54]. Finally, given the huge impact of governance on mine closure outcomes, a governance risk component could be added to the risk rating system. This would look at corporate governance, local government capacity and provincial policy. If the rating systems were expanded to other countries, then national indicators would become important. Additional research could include correlation analysis between commodities, provinces and regions to better understand the risks and implications for different areas and sectors.

4.3. Communication, Stakeholder Engagement and Decision-Making

As social, environmental and governance aspects of mine closure are intertwined, the weightings need to be reviewed, and the local contexts must be considered. The weighting of factors may differ across local contexts, and therefore, critical engagement by mine closure stakeholders with the risk ratings is essential. Mine closure decision-making tools should be used in conjunction with efforts to build relations in or among communities potentially impacted by mine closure, and time should be given to the participatory process to ensure as wide an input as possible.

The risk rating maps described in this paper have been incorporated into an online Mine Closure Risk and Opportunity Atlas that can operate on a computer, tablet or smartphone to allow them to be communicated to a wide audience [76]. The Atlas has all the datasets used to calculate the risk ratings so that a user can interrogate the individual risk factors and make their own assessments to support mine closure planning and aid decision-making processes at all scales. Making information about mine closure risks and opportunities accessible to mining host communities, particularly through the Atlas being supported on smartphones, empowers them to engage in and influence decision-making processes. Showing the evidence for socio-economic deprivation in mining host communities to mining companies gives them the understanding and motivation to do more to support these communities when mine closure occurs.

There are numerous areas in the South African mining policy and planning arena where this risk rating system can be applied. It can be used in developing mine closure plans, undertaking environmental risk assessments and assessing financial provisions for rehabilitation of negative environmental impacts, all required by mining companies in South Africa under Regulation 11 (1) of Government Notice R1147. It can also be used by Future Forums designated in SLPs for engaging host communities in mine closure planning [77], in the development of Integrated Development Plans (IDPs) by local municipalities, and in regional spatial planning by district municipalities and provincial governments. Finally, it could play a supporting role in the implementation of the National Mine Closure Strategy by the national DMRE [78], which is being gazetted for another round of public participation in early 2024. There are several options for post-closure land use and economic diversification and job creation for different types of mines. These involve agriculture, forestry, tourism and recreation, conservation, energy supply, water storage, community and culture facilities, and research [79]. The risk ratings and their underlying datasets can contribute to the selection of viable opportunities as risk assessments are critical to mine closure planning [80].

5. Conclusions

This paper has described the development of a national mine closure risk rating system for South Africa to aid management and policy-making regarding mine closure in the country. The risk rating system identifies mines and regions where mine closure is highly likely and needs immediate attention, and provincial, district and local governments should take note of how this may affect their planning and budgeting in the short and medium term. Currently, 48 operating mines have a life of mine of 10 years or less, and the coal mines in Mpumalanga province and the gold mines in Gauteng stand out as areas requiring urgent attention. The rating system ranks mines by the risk of negative social and environmental impacts potentially resulting from closure, enabling the prioritisation of mitigation and intervention by mining companies and government. The social risks are highest in the most vulnerable communities, highlighting the need for a broad response by national and provincial governments that alleviate poverty and support economic growth. The environmental risks are highest for coal and gold mining regions, which are also where the best arable land is found, highlighting the importance of rehabilitation in the mine closure process to enable post-closure land use and economic diversification.

The risk ratings proposed here are a first attempt at a national comparison of likelihood of mine closure, social risk and environmental risk, and will require further testing and analysis in the future. Additional datasets will be required to fill gaps and ensure that the risk ratings capture all of the influencing factors. The results of the risk ratings could promote deeper discussions on mine closure management and planning among a diverse group of stakeholders and support evidence-based decision-making. Finally, while the risk rating system has been developed for South Africa, its concepts, design and insights could be applied to any mining country in the world.

Supplementary Materials: The following supporting information can be downloaded at: https://www.mdpi.com/article/10.3390/mining4010005/s1, Table S1: Data and risk ratings for all operating mines in South Africa.

Funding: This research was funded by the Water Research Commission, grant number C2021_2023-00475.

Data Availability Statement: Data supporting reported results will be made available in the online South African Mine Closure Risk and Opportunity Atlas, including links to publicly archived datasets analysed or generated during the study.

Acknowledgments: The author would also like to thank all the experts and stakeholders who gave of their time and expertise, and the wider project team for their comments and insights.

Conflicts of Interest: The author declares no conflict of interest.

Appendix A

Table A1. Indicators of social well-being for mining host communities [51].

Dimension and SDG Target	Indicators
Income (SDG 1.2)	% households with income more than R19,600/year (est. national poverty line)
Household goods	% households that own a refrigerator
Health	% population without a disability
Education (SDG 4.1)	% adults (≥20 years) with NQF4 qualification (Grade 12, NTC3) or better
Gender representation (SDG 5.5)	% of female councillors in local government
Water access (SDG 6.1)	% population with piped water in their dwelling or yard
Sanitation (SDG 6.2)	% population with access to a flush toilet or chemical toilet
Electricity access (SDG 7.1)	% population with electricity as main source of lighting
Clean cooking fuel (SDG 7.1)	% population using clean cooking fuel
Employment (SDG 8.5)	% labour force (including discouraged jobseekers) employed
Housing (SDG 11.1)	% population in formal housing
Waste management (SDG 11.6)	% population with refuse removal
Internet access (SDG 17.8)	% population with access to internet

References

1. Cole, M.J.; Broadhurst, J.L. Mapping and Classification of Mining Host Communities: A Case Study of South Africa. *Extr. Ind. Soc.* **2020**, *7*, 954–964. [CrossRef]
2. Minerals Council South Africa. *Facts & Figures Pocketbook 2022*; Minerals Council South Africa: Johannesburg, South Africa, 2023.
3. USGS. *Platinum-Group Metals Mineral Commodity Summary*; United States Geological Survey: Reston, VA, USA, 2023.
4. USGS. *Manganese; Mineral Commodity Summary*; United States Geological Survey: Reston, VA, USA, 2023.
5. USGS. *Zirconium and Hafnium; Mineral Commodity Summary*; United States Geological Survey: Reston, VA, USA, 2023.
6. USGS. *Vanadium; Mineral Commodity Summary*; United States Geological Survey: Reston, VA, USA, 2023.
7. SADPMR. *Annual Report 2021/22*; SADPMR: Kempton Park, South Africa, 2022.
8. USGS Fluorspar. *Mineral Commodity Summary*; United States Geological Survey: Reston, VA, USA, 2023.
9. USGS Gold. *Mineral Commodity Summary*; United States Geological Survey: Reston, VA, USA, 2023.
10. BP. *Statistical Review of World Energy 2022*, 71st ed.; BP: London, UK, 2022.
11. Yager, T.R. *2016 Minerals Yearbook the Mineral Industry of South Africa*; United States Geological Survey: Reston, VA, USA, 2021.
12. DMR. *South African Mineral Industry 2016/2017 SAMI*, 34th ed.; Department of Mineral Resources, Government of South Africa: Pretoria, South Africa, 2018; ISBN 9780621450088.
13. StatsSA. *Census 2022: Provinces at a Glance*; Statistics South Africa: Pretoria, South Africa, 2023; ISBN 9780621413908.

14. Krause, R.; Snyman, L. Rehabilitation and Mine Closure Liability: An Assessment of the Accountability of the System to Communities. In Proceedings of the 9th International Conference on Mine Closure, Sandton, South Africa, 1–3 October 2014; Fourie, A., Tibbett, M., Weiersbye, I., Eds.; Australian Centre for Geomechanics: Johannesburg, South Africa, 2014.
15. Marais, L.; Nel, E. The Dangers of Growing on Gold: Lessons for Mine Downscaling from the Free State Goldfields, South Africa. *Local Econ.* **2016**, *31*, 282–298. [CrossRef]
16. Watson, I.; Olalde, M. The State of Mine Closure in South Africa—What the Numbers Say. *J. S. Afr. Inst. Min. Metall.* **2019**, *119*, 639–645. [CrossRef]
17. Auditor-General South Africa. *Report of the Auditor-General to Parliament on a Performance Audit of the Rehabiliation of Abandoned Mines at the Department of Minerals and Energy*; Auditor-General South Africa: Pretoria, South Africa, 2009.
18. SAHRC. *SAHRC National Hearing on the Underlying Socio-Economic Challenges of Mining-Affected Communities in South Africa*; South Africa Human Rights Commission: Johannesburg, South Africa, 2016; ISBN 9780080453705.
19. IHRC. *The Cost of Gold: Environmental, Health and Human Rights Consequences of Gold Mining in South Africa's West and Central Rand*; IHRC: London, UK, 2016.
20. Owen, J.; Kemp, D.; Marais, L. The Cost of Mining Benefits: Localising the Resource Curse Hypothesis. *Resour. Policy* **2021**, *74*, 102289. [CrossRef]
21. Brueckner, M.; Spencer, R.; Knowles, S.; Paull, M. Mining Legacies—Broadening Understandings of Mining Impacts. *Extr. Ind. Soc.* **2021**, *8*, 100950. [CrossRef]
22. Marais, L.; Nel, V.; Rani, K.; van Rooyen, D.; Sesele, K.; van der Watt, P.; du Plessis, L. Economic Transitions in South Africa's Secondary Cities: Governing Mine Closures. *Polit. Gov.* **2021**, *9*, 381–392. [CrossRef]
23. Cole, M.J.; Mthenjane, M.; Zyl, A.T. Van Assessing Coal Mine Closures and Mining Community Profiles for the 'Just Transition' in South Africa. *J. S. Afr. Inst. Min. Metall.* **2023**, *123*, 329–341. [CrossRef]
24. MISTRA. *The Future of Mining in South Africa: Sunset or Sunrise?* MISTRA: Johannesburg, South Africa, 2018.
25. Boggs, G.; Measham, T.; D'Urso, J. What Are We Transitioning into? Re-Thinking the Model of Mine Closure. In Proceedings of the Mine Closure 2023, Reno, NV, USA, 2–5 October 2023; Australian Centre for Geomechanics: Crawley, Australia, 2023.
26. Dzakpata, I.; Qureshi, M.; Kizil, M.; Maybee, B. *Exploring the Issues in Mine Closure Planning*; CRCTiME: Perth, Australia, 2021; ISBN 978-1-922704-05-4.
27. Everingham, J.; Svobodova, K.; Mackenzie, S.; Witt, K. *Participatory Processes for Mine Closure and Social Transitions*; Centre for Social Responsibility in Mining, University of Queensland: Brisbane, Australia, 2020.
28. Cole, M.J. ESG Risks to Global Platinum Supply: A Case Study of Mogalakwena Mine, South Africa. *Resour. Policy* **2023**, *85*, 104054. [CrossRef]
29. Laurence, D. Optimisation of the Mine Closure Process. *J. Clean. Prod.* **2006**, *14*, 285–298. [CrossRef]
30. Cairncross, B. The Okiep Copper District Namaqualand, Northern Cape Province South Africa. *Mineral. Rec.* **2004**, *35*, 289–317.
31. Binns, T.; Nel, E. The Village in a Game Park: Local Response to the Demise of Coal Mining in KwaZulu-Natal, South Africa. *Econ. Geogr.* **2003**, *79*, 41–66. [CrossRef]
32. Falcon, L.M. Tin in South Africa. *J. S. Afr. Inst. Min. Metall.* **1985**, *85*, 333–345.
33. Godsell, S. Rooiberg: The Little Town That Lived. *S. Afr. Hist. J.* **2011**, *63*, 61–77. [CrossRef]
34. Abratt, R.P.; Vorobiof, D.A.; White, N. Asbestos and Mesothelioma in South Africa. *Lung Cancer* **2004**, *45*, 3–6. [CrossRef] [PubMed]
35. Brock, D. ICMM Guidance and Resources for Integrating Closure into Business Decision Making Processes. In Proceedings of the International Conference on Mine Closure 2021, Perth, Australia, 26–28 November 2021; Fourie, A., TIbbett, M., Sharkuu, A., Eds.; Australian Centre for Geomechanics: Crawley, Australia, 2021.
36. ICMM. *Integrated Mine Closure Good Practice Guide*, 2nd ed.; International Council for Mining and Metals (ICMM): London, UK, 2019.
37. Government of Western Australia. *Statutory Guidelines for Mine Closure Plans*; Government of Western Australia: Perth, Australia, 2020.
38. Coppin, N. An Ecologist in Mining—A Retrospective of 40 Years in Mine Closure and Reclamation. In Proceedings of the Eighth International Seminar on Mine Closure, St. Austell, UK, 17–22 September 2013; Tibbett, M., Fourie, A., Digby, C., Eds.; Australian Centre for Geomechanics: Cornwall, Australia, 2013; pp. 295–309.
39. Kabir, S.M.Z.; Rabbi, F.; Chowdhury, M.B. Mine Closure Planning and Practice in Canada and Australia: A Comparative Review. *World Rev. Bus. Res.* **2015**, *5*, 140–159.
40. Manero, A.; Kragt, M.; Standish, R.; Miller, B.; Jasper, D.; Boggs, G.; Young, R. A Framework for Developing Completion Criteria for Mine Closure and Rehabilitation. *J. Environ. Manag.* **2020**, *273*, 111078. [CrossRef] [PubMed]
41. Bainton, N.; Holcombe, S. A Critical Review of the Social Aspects of Mine Closure. *Resour. Policy* **2018**, *59*, 468–478. [CrossRef]
42. Beckett, C.; Dowdell, E.; Monosky, M.; Keeling, A. *Integrating Socio-Economic Objectives for Mine Closure and Remediation into Impact Assessment in Canada*; Social Sciences and Humanities Research Council: Ottawa, ON, Canada, 2020.
43. Luckeneder, S.; Giljum, S.; Schaffartzik, A.; Maus, V.; Tost, M. Surge in Global Metal Mining Threatens Vulnerable Ecosystems. *Glob. Environ. Chang.* **2021**, *69*, 102303. [CrossRef]
44. Macklin, M.G.; Thomas, C.J.; Mudbhatkal, A.; Brewer, P.A.; Hudson-Edwards, K.A.; Lewin, J.; Scussolini, P.; Eilander, D.; Lechner, A.; Owen, J.; et al. Impacts of Metal Mining on River Systems: A Global Assessment. *Science* **2023**, *381*, 1345–1350. [CrossRef]
45. Meißner, S. The Impact of Metal Mining on Global Water Stress and Impact Assessment. *Resources* **2021**, *10*, 120. [CrossRef]

46. Owen, J.; Kemp, D.; Lèbre, E.; Harris, J.; Svobodova, K. A Global Vulnerability Analysis of Displacement Caused by Resource Development Projects. *Extr. Ind. Soc.* **2021**, *8*, 100877. [CrossRef]
47. Marais, L. The Impact of Mine Downscaling on the Free State Goldfields. *Urban Forum* **2013**, *24*, 503–521. [CrossRef]
48. Lèbre, É.; Owen, J.R.; Stringer, M.; Kemp, D.; Valenta, R.K. Global Scan of Disruptions to the Mine Life Cycle: Price, Ownership, and Local Impact. *Environ. Sci. Technol.* **2021**, *55*, 4324–4331. [CrossRef]
49. Schers, J.; Burton, J. *Managing the Coal Transition for Workers in South Africa: A Scenario Analysis of Age Skills Profiles of the Coal Mining Workforce*; International Association of Exhibitions and Events: Dallas, TX, USA, 2021.
50. Cole, M.J.; Broadhurst, J.L. Sustainable Development in Mining Communities: The Case of South Africa's West Wits Goldfield. *Front. Sustain. Cities* **2022**, *4*, 895760. [CrossRef]
51. Cole, M.J.; Broadhurst, J.L. Measuring the Sustainable Development Goals (SDGs) in Mining Host Communities: A South African Case Study. *Extr. Ind. Soc.* **2021**, *8*, 233–243. [CrossRef]
52. SAMREC. *The South African Code for the Reporting of Exploration Results, Mineral Resources and Mineral Reserves (THE SAMREC CODE)*; SAMCODE: Johannesburg, South Africa, 2016.
53. Valenta, R.K.; Kemp, D.; Owen, J.R.; Corder, G.D. Re-Thinking Complex Orebodies: Consequences for the Future World Supply of Copper. *J. Clean. Prod.* **2019**, *220*, 816–826. [CrossRef]
54. Koovarjee, B.; Becker, M.; Von Holdt, J.R.; Petersen, J.; Cole, M. Developing a Mine Waste Atlas for the Northern Cape, South Africa. In *Next Generation Tailings—Opportunity or Risk*; SAIMM: Johannesburg, South Africa, 2023.
55. Hund, K.; La Porta, D.; Fabregas, T.; Laing, T.; Drexhage, J. *Minerals for Climate Action: The Mineral Intensity of the Clean Energy Transition*; World Bank: Washington, DC, USA, 2020.
56. Elshkaki, A.; Graedel, T.E.; Ciacci, L.; Reck, B.K. Resource Demand Scenarios for the Major Metals. *Environ. Sci. Technol.* **2018**, *52*, 2491–2497. [CrossRef] [PubMed]
57. *Fortune Business Insights Diamond Market Analysis and Regional Forecast 2023–2030*; Fortune Business Insights: Maharashtra, India, 2023.
58. Lonergan, W. *The Valuation of Mineral Assets*; Sydney University Press: Sydney, Australia, 2006; ISBN 9781920898267.
59. Rudenno, V. *The Mining Valuation Handbook*; John Wiley and Sons Australia: Milton, Australia, 2012; ISBN 9780730377078.
60. Tholana, T. *Industry Cost Curves as a Tool to Analyse Cost Performance of South African Mining Operations: Gold, Platinum, Coal and Diamonds*; University of the Witwatersrand: Johannesburg, South Africa, 2012.
61. Kriel, H. *The Extent, Nature and Economic Impact of the Junior and Emerging Mining Sector in South Africa*; Minerals Council South Africa: Johannesburg, South Africa, 2019.
62. Wright, G.; Noble, M. *The South African Index of Multiple Deprivation 2007 at Municipality Level*; Department of Social Development, Government of South Africa: Pretoria, South Africa, 2009.
63. UN ESA. Dependency Ratio. Available online: http://www.un.org/esa/population/unpop.htm (accessed on 18 October 2023).
64. STEPSA. CSIR SA CSIR MesoZone 2018v2; 2018. Available online: https://www.dpme.gov.za/keyfocusareas/gwmeSite/The%20PME%20Forum%202018/Annexure-Draft-NSDF-2018_September-2018-3.pdf (accessed on 22 January 2023).
65. Auditor-General South Africa. *Consolidated General Report on Local Government Audit Outcomes MFMA 2020-21*; Auditor-General South Africa: Pretoria, South Africa, 2022; ISBN 9788578110796.
66. Skowno, A.L.; Poole, C.J.; Raimondo, D.C.; Sink, K.J.; Van Deventer, H.; Van Niekerk, L.; Harris, L.R.; Smith-Adao, L.B.; Tolley, K.A.; Zengeya, T.A.; et al. *National Biodiversity Assessment 2018: The Status of South Africa's Ecosystems and Biodiversity—Synthesis Report*; South African National Biodiversity: Pretoria, South Africa, 2019; ISBN 978-1-928224-34-1.
67. *DFFE Protected and Conservation Areas Database 2023*; Department of Forestry, Fisheries and the Environment, Government of South Africa: Pretoria, South Africa, 2023.
68. Le Maitre, D.; Seyler, H.; Holland, M.; Smith-Adao, L.; Maherry, A.; Nel, J.; Witthuser, K. *Identification, Delineation and Importance of the Strategic Water Source Areas of South Africa, Lesotho and Swaziland for Surface Water and Groundwater*; Water Research Commission: Pretoria, South Africa, 2018; ISBN 9780639200064.
69. WRC. *The South African Mine Water Atlas*; Water Research Commission: Pretoria, South Africa, 2018; ISBN 9781431208067.
70. Collett, A. The Impact of Effective (Geo-Spatial) Planning on the Agricultural Sector. In Proceedings of the South African Surveying and Geomatics Indaba, Kempton Park, South Africa, 22–24 July 2013.
71. Schoeman, J.L.; van der Walt, M.; Monnik, K.A.; Thackrah, A.; Malherbe, J.; le Roux, R.E. *Development and Application of a Land Capability Classification System for South Africa*; Institute for Soil, Climate and Water: Pretoria, South Africa, 2002.
72. Owen, J.R.; Kemp, D.; Lèbre, E.; Svobodova, K.; Pérez Murillo, G. Catastrophic Tailings Dam Failures and Disaster Risk Disclosure. *Int. J. Disaster Risk Reduct.* **2020**, *42*, 101361. [CrossRef]
73. Lèbre, É.; Stringer, M.; Svobodova, K.; Owen, J.R.; Kemp, D.; Côte, C.; Arratia-Solar, A.; Valenta, R.K. The Social and Environmental Complexities of Extracting Energy Transition Metals. *Nat. Commun.* **2020**, *11*, 4823. [CrossRef] [PubMed]
74. GRID-Arendal Global Tailings Portal. Available online: https://tailing.grida.no/ (accessed on 6 July 2023).
75. Cole, M.; Chimbganda, T.; Esau, M.; Abrams, A.; Broadhurst, J. *Developing National Mine Closure Risk and Opportunity Rating Systems for South Africa*; Draft Final Report to the Water Research Commission; Water Research Commission: Cape Town, South Africa, 2023.
76. Esau, M.; Cole, M.; Broadhurst, J.; Chimbganda, T.; Abrams, A. Developing a National Mine Closure Risk and Opportunity Atlas for South Africa. In Proceedings of the Mine Closure 2023, Reno, NV, USA, 2–5 October 2023; Abbasi, B., Parshley, J., Fourie, A., Tibbett, M., Eds.; Australian Centre for Geomechanics: Crawley, Australia, 2023; pp. 1–14.

77. CALS. *Social and Labour Plan Mining Community Toolkit*; Centre for Applied Legal Studies (CALS): Johannesburg, South Africa, 2017.
78. DMRE. *Publication of the Draft Mine Closure Strategy 2021 for Public Comments*; Department of Mineral Resources and Energy: Pretoria, South Africa, 2021.
79. Keenan, J.; Holcombe, S. Mining as a Temporary Land Use: A Global Stocktake of Post-Mining Transitions and Repurposing. *Extr. Ind. Soc.* **2021**, *8*, 100924. [CrossRef]
80. Maybee, B.; Lilford, E.; Hitch, M. Environmental, Social and Governance (ESG) Risk, Uncertainty, and the Mining Life Cycle. *Extr. Ind. Soc.* **2023**, *14*, 101244. [CrossRef]

minerals

MDPI

Article

Study of the Critical Safe Height of Goaf in Underground Metal Mines

Qinli Zhang [1], Peng Zhang [1], Qiusong Chen [1,*], Hongpeng Li [2], Zian Song [1] and Yunbo Tao [1]

[1] School of Resources and Safety Engineering, Central South University, Changsha 410083, China; zhangqinlicn@126.com (Q.Z.); 215512116@csu.edu.cn (P.Z.); 225512120@csu.edu.cn (Z.S.); yunbo.tao@csu.edu.cn (Y.T.)

[2] Jiangxi Copper Industry Group Yinshan Mining Industry Co., Ltd., Shangrao 334200, China; lihongpeng2024@163.com

* Correspondence: qiusong.chen@csu.edu.cn; Tel.: +86-199-1150-0066

Abstract: The empty-space subsequent filling mining method is the main mining scheme for underground metal mines to achieve large-scale mechanized mining. The stage height, one of the main parameters of this method, affects the various production process aspects of the mine and influences the stability of the goaf. In order to determine the stage height scientifically and rationally in the empty-space subsequent filling mining method, a formula for the stabilized critical safe height of a high goaf in an underground metal mine was derived based on Pu's arch equilibrium theory, Bieniawski's pillar strength limit theory, and the Kastner equation and combined with the results of an orthogonal analysis to rank the importance of the main factors in the formula. A copper mine in Jiangxi Province was used as a case study, with the reliability of the formula verified by numerical simulation and industrial testing. The factors in the formula influencing the critical stabilized safe height of the goaf were, in descending order, the compressive strength of the rock body, the width of the two-step mining pillar, the width of the one-step mining room, the mining height, and the depth of mining. Based on the calculation results, the recommended stage heights are 30 m (−378 m middle section) and 25 m (−478 m middle section) in the area of poor rock body stability and 50 m in the area of better rock body stability. The simulation results show that the goaf is significantly affected by the compressive stress under the condition of a certain rock body stability and that the compressive stress increases with increasing goaf height. The minimum recommended values of the sidewall safety coefficients in areas of poor and better rock stability are 1.04 and 1.06, respectively. The volume deviation coefficients of the three industrial test mines were all controlled within 3%, indicating that no obvious collapse and destabilization phenomenon occurred in the goaf. This paper provides some theoretical and applied guidance for the stage height design of similar underground metal mines using the empty-space subsequent filling mining method.

Keywords: critical height of goaf; the empty-space subsequent filling mining method; rock stability; numerical simulation; industrial tests

1. Introduction

In underground metal mining, the empty-space subsequent filling mining method has become one of the main mining schemes because of its advantages of high production capacity, low cost, high recovery rate, and protection of underground and surface environments [1–7]. The empty-space subsequent filling mining method, whether it is staged open-field subsequent filling or segmented open-field subsequent filling, generally divides the ore block into a one-step mining room and a two-step ore pillar during the mining process and carries out the mining in two steps. After the completion of stoping, the stage height of the goaf can reach tens or even hundreds of meters. Once the goaf height becomes unstable, it poses a threat to the equipment and personnel involved in underground production and increases the difficulty of mining in adjacent areas, resulting in resource waste [8–10].

Many hydrogeological, rock stress-strain behavior, geometric, and engineering factors affect the safe proportions of goaf [11–15]. Extensive research has been conducted on the ultimate exposed area and ultimate span of the goaf roof in underground mining. Based on a summary of numerous mining engineering practices, Matthews proposed a diagrammatic method for assessing the relationship between the stability index N of a rock mass and the shape coefficient of the exposed mining area [16]. Jang et al. proposed a stability stress mechanics model for goaf and discussed the relationship between the ultimate exposed area of goaf and the span ratio, joint density of the fracture surface, and joint diameter [17]. Swift et al. used an elastic beam model and a pillar strength formula to study the relationship between the exposed area of goaf, the pillar stability, and a safety factor [18]. Gao et al. simulated the distribution and variation patterns of stress, displacement, and the plastic zone in goaf for different span ratios using the PLAXIS 2D finite element analysis software [19]. Zhienbayev, A. et al. performed the calculations, modelling, and statistic analysis of factual rock falls from the roof to ensure the roof span stability in terms of room-and-pillar system of mining [20]. Hosseini, M. et al. conducted a sensitivity analysis using the numerical, squat pillar, and Mathews stability methods through the Taguchi technique to properly understand the influence of geometric parameters and stress on stope stability according to Sormeh underground mine data [21]. Although the ultimate exposed area and ultimate span of the goaf roof have been studied extensively, the influence of goaf height on goaf stability has not been adequately examined.

Bagde evaluated the impact of mining height on the stability of goaf through empirical methods, numerical simulations, and on-site rock deformation monitoring [22]. Soni et al. used numerical simulation to study the variation of pillar strength with height during the implementation of the room-and-pillar mining method in mines [23]. Zhang et al. used an orthogonal experimental design method to investigate the effects of the height, length, and width of mining pillars and the length and width of mining rooms on the stability of goaf [24]. Qiu et al. derived a limit equilibrium equation for mining pillars based on the area-bearing theory and used this equation to optimize the structural parameters of the mining area, such as the mining height, pillar spacing, and row spacing [25]. The research results obtained to date are of great significance in optimizing the mining height using the room column method. However, there has been relatively little research on the stability of high goaf using the subsequent filling mining method. Research on the safe critical height for maintaining self-stability in high goaf is still at an early stage.

A copper mine in Jiangxi was taken as a case study in this research to examine this issue. Using Bieniawski's ultimate pillar strength theory [26], Pu's arch equilibrium theory [27], and the Kastner equation [28], a formula was derived for the critical stable height of goaf. A sensitivity analysis of the factors affecting the critical safe height of goaf was conducted using the orthogonal range analysis method. On the basis of rock mechanics experiments and reduction of rock mechanics parameters, the critical safe height of the goaf of the copper mine was calculated. The value obtained was verified through numerical simulation analysis and on-site experiments. The research results provide a theoretical basis and engineering reference for designing reasonable mining stage heights for similar underground metal mines under various geological and mining conditions.

2. Methods

2.1. Study Site

The Jiangxi copper mine is located in shangrao City, Jiangxi Province. The mine area is about 2.7 km long from north to south and 2.5 km wide from east to west, covering an area of about 6.75 km^2. The average surface elevation of the mine is +50 m, and the elevations of the first and second mining sections are −308 m to −378 m and −408 m to −478 m, respectively. There are a total of 11 large ore bodies in the copper–sulfur–gold section, with S1, N1, N2, and N3 being the main ore bodies. The dip of the ore body ranges from 76°

to 88°, with an average of 83°; the ore body mainly strikes EW-NEE, S dips; distributed between exploration lines 03 and 14; and the length of the ore body along the strike ranges from 700 to 1300 m; the thickness of the single project ranges from 1.00 m to 66.70 m, with an average true thickness of 12.5 m. The ore body mainly exists in the contact zone with altered phyllite, with a small amount occurring in blast breccia, altered quartz diorite, and quartz porphyry. Tectonic development of folds and fractures results in poor stability of the ore rocks. The empty-space subsequent filling mining method is the main mining method used in the mine, accounting for 66% of the total. Mining is conducted in intervals. The current mining structure parameters used in the mine have poor compatibility with geological and mining conditions, resulting in frequent collapse events of the one-step mining room during the mining process.

2.2. Experiments Rock Mechanics Parameter and Their Reduction

The rock samples were taken from the underground mining stope of a lead-zinc copper mine in Jiangxi Province. In accordance with the requirements of the rock mechanics parameter experiments, the specimens were drilled using a rock drilling core machine with a diameter of φ = 50 mm. The specimen is cut and sanded into a cylindrical shape. For uniaxial and triaxial compression tests, the recommended ratio of specimen height to diameter is 2:1, which is 100 mm in height and 50 mm in diameter. The actual error is controlled within ±3 mm. For the Brazilian splitting test, the ratio of specimen height to diameter is controlled at 1:1. The actual error is controlled within ±2 mm.

Rock mechanical parameter experiments were conducted using MTS 322 rock mechanical press and MTS 815 rock mechanical press. The experiment was conducted using the transverse isobaric triaxial method at room temperatures in the range of 24–28 °C and 55%–79% relative humidity. The displacement loading method was used to maintain a loading rate of 0.045 mm/min throughout the experiment. The loading process stops automatically when the maximum load is reached and the specimen fails.

Since the rock strength obtained from the indoor tests is actually the rock strength of the intact rock mass, it differs significantly from the rock strength in the field. Therefore, it is necessary to discount its strength. The empirical discounting method has the advantages of being simple to apply and dealing effectively with some uncertain factors. It has become the most important method for studying the strength parameters of rock masses. The Hoek–Brown strength criterion for discounting the strength of a rock mass based on certain indoor and field test data has become the most commonly used discounting method. Empirical formulas for discounting the strength of the rock mass based on the geological description of the rock mass in the field and the statistical data from the investigation of the joints and fractures on the structural surface are considered.

2.3. Theoretical Modelling

After excavation of the rock mass, the stresses in the rock mass are redistributed. The pillar is pressurized by the overlying rock layer, and the top rock layer shows large compressive deformation, forming a small pressure-free arch. It is worth noting that once formed, the pressure-free arch will not disappear with the backfilling of the mining space and is an irreversible deformation phenomenon. Therefore, the smaller pressure-free arches above adjacent spaces gradually merge to form a large pressure-free arch. As shown in Figure 1, a plastic zone with radius R_p will be formed above the goaf. The size of the plastic zone is not only related to the nature of the rock body and the structural parameters of the stope, but is also affected by the mining burial depth. The load on the pillar is the self-weight of the rock body within the plastic zone above its roof. As long as the pillar can withstand the gravity of the rock body within the plastic zone of the top plate surrounding the rock, the long-term stability of the pillar can be ensured.

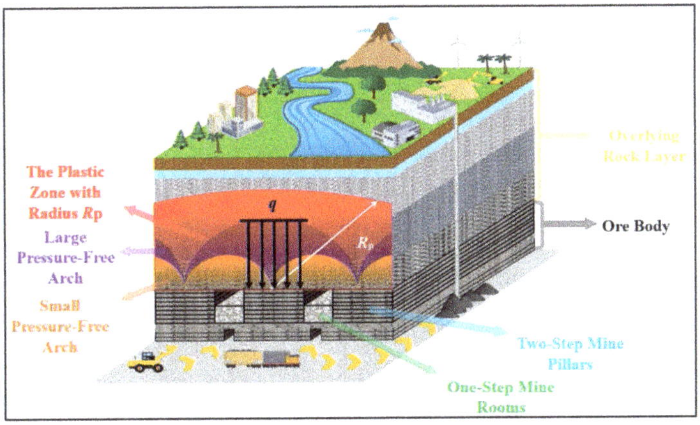

Figure 1. Theoretical modelling of the critical safe height of goaf.

According to the bearing mechanism of the pillar, the load carried by the pillar is the self-weight of all the rock bodies within the plastic zone above its roof. The radius of the plastic zone can be solved using the Kastner equation:

$$R_p = R_0\left[\frac{(P_0 + c \cdot \cot\varphi)(1-\sin\varphi)}{c \cdot \cot\varphi}\right]^{\frac{1-\sin\varphi}{2\sin\varphi}} \tag{1}$$

where R_p is the radius of the plastic zone, m; R_0 is the radius of excavation, m; P_0 is the vertical self-weight stress at the excavation, MPa, $P_0 = \gamma$ H; c is the cohesive force of the rock mass, MPa; φ is the angle of internal friction of the rock mass, °; γ is the capacitance of the top plate surrounding rock, kN/m^3; and H is the mining depth, m.

In order to ensure the stability of the pillar, the maximum range is used to calculate the roof pressure of the stope. Practice has shown that the radius of the plastic zone is less affected by the shape of the excavation section. Therefore, the excavation radius in the calculation is replaced by the equivalent excavation radius—that is, the radius of the external circle with different shapes is used instead. The excavation radius R_0 can be expressed as follows:

$$R_0 = \sqrt{\left(\frac{L}{2}\right)^2 + \left(\frac{H_1}{2}\right)^2} \tag{2}$$

where L is the mining space span, m, and H_1 is the height of the mining pillar, m.

The principle of calculating a rectangular mining pillar is shown in Figure 1. According to the theory of area-bearing capacity, the bearing capacity of a mining pillar is the gravity of the rock body in the overlying plastic zone, and the area of bearing is the sum of the area of the pillar itself and the mining area apportioned by the pillar, so that the following equilibrium equations can be obtained:

$$\sigma_p W_p L_p = (W_0 + W_p) L_p P_{zz} \tag{3}$$

where σ_p is the axial average stress of the ore column, MPa; P_{zz} is the vertical stress in the thickness of the plastic zone of the overlying rock layer, MPa; W_0 the width of the mining room, m; W_p the width of the mine pillar, m; L_p the length of the stope, m.

Knowing that $P_{zz} = \gamma R_p$ and substituting it into the above equation, the average axial stress of the ore column can be obtained as follows:

$$\sigma_p = \frac{(W_0 + W_p)\gamma R_0\left[\frac{(\gamma H + c\cdot\cos\varphi)(1-\sin\varphi)}{c\cdot\cot\varphi}\right]^{\frac{1-\sin\varphi}{2\sin\varphi}}}{W_p} \tag{4}$$

The bearing strength of the pillar depends mainly on the compressive strength of the rock body (obtained by discounting based on the uniaxial compressive strength of the rock) and is related to the shape and width of the pillar. Researchers have proposed a variety of theoretical and empirical formulas for the bearing strength of mining pillars. The Bieniawski strength formula for mining pillars is one of the more widely used. In this study, the Bieniawski formula was used to calculate the strength of the mine pillar as follows:

$$S_p = \sigma_c \left[0.64 + 0.36\left(\frac{W_p}{h}\right)\right]^{\alpha} \tag{5}$$

where S_p is the strength of the ore column, MPa; σ_c is the average compressive strength of the ore column, MPa; h is the height of the ore column, m; α is a constant = 1.0 when the width-to-height ratio of the ore column is less than 5 and 1.4 when the width-to-height ratio of the ore column is greater than 5.

In order to simplify the problem, only the main influences that can be quantified are considered. From Equations (4) and (5), a factor of safety can be derived for rectangular pillars, $K = \sigma_p/S_p$. When $K > 1$, the ore column is in a stable state; when $K > 1$, the ore column is in an unstable state; when $K = 1$, the ore column is in a critical state. When the column is in the critical state, the critical height of the column instability meets the relationship equation:

$$h = \sqrt[\alpha]{\frac{\sigma_c W_p (0.36 W_p)^{\alpha}}{(W_0 + W_p)\gamma R_0 \left[\frac{(\gamma H + c\cdot\cot\varphi)(1-\sin\varphi)}{c\cdot\cot\varphi}\right]^{\frac{1-\sin\varphi}{2\sin\varphi}} - 0.64\sigma_c W_p}} \tag{6}$$

2.4. Numerical Simulation Tests

FLAC 3D (Fast Lagrangian Analysis of Coutinua) uses the "Explicit Lagrangian" algorithm and the "Hybrid-Discrete Partitioning" technique to simulate the plastic damage and flow of materials very accurately. It has a good pre-processing function. When calculating, the program automatically dissects the model into hexahedral cells, and each cell can have its own material model, and the material can be yielded and flowed under the action of external force and stress field. Its post-processing functions are also very powerful. Users can print or plot their own data or graphs with the appropriate commands according to their needs. Therefore, FLAC 3D is one of the most ideal tools for solving the goaf stability problem. In this study, the software was used to simulate and analyze the safety critical heights of the goaf. In addition, the stability of the mining pillars is also discriminated using the Mohr-Coulomb strength criterion [29,30]. The principal stress is the main parameter of rock mechanics after discounting, and its relation is as follows [31,32]:

$$\sigma_3 = \sigma_1 \tan^2\left(45^\circ - \frac{\varphi}{2}\right) - 2c \cdot \tan\left(45^\circ - \frac{\varphi}{2}\right) \tag{7}$$

$$f = \frac{\sigma_3}{\sigma\prime_3} \tag{8}$$

2.4.1. Modeling

The model created in Rhino 6.0 was meshed and imported into Flac 3D, as shown in Figure 2. The length of the model along the strike of the ore body was set to 300 m, the length and width of the one-step mining room were 50 m and 12 m, respectively, and the length and width of the two-step mining pillar were 50 m and 18 m, respectively. To minimize the influence of the boundary conditions on the stresses in the mining operation area, the left and right sides of the back-mining area were enlarged by 150 m of bedrock to ensure that the boundary conditions were applied to the bedrock and were transmitted to the mining area through the bedrock. The height of the model was set to 800 m, and the average thickness of bedrock at the top was set to 400 m to avoid the top self-gravitating stress acting directly on the goaf. The average thickness of bedrock at the bottom was set to 400 m to avoid the bottom boundary condition acting directly on the mining area, which would have caused stress concentration.

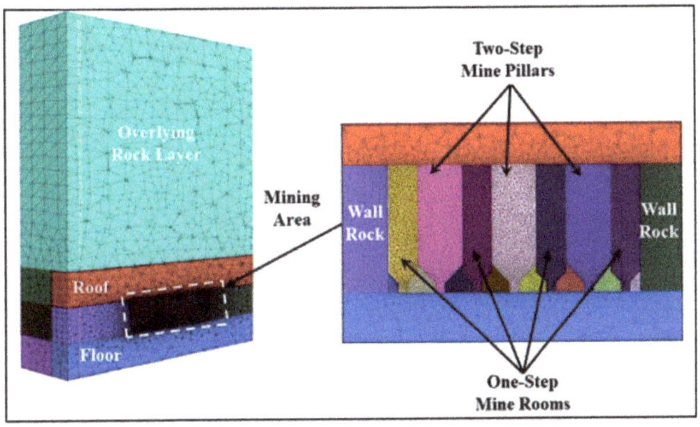

Figure 2. Numerical modeling of the critical safe height of goaf.

Based on the hosting conditions of the ore body, the model boundary was constrained from horizontal displacement, i.e., the boundary displacements in the x and y directions were zero.

2.4.2. Basic Assumption

In view of the complexity and variability of the actual engineering geological environment of the mine and the characteristics of the ore body endowment, the following assumptions need to be made in the numerical simulation process:

(1) The horizontal thickness and inclination of the excavated ore body are fixed values;
(2) The ore body is assumed to be an ideal elastic-plastic body, and the strength and volume of the material do not change with plastic deformation after the yield point;
(3) The ore body and peripheral rock are locally homogeneous and isotropic materials, and plastic deformation does not change the material isotropy;
(4) Considering the brittleness of the rock, all physical quantities involved in the analysis are independent of time;
(5) Strain hardening (or softening) is not considered;
(6) The sides of the model limit horizontal displacements and the bottom limits vertical displacements;
(7) Considering the limitations of the finite element program, it is assumed that there is no influence of tectonic activities in the site, the original rock ground stress is of geostrophic field type, there is an integrated contact between the rock layers, the interior of the rock layer is a continuous medium, the influence of groundwater activities is not taken into account in the model, and the existence and influence of structural surfaces, fissures, and weak layers in the rock layer and ore body are not taken into account;
(8) The loads selected for the calculations will not change with the orientation of the units, but will always maintain their original orientation, with the surface loads acting normal to the surface of the deformed units and can be used to simulate "following" forces, the magnitude of which is the weight of the overlying rock formation;
(9) Simulation of the most hazardous scenario, i.e., mining in intervals.

2.4.3. Parameters

Based on the numerical simulation test scheme shown in Table 1, combined with the results of the rock body mechanical parameter discounting in Table 3 in Section 3.1, the corresponding mechanical parameters were set for the overlying rock layer, the ore body, and the bottom plate of the model.

Table 1. Test Programs.

Program Number	Rock Stability	Compressive Strength (σ_t/MPa)	Width of the Stope One-Step Mine Room (m)	Width of the Stope Two-Step Mine Pillar (m)	Length of the Stope (m)	Height of the Stope (m)
1	Poor	10.93	12	18	50	25
2	Poor	10.93	12	18	50	30
3	Poor	10.93	12	18	50	35
4	Better	22.97	12	18	50	45
5	Better	22.97	12	18	50	50
6	Better	22.97	12	18	50	55

2.4.4. Simulation Schemes

The two-step mining process was used to analyze the critical height of the pillar instability for the two types of areas with better and worse surrounding rock stability. The length and width parameters of the one-step mining room and the two-step pillar were unchanged; only the mining height of the stope was changed. Parameterization of the mining height in the field of values in the vicinity of the theoretically calculated values of the critical safe height of stope in different stability area was conducted. The specific parameters of the mining height are shown in Table 1.

3. Results and Discussion

3.1. Results of Experiments Rock Mechanics Parameter and Their Reduction

Comparison of the results in Tables 2 and 3 reflects that:

Table 2. Experiment results of rock mechanical parameters.

Materials	Compressive Strength (σ_t/MPa)	Tensile Strength (σ_c/Mpa)	Elasticity Modulus (E_m/Gpa)	Density (ρ/t·m^{-3})	Poisson's Ratio (μ)	Cohesion (C_m/Mpa)	Internal Friction Angle (φ_m/°)
Overburden Rock	68.54	8.89	39.29	2.84	0.24	10.31	50.41
Phyllite ore body	33.80	6.28	27.40	2.97	0.34	3.33	43.39
Floor	46.69	6.05	24.63	2.88	0.32	7.14	50.43

Table 3. Summary of discounted rock mechanics parameters results of rock mechanical parameters.

Materials	Compressive Strength (σ_t/Mpa)	Tensile Strength (σ_c/Mpa)	Elasticity Modulus (E_m/Gpa)	Density (ρ/t·m^{-3})	Poisson's Ratio (μ)	Cohesion (C_m/Mpa)	Internal Friction Angle (φ_m/°)
Overburden Rock	29.99	3.89	25.25	2.84	0.24	10.31	50.41
Phyllite ore body (better)	22.97	4.27	24.64	2970	0.34	3.33	43.39
Phyllite ore body (poor)	10.93	2.03	9.69	2970	0.34	3.33	43.39
Floor	18.23	3.75	20.69	2880	0.32	7.14	50.43

The gap between the mechanical strength of rocks before and after discounting is significant. For example, the compressive strength, tensile strength, and elasticity modulus of the Overburden rock are 68.54 Mpa, 8.89 Mpa, and 39.29 Gpa before reduction, and 22.99 Mpa, 3.89 Mpa, and 25.25 Gpa after reduction.

The gap between the same kind of rocks in the areas with different stability is significant after reduction. For example, the compressive strength, tensile strength, and elasticity modulus of the Phyllite ore body are 22.97 Mpa, 4.27 Mpa, and 24.64 Gpa, respectively, in the region with good stability, and 10.93 Mpa, 2.03 Mpa, and 9.69 Gpa, respectively, in the region with poor stability.

Other rock mechanical parameters remain unchanged before and after the reduction.

3.2. Quantitative Analysis of Each Influencing Factor

The analysis of the critical safe height of the goaf showed that the factors affecting the stability of the goaf were the rock body compressive strength, mine room width, pillar width, mining height, bearing capacity of the overlying rock layer, mining depth, and mining radius. Together with the geological conditions and mining technology conditions of the mine, the main influencing factors were identified as the rock body compressive strength, mine room width, mine pillar width, mining height, and mining depth. The sensitivity of each of these factors was analyzed quantitatively using orthogonal extreme difference analysis. The values of the main influencing factors were established in the range of five-factor, five-level-structure orthogonal tests. Table 4 lists the parameters of the critical safe height of the goaf and the results of the orthogonal experiments.

Table 4. Parameters and results of orthogonal experiments.

Number of Tests	Mining Depth (m)	Compressive Strength (MPa)	Width of Mining Room (m)	Width of Mining Pillar (m)	Height of Stope (m)	Critical Height for Pillar Destabilization (m)
1	300	11	10	12	20	20.7
2	300	12	12	14	30	32.1
3	300	13	14	16	40	43.7
4	300	14	16	18	50	55.4
5	300	15	18	20	60	67.2
6	400	11	10	16	50	33.7
7	400	12	12	18	60	45.1
8	400	13	14	20	20	67.7
9	400	14	16	12	30	44.9
10	400	15	18	14	40	16.0
11	500	11	10	20	30	57.3
12	500	12	12	12	40	49.6
13	500	13	14	14	50	39.5
14	500	14	16	16	60	17.1
15	500	15	18	18	20	31.6
16	600	11	10	14	60	42.8
17	600	12	12	16	20	66.6
18	600	13	14	18	30	22.4
19	600	14	16	20	40	33.7
20	600	15	18	12	50	23.5
21	700	11	10	18	40	71.2
25	700	12	12	20	50	23.5
h_1	42.02	43.82	45.14	19.94	31.4	
h_2	40.04	41.48	43.38	29.88	32.68	
h_3	37.16	39.02	38.32	40.52	40.82	
h_4	35.84	37.8	36.82	51.7	45.14	
h_5	36.24	37.8	36.26	57.88	49.88	
R	6.02	8.88	37.94	18.48	8.80	

Based on the results of the orthogonal experiment polar analysis shown in Table 4, the extreme difference R of the rock body discounted compressive strength, and the width of the pillar were 37.94 m and 18.48 m, respectively. The extreme difference of the width of the mine room, the height of the stope, and the depth of the mining were 8.88 m, 8.80 m, and 6.02 m, respectively. The results show that the critical height of the pillar instability in this copper mine results in the most significant factor being the rock body discounted compressive strength, followed by the width of the pillar, the width of the mine room, and the height of the stope. It is worth noting that the width of the mine room and the height of the stope have almost the same level of influence on the critical height of pillar instability. Therefore, the structural parameters of the stope should be adjusted according to the geological conditions of the mine in the process of ore body recovery. The width of the ore column should be increased as much as possible to ensure that the critical height of the ore column instability is greater than the height of the stope while ensuring the maximum width permitted for the limited exposure area of the empty area and the minimum width permitted for the self-supporting height of the filling body of the one-step stope.

One factor at a time was studied to analyze the quantitative relationship between the critical safe height of the goaf and the main influencing factors. Four model forms, a linear function, exponential function, power function, and logarithmic function, were used for fitting. The initial values of each influencing factor were set as follows: γ = 2.97 kN·m^{-3}, H = 400 m, W_0 = 12 m, W_p = 18 m, σ_c = 10.93 MPa, C = 3.33 MPa, $L = 3(W_0 + W_p)$ = 90 m, H_1 = 50 m, and φ = 43.39°. The fitting results are shown in Table 5.

Table 5. Functional Relationship Fitting Results.

Main Influencing Factors	Fitting Methods	Fitting Formula	Correlation Coefficient
Depth of Mining	Linear Function	$y = -0.0097x + 29.91$	0.98279
	Exponential Function	$y = 30.3933e^{-0.0004x}$	0.98896
	Power Function	$y = 77.0907x^{-0.1823}$	0.99827
	Logarithmic Function	$y = 56.9144 - 5.0834\ln(x + 45.6918)$	0.99976
Width of Mine Room	Linear Function	$y = -1.42x + 43.0000$	0.99683
	Exponential Function	$y = 51.9860 \times 0.9436^x$	0.99962
	Power Function	$y = 163.8241x^{-0.7431}$	0.99853
	Logarithmic Function	$y = 68.4447 - 17.482\ln(x - 0.5964)$	0.99983
Compressive Strength	Linear Function	$y = 2.404x - 1.06$	0.98682
	Exponential Function	$y = 15.7814e^{0.0514x}$	0.98707
	Power Function	$y = 2.0802x^{1.0394}$	0.98765
	Logarithmic Function	$y = -14343.2765 + 2228.6499\ln(x + 611.5214)$	0.87614
Width of Mine Pillar	Linear Function	$y = 1.45x - 0.3$	0.99778
	Exponential Function	$y = 8.27504 \times 1.0646^x$	0.98247
	Power Function	$y = 1.3922x^{1.0099}$	0.99769
	Logarithmic Function	$y = -119.4666 + 42.2206\ln(x - 13.2732)$	0.99999
Height of the Stope	Linear Function	$y = -0.108x + 31.24$	0.98822
	Exponential Function	$y = 31.5312e^{-0.0039x}$	0.98161
	Power Function	$y = 44.4409x^{-0.1389}$	0.90038
	Logarithmic Function	$y = 755.4333 + 105.7652\ln(x + 940.5610)$	0.98000

1. The highest correlation coefficient for the function fit between the mining depth and the critical safe height of the goaf was 0.99976 for the logarithmic function fit, i.e., it follows the decreasing law of the logarithmic function $y = a_1 + b_1\ln(x + c_1)$ (in which a_1, b_1, and c_1 mainly depend on the mining depth, the rock body compressive strength, and other factors). Figure 3a shows the logarithmic function of the mining depth and the critical safe height of the goaf fitting curve. With increasing mining

depth, the critical safety height of the goaf followed a logarithmic function with the rate of reduction gradually decreasing.

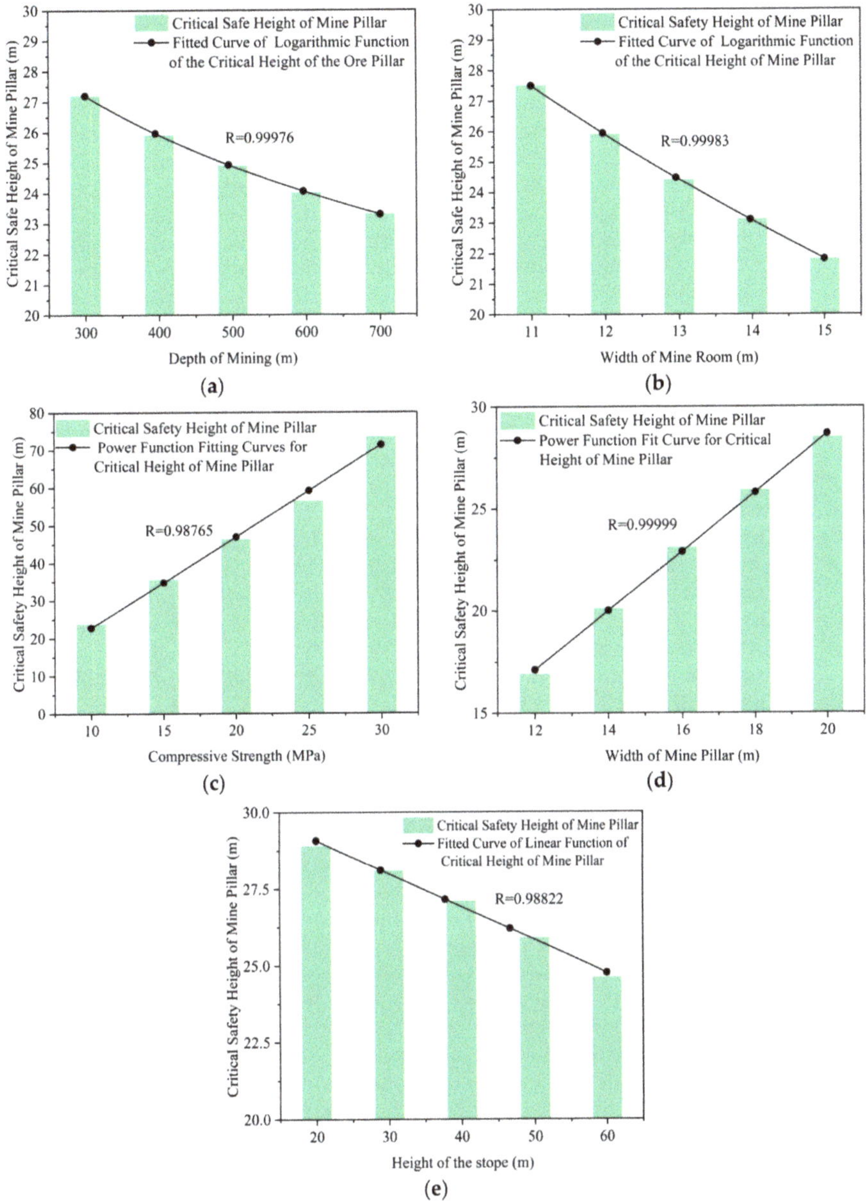

Figure 3. Function fitting curve. (**a**) Fitted curve of the logarithmic function of the critical safe height of mine pillar and the mining depth; (**b**) Fitted curve of the logarithmic function of the critical safe height of mine and the width of mine room; (**c**) Fitted curve of the logarithmic function of the critical safety height of mine pillar and the compressive strength of the rock body; (**d**) Fitted curve of the logarithmic function of the critical safe height of mine pillar and the width of mine pillar; (**e**) Fitted curve of logarithmic function of the critical safe height of mine pillar and the height of stope.

2. The highest correlation coefficient between the width of the mining room and the critical safety height of the goaf function fit was the logarithmic function fit of 0.99983, that is, following a logarithmic function of the form $y = a_2 - b_2 \ln(x + c_2)$, where a_2, b_2, and c_2 mainly depend on the width of the mining room, the rock body compressive strength, and other factors. Figure 3b shows the fitting curve of the logarithmic function between the width of the mine room and the critical safe height of the goaf. As the width of the mine room increases, the critical safe height of the goaf decreases according to a logarithmic function with a gradually decreasing rate of decrease.
3. The highest correlation coefficient between the rock compressive strength and the critical safety height of the goaf was 0.98765 for the power function fit, that is, following a power function $y = a_3 x^{b_3}$, in which a_3 and b_3 mainly depend on the rock compressive strength and other factors. Figure 3c shows the rock body compressive strength and the critical safe height of the goaf power function fitting curve. As the rock body compressive strength increases, the critical safe height of the goaf increases gradually according to a power function, with the rate of increase gradually increasing.
4. The highest correlation coefficient between the width of the mine pillar and the critical safety height of the goaf function fit as the logarithmic function fit of 0.99999, that is, following the logarithmic function $y = a_4 - b_4 \ln(x + c_4)$ of the incremental law (where a_4, b_4, and c_4 mainly depend on the width of the mine pillar, the rock body compressive strength, and other factors). Figure 3d shows the logarithmic function fitting curve of the mine pillar width and critical safety height of the goaf. As the mine pillar width increases, the critical safe height of the goaf increases gradually according to a logarithmic function, with the rate of increase gradually increasing.
5. The highest correlation coefficient of the function fit between the mining height and the critical safety height of the goaf was 0.98822, following a decreasing linear function of the form $y = a_5 + b_5 x$ (in which a_5 and b_5 mainly depend on the mining height and other factors). Figure 3e shows the fitting curve of the logarithmic function between the mining height and the critical safe height of the goaf. As the mining height increases, the critical safe height of the goaf decreases linearly.

3.3. Theoretical Analysis of Project Examples

The results in Table 2 show a large gap between the results of rock body compressive strength discounting in different areas of the mine. The mine is strictly in accordance with the established structural parameters, i.e., the one-step mining room being 12 m, the two-step mining pillar being 18 m divided into blocks, and the existing middle section of the division of the stopes being carried out in accordance with engineering construction specifications. Therefore, the critical safe height of the goaf depends only on the rock body compressive strength and the mining height. The calculation results are shown in Figure 4.

According to the results shown in Figure 4a, in the section with poor stability, when the design mining height is in the range of 10–70 m, the critical safe height of the goaf is between 20 and 30 m. When the design mining height is 25 m, the critical safe heights of the goaf are 29.08 m (−378 m middle section) and 27.84 m (−478 m middle section). When the design mining height is 30 m, the critical safe heights of the goaf are 28.63 m (−378 m middle section) and 27.41 m (−478 m middle section). Therefore, in order to achieve higher recycling efficiency with the goal of guaranteeing safety, the height of the stope should not exceed 28 m (in the −378 m middle section) or 27 m (in the −478 m middle section).

According to the results shown in Figure 4b, for the section with better stability, when the design mining height is in the range of 10–70 m, the critical safe height of the goaf is between 45 and 65 m. When the design mining height is 50 m, the critical safe heights of the goaf are 55.43 m (−378 m middle section) and 53.08 m (−478 m middle section). When the design mining height is 55 m, the critical safe heights of the stope are 54.11 m (−378 m middle section) and 51.82 m (−478 m middle section). Therefore, in order to achieve higher recovery efficiency with the goal of guaranteeing safety, the heights of the goaf should not exceed 54 m (−378 m middle section) and 51 m (−478 m middle section).

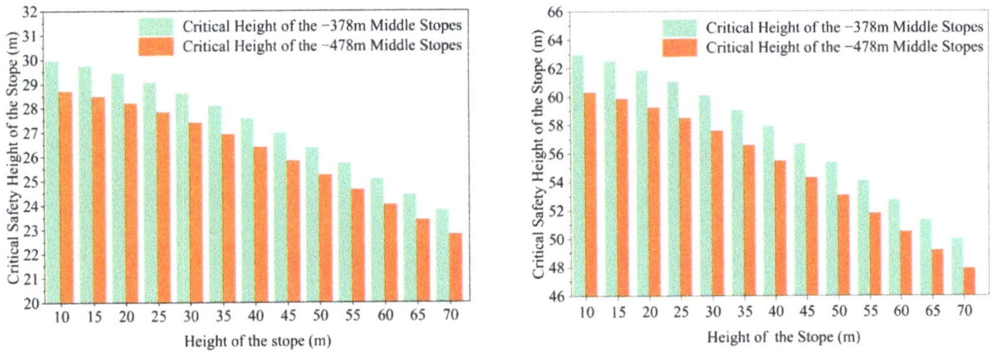

(**a**) (**b**)

Figure 4. Correspondence between the design mining height and mining depth of stopes in different regions and the safety critical height of goaf. (**a**) Correspondence between the design mining height of the stopes and the safety critical height of the goaf in the area of poor rock stability in the middle section of −378 m and the middle section of −478 m; (**b**) Correspondence between the design mining height of the stopes and the safety critical height of the goaf in the area of better rock stability in the middle section of −378 m and the middle section of −478 m.

3.4. Numerical Simulation Testing of Project Examples

The analysis of the results is aimed at the empty field state in which the back mining of the one-step mining room is completed and unfilled. Figure 5 shows maps of the stress and displacement during the mining process obtained by taking the example of the mining site in the area with better rock stability and the designed height of the mining site of 50 m; the mining site in the area with poorer rock stability and the designed height of the mining site of 30 m. The maximum compressive stress, maximum tensile stress, maximum displacement, and maximum compressive stress safety coefficient within the roof and sidewall (two-step mining pillars) of the one-step mining room were counted in Figure 6 under different simulation scenarios, respectively.

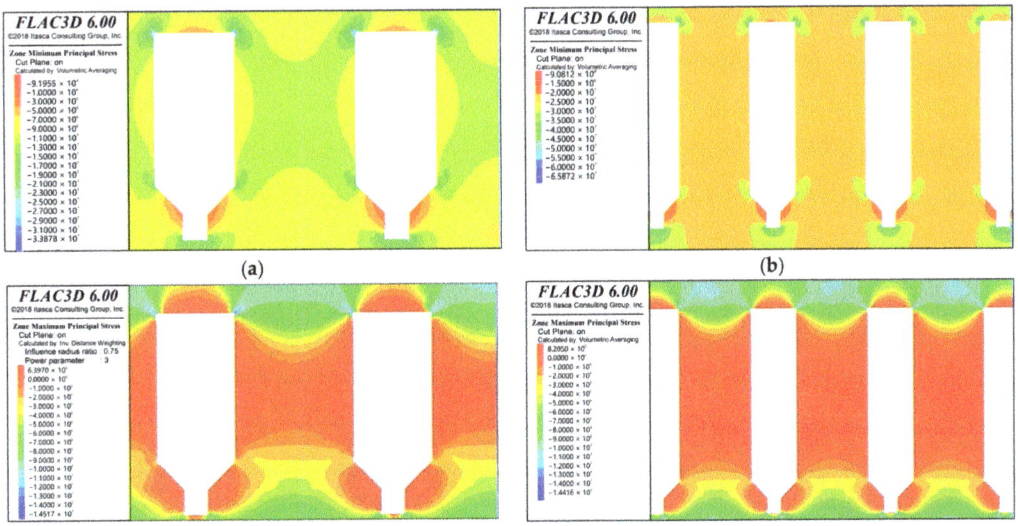

(**a**) (**b**) (**c**) (**d**)

Figure 5. *Cont.*

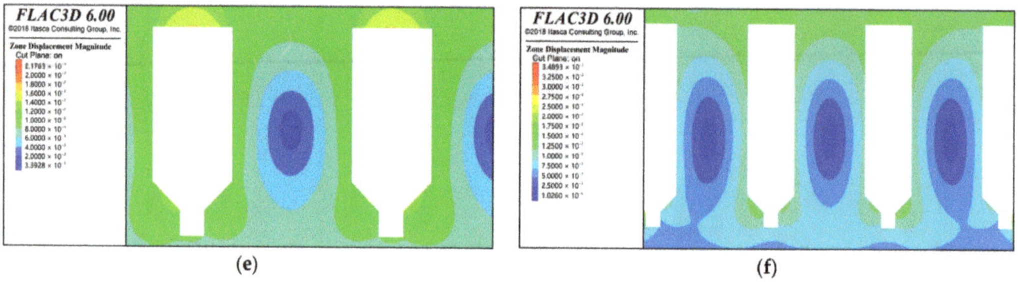

Figure 5. Stress and displacement maps of the mining process. (**a**) Compressive stress cloud at 30 m height of the goaf; (**b**) Compressive stress cloud at 50 m height of the goaf; (**c**) Tensile stress cloud at 30 m height of the goaf; (**d**) Tensile stress cloud at 50 m height of the goaf; (**e**) Displacement cloud at 30 m height of the goaf; (**f**) Displacement cloud at 50 m height of the goaf.

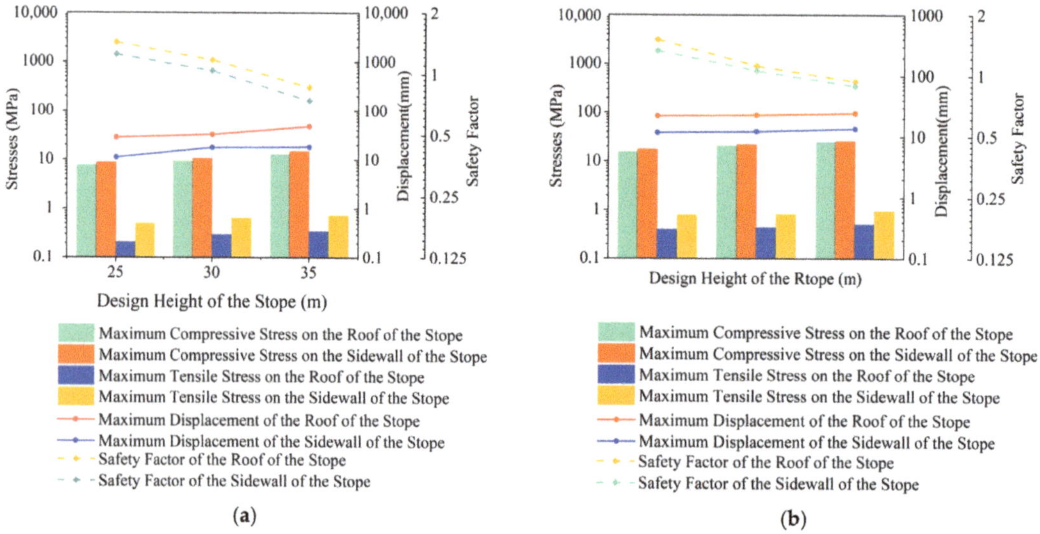

Figure 6. Trends in the results of the numerical simulation schemes. (**a**) Trends in stress, displacement and factor of safety in areas of poor rock stability; (**b**) Trends in stress, displacement and factor of safety in areas of better rock stability.

3.4.1. Compressive Stress Analysis

From Figures 5a,b and 6a, it can be seen that the compressive stress distribution of both the roof of the goaf and the sidewall are wider, and the values are generally larger. This is due to the stress release at the mining location after mining back and the stress concentration in the surrounding rock around the mining area. It is worth noting that the stress concentration phenomenon occurs at the junction of two or three faces. Therefore, the corners of the goaf in the actual mining process are generally curved instead of forming a right angle as in the design.

Under the conditions of the same rock stability level and the same height of the goaf, the pressure of the sidewall in the goaf is slightly larger than the pressure of the roof, but the difference is not large. For example, in the area of poor rock stability, under the condition that the height of the stope is 30 m, and the maximum compressive stresses of the side gang and the roof plate are 10.51 MPa and 9.36 MPa, respectively. Under the condition of the same rock stability level, as the height of the goaf increases, the compressive stresses on the roof plate and sidewalls of the goaf increase gradually. For example, in the area of poor stability, when the height of the goaf is increased from 25 m to 35 m, the maximum compressive stress of the roof plate of the goaf is increased from 7.65 MPa to 12.77 MPa, and the maximum compressive stress of the sidewalls is increased from 8.74 MPa to 14.85 MPa. Therefore, when the stability of the surrounding rock is poor, it is appropriate to reduce the height of the goaf to ensure the safety of the back mining.

3.4.2. Tensile Stress Analysis

From Figures 5a,b and 6b, it can be seen that the tensile stresses on the roof and sidewalls of the goaf are distributed over a wide range and that the maximum tensile stresses on the sidewalls run through the entire two-step pillar. On the one hand, the tensile stress is dispersed from the center of the pillar to the left and right sides in a symmetric trumpet shape, and the magnitude of the tensile stress is unchanged. On the other hand, it decreases gradually from the center of the pillar to the upper and lower ends and decreases gradually.

For a given rock stability level and mining height, the tensile stress of the sidewalls in the mining area is slightly larger than that of the roof, but they are not very different. For example, in the better stability area, the maximum tensile stresses of the sidewalls and top plate are 0.82 MPa and 0.84 Mpa, respectively, when the height of the stope is 50 m.

For the same rock stability grading conditions, as the height of the stope increases, the tensile stresses on the roof plate and sidewall of the goaf are gradually increased. For example, in the area of better rock stability, when the height of the stope increases from 45 m to 55 m, the maximum tensile stress of the roof plate of the stope increases from 0.4 MPa to 0.51 MPa, and the maximum tensile stress of the sidewall increases from 0.78 MPa to 0.95 MPa. Thus, at greater goaf heights, the tensile stresses of the rock around the mining area are much smaller than the ultimate tensile strength of the rock. Therefore, the maximum compressive stress should be mainly considered in assessing the potential for extrusion damage to the top plate and sidewall.

3.4.3. Displacement Analysis

Figure 6a,b show that the displacements of the roof plate and sidewalls of the stope have large distribution ranges, mainly concentrated in the upper part of the mined area, with a funnel-shaped distribution. In particular, the maximum displacement generally occurs in the top plate of the mining hollow area, mainly due to the loss of support for the top plate in the hollow area, which ultimately makes the top plate appear larger deformation, and the side gang deformation is smaller.

Under the same rock stability grading conditions, as the height of the stope increases, the displacement of the roof plate of the stope gradually increases. For example, in the area of poor stability, when the height of the stope increases from 25 m to 35 m, the maximum displacement of the roof plate of the stope increases from 28.23 mm to 47.52 mm. In the area of good stability, when the height of the stope increases from 45 m to 55 m, the maximum displacement of the roof plate of the stope increases from 21.85 mm to 24.36 mm. It is worth noting that the maximum displacement of the stope roof in the less stable area is greater than the maximum displacement of the stope roof in the more stable area. The goaf roof displacement is strongly influenced by the rock stability level.

In summary, although the height of the stope and the stability of the regional rock body in the tensile stress have a certain impact on the roof of the stope and the sidewall, they are not beyond the ultimate tensile strength of the rock body. In terms of displacement, the main influence is the deformation of the roof of the stope; the amount of deformation is relatively small. Therefore, in order to determine the reasonable height of the stope accurately, it is also necessary to determine the maximum compressive stress safety coefficient from the goaf roof and sidewall. As Figure 6a,b show, when the rock stability is poor, the height of the stope is 30 m. The roof plate and sidewall safety coefficients are 1.17 and 1.04, so the height of the stope in this section should not be more than 30 m. In the region of good rock stability, the height of the stope is 50 m, and the top plate and sidewall safety coefficients are 1.12 and 1.06, so the height of the stope in the section should not be more than 50 m.

4. Industrial Tests

4.1. Overview of the Experimental Stope

Three stopes, numbered 1, 2, and 3, were selected for industrial testing. The #1 test stope was located between exploration lines 4 and 5 of the N1 ore body in the middle section of −478 m, the #2 test stope was located between exploration lines 6 and 7 of the N3 ore body in the middle section of −378 m, and the #3 test stope was located between exploration lines 8 and 9 of the S1 ore body in the middle section of −378 m. The field investigation and rock stability evaluation results show that the #1 test stope and #2 test stope area rock body is poor and stability is poor; the #3 test stope area rock body is better.

Each test stope adopts the method of filling the empty field after the rock drilling stage, and does not change the width of the one-step mining room and the two-step mining pillar. The length of the stope is arranged according to the standard stope. However, according to the different geological conditions and mining conditions of the areas where the test stopes are located, different heights of the stopes are designed. Specifically, the height of the #1 test stope was 25 m, the height of the #2 test stope was 30 m, and the height of the #3 test stope was 50 m.

4.2. Results of Industrial Experiments

Immediately after the completion of the last ore discharge from each test stope, the GOSLAM RS100 scanner was used to scan the voided area, a 3D voided area scanning model was established, and the scanning boundary profile of the voided area was plotted, as shown in Figure 7. In order to more intuitively reflect the collapse of the hollow zone, the volume of the hollow zone was compared with the design volume, as shown in Table 6.

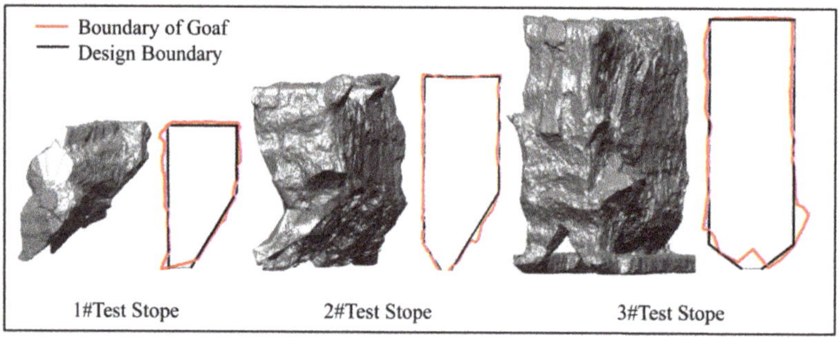

Figure 7. Goaf model and boundary profile.

Table 6. Statistical analysis of empty space volume in test stope.

Test Stope	Design Parameters of the Stope Length (m)	Width (m)	Height (m)	Volume (V_1/m^3)	Volume of the 3D Model of the Goaf(V_2/m^3)	Deviation Factor * (f/%)
1#	50	12	25	12,511	12,787	2.21
2#	50	12	30	15,405.5	15,752	2.25
3#	50	12	50	27,292.5	27,665.5	1.37

Note: Deviation Factor * $f = \frac{V_2 - V_1}{V_1} \times 100\%$. If the stope deviation "f" factor is "+", it indicates that the goaf is generally over-excavated; conversely, if the goaf deviation factor "f" is "−", it indicates that the stope is generally under-excavated.

The over-excavation amounts for test stopes 1, 2, and 3 were 2.21%, 2.25%, and 1.37%, respectively, which are relatively small amounts. The over-excavation was mainly concentrated in the corners of the roof slab of the mining area and the bottom structure of the stope, which may be affected by many factors such as stress concentration, skewed boreholes, fault development, poor charging accuracy, and blasting construction. The actual mining operation in the two-step mining room increases the difficulty of retaining the shoring wall pillars, the ore depletion rate, and the difficulty of constructing the mine access road.

On the whole, the actual boundaries of the three test stopes under the empty field condition are basically consistent with the design boundaries, and there is no obvious instability and collapse phenomenon. The industrial test results show that the mine does not change the one-step and two-step width of the stope according to the stability of the regional rock body and mining depth. The height of the stopes, adjusted for the stope safety below the critical height of the stope, can effectively reduce the phenomenon of instability and collapse of the stope. Specifically, in the middle section of −378 m, the height of the stope in the area of better rock stability is 50 m, and the height of the stope in the area of poorer rock stability is 30 m; in the middle section of −478 m, the height of the stope in the area of better rock stability is 50 m, and the height of the stope in the area of poor rock stability is 25 m.

5. Conclusions

In this study, a formula for the safe critical goaf height for the stage airfield filling mining method was derived based on rock body strength discounting, Pu's arch equilibrium theory, and the ultimate strength formula of the Bieniawski pillar. The main influencing factors were analyzed by means of an extreme quadrature experiment. In addition, the reliability of the theoretical calculation results was verified by numerical simulation tests and industrial tests. Based on this study, the following conclusions can be drawn.

(1) According to a comprehensive analysis of polar–orthogonal experiments, among the main factors affecting the critical safety height of the goaf, the impact of the compressive strength of the rock body is the most significant; followed by the width of the two-step mining pillar; and finally, the height of the stope, the depth of mining, and the width of the one-step mining room, the effect of which are almost the same. The effects of the mining depth, the width of the one-step mining room, and the critical safe height of the goaf follow decreasing logarithmic functions; the effects of the compressive strength of the rock body and the critical safe height of the goaf follow increasing power functions; the effects of the width of the two-step mining pillar and the critical safe height of the goaf follow increasing logarithmic functions; and the height of the goaf and the critical safe height of the goaf follow decreasing linear functions.

(2) Based on the geological conditions of the mine and the current mining situation, the permissible critical height of mining safety was calculated for each area of the mine. The calculation results show that: in the area with poor rock stability, the heights of the stope should not exceed 28 m (−378 m middle section) and 27 m (−478 m middle section); in the area with better rock stability, the heights of the stope should not exceed 54 m (−378 m middle section) and 51 m (−478 m middle section).

(3) Numerical simulation tests were carried out using FLAC 3D, and the results show that the stope is significantly affected by compressive stress and increases with increasing stope height; the tensile stress does not exceed the ultimate tensile strength of the rock body; and the displacement and deformation are relatively small and do not exhibit sudden changes. In the area of poor rock stability, when the height of the stope is 25 m, 30 m, and 35 m, respectively, the safety coefficients of the roof plate in the goaf is 1.43, 1.17, and 0.86; the safety coefficients of the sidewall in the goaf is 1.25, 1.04, and 0.74. In the area of better rock stability, when the height of the stope is 45 m, 50 m, and 55 m, respectively, the safety coefficients f of the roof plate in the goaf is 1.51, 1.12, and 0.94; the safety coefficients f of the r sidewall in the goaf is 1.33, 1.06, and 0.89. The former heights are slightly greater than the results suggested by the theoretical calculations, and the latter are basically the same as the theoretical calculation results.

(4) The industrial test results show that the three test stopes are in a state of over-excavation but that the amount of over-excavation is relatively small, within the range of 1%–3%. The actual boundary of the open area is relatively regular and is basically consistent with the design boundary. On the whole, the boundary meets the basic requirements of safe production and does not affect two-step pillar mining.

The method provides a reference and a practical basis for determining a reasonable stage height in similar metal mines using the empty-space subsequent filling mining method. It is worth noting that the stope length does not contribute to the use of the formula in the empty-space subsequent filling mining method. Therefore, in the future, the stope length can be included as a key factor in the formula to adapt to different mining methods. Meanwhile, in addition to the influencing factors considered in this study, with the increase of the mining depth, the change of the plastic zone in the upper part of the goaf, the effect of the backfilling body or goaf group in the upper part of the goaf, and the vibrational inertial force generated by the blasting vibration on the stability of the goaf are the main aspects to be considered to improve its accuracy and predictive ability.

Author Contributions: Writing—original draft preparation, Q.Z. and P.Z.; data curation, P.Z. and H.L. conceptualization, Q.Z. and P.Z.; methodology, Q.C., H.L. and Z.S.; software, Q.Z., Z.S. and Y.T.; validation, P.Z. and Q.C.; writing—review and editing, Q.Z. and P.Z.; supervision, Q.C.; funding acquisition, Q.Z. All authors have read and agreed to the published version of the manuscript.

Funding: This research was funded by the National Natural Science Foundation of China grant number 52274151 the Science and Technology Innovation Program of Hunan Province grant number 2021RC3125,and the APC was funded by Qiusong Chen.

Data Availability Statement: All data are shown in the article.

Acknowledgments: The authors also would like to thank Jiaduo Ding, and Zhiyue Lin for their support with industrial tests during the research process.

Conflicts of Interest: Author H.L. was employed by the Jiangxi Copper Industry Group Yinshan Mining Industry Co., Ltd. The remaining authors declare that the research was conducted in the absence of any commercial or financial relationships that could be construed as a potential conflict of interest.

References

1. Chen, Q.; Zhou, H.; Wang, Y.; Wang, D.; Zhang, Q.; Liu, Y. Erosion wear at the bend of pipe during tailings slurry transportation: Numerical study considering inlet velocity, particle size and bend angle. *Int. J. Miner. Metall. Mater.* **2023**, *30*, 1608–1620. [CrossRef]
2. Guo, L.; Wu, Y.; Zhang, Q.; Chen, Q. Stability evaluation of layered backfill considering filling interval, backfill strength and creep behavior. *Minerals* **2022**, *12*, 271. [CrossRef]
3. Liu, Y.; Wang, Y.; Chen, Q. Using cemented paste backfill to tackle the phosphogypsum stockpile in china: A down-to-earth technology with new vitalities in pollutants retention and CO_2 abatement. *Int. J. Miner. Metall. Mater.* **2023**. Available online: http://ijmmm.ustb.edu.cn/en/article/id/e87e96bf-4935-4a0f-bbf2-0033535b0b9b (accessed on 1 December 2023).
4. Liu, Y.; Molinari, S.; Dalconi, M.C.; Valentini, L.; Bellotto, M.P.; Ferrari, G.; Pellay, R.; Rilievo, G.; Vianello, F.; Salviulo, G.; et al. Mechanistic insights into pb and sulfates retention in ordinary portland cement and aluminous cement: Assessing the contributions from binders and solid waste. *Hazard. Mater.* **2023**, *458*, 131849. [CrossRef]
5. Wang, R.; Zeng, F.; Li, L. Stability analyses of side-exposed backfill considering mine depth and extraction of adjacent stope. *Int. J. Rock. Mech. Min. Sci.* **2021**, *142*, 104735. [CrossRef]
6. Qi, C.; Wu, M.; Liu, H.; Liang, Y.; Liu, X.; Lin, Z. Machine learning exploration of the mobility and environmental assessment of toxic elements in mining-associated solid wastes. *J. Clean. Prod.* **2023**, *401*, 136771. [CrossRef]
7. Wu, M.; Qi, C.; Derrible, S.; Choi, Y.; Fourie, A.; Ok, Y.S. Regional and global hotspots of arsenic contamination of topsoil identified by deep learning. *Commun. Earth Environ.* **2024**, *5*, 10. [CrossRef]
8. Yuan-hui, L.; Gang, L.; Shi-da, X.; Da-wei, W. The spatial-temporal evolution law of microseismic activities in the failure process of deep rock masses. *J. Appl. Geophys.* **2018**, *154*, 1–10. [CrossRef]
9. Golik, V.I.; Klyuev, R.V.; Martyushev, N.V.; Kondratiev, V.V.; Tynchenko, V.S.; Gladkikh, V.A.; Iushkova, L.V.; Brigida, V. Reuse and Mechanochemical Processing of Ore Dressing Tailings Used for Extracting Pb and Zn. *Materials* **2023**, *16*, 7004. [CrossRef] [PubMed]

10. Uzarowicz, Ł.; Wolińska, A.; Błońska, E.; Szafranek-Nakonieczna, A.; Kuźniar, A.; Słodczyk, Z.; Kwasowski, W. Technogenic soils (technosols) developed from mine spoils containing fe sulphides: Microbiological activity as an indicator of soil development following land reclamation. *Appl. Soil. Ecol.* **2020**, *156*, 103699. [CrossRef]
11. Kongar-Syuryun, C.B.; Kovalski, E.R. Hardening backfill at potash mines: Promising materials regulating stress-strain behavior of rock mass. *Geol. I Geofiz. Yuga Ross. Geol. Geophys. Russ. South.* **2023**, *13*, 177–187. [CrossRef]
12. Kovalski, E.R.; Kongar-Syuryun, C.B.; Petrov, D.N. Challenges and prospects for several-stage stoping in potash minining. *Sustain. Dev. Mt. Territ.* **2023**, *15*, 349–364. [CrossRef]
13. Khayrutdinov, M.M.; Golik, V.I.; Aleksakhin, A.V.; Trushina, E.V.; Lazareva, N.V.; Aleksakhina, Y.V. Proposal of an Algorithm for Choice of a Development System for Operational and Environmental Safety in Mining. *Resources* **2022**, *11*, 88. [CrossRef]
14. Ganapathy, G.P.; Zaalishvili, V.B.; Chandrasekaran, S.S.; Melkov, D.A. Integrated monitoring of slope process in India and Russia. *Sustain. Dev. Mt. Territ.* **2020**, *12*, 572–581. [CrossRef]
15. Driouch, A.; Ouadif, L.; Lahmili, A.; Belmi, M.A.; Benjmel, K. Geotechnical modeling of the method for mining cobalt deposits at the Bou Azzer Mine, Morocco. *Min. Miner. Depos.* **2023**, *17*, 51–58. [CrossRef]
16. Obert, L.; Duvall, W.I. *Rock Mechanics and Design of Structures in Rock*; Jonh Wiley & Sons: New York, NY, USA, 1967.
17. Jiang, L.; Yang, C.; Jiao, H. Ultimately exposed roof area prediction of bauxite deposit goaf based on macro joint damage. *Int. J. Min. Sci. Technol.* **2020**, *30*, 699–704. [CrossRef]
18. Swift, G.; Reddish, D. Stability problems associated with an abandoned ironstone mine. *Bull. Eng. Geol. Environ.* **2002**, *6*, 227–239. [CrossRef]
19. Peng, G.; Gaoyi, D.; Jingsong, C.; Zhou, C.; Manqing, L.; Weizhong, Z.; Yang, S. Study on optimization of stope structural parameters and filling scheme of wawu phosphate mine in yichang city, china. *Front. Earth Sci.* **2022**, *10*, 883572. [CrossRef]
20. Zhienbayev, A.; Balpanova, M.; Asanova, Z.; Zharaspaev, M.; Nurkasyn, R.; Zhakupov, B. Analysis of the roof span stability in terms of room-and-pillar system of ore deposit mining. *Min. Miner. Depos.* **2023**, *17*, 129–137. [CrossRef]
21. Hosseini, M.; Azhari, A.; Lotfi, R.; Baghbanan, A. Safety analysis of Sormeh underground mine to improve sublevel stoping stability. *Deep. Undergr. Sci. Eng.* **2023**, *2*, 173–187. [CrossRef]
22. Bagde, M.; Sangode, A.; Jhanwar, J. Evaluation of Stoping Parameters through Instrumentation and Numerical Modelling in Manganese Mine in India: A Case Study. *Procedia Eng.* **2017**, *191*, 10–19. [CrossRef]
23. Soni, A.; Monsalve, J.J.; Bishop, R.; Ripepi, N. Modified design of pillar based on estimated stresses and strength of pillar in an underground limestone mine. *Min. Metall. Explor.* **2023**, *40*, 529–541. [CrossRef]
24. Zhang, Y.F.; Ni, P.P. Design optimization of room and pillar mines: A case study of the xianglushan tungsten mine. *Q. J. Eng. Geol. Hydrogeol.* **2018**, *51*, 352–364. [CrossRef]
25. Qiu, H.Y.; Huang, M.Q.; Weng, Y.-J. Stability evaluation and structural parameters optimization of stope based on area bearing theory. *Minerals* **2022**, *12*, 808. [CrossRef]
26. Bieniawski, Z.T. A method Revisited: Coal Pillar Strength Formula Based on Field Investigations. In *Proceedings of the Workshop on Coal Pillar Mechanics and Design*; US Department of the Interior, US Bureau of Mines: Washington, DC, USA, 1992; pp. 158–165.
27. Fan, H.; Li, L.; Liu, H.; Hu, J.; Zhang, M.; Zhou, S.; Yang, G. Improvement to the calculating model of the pressure arch's height considering the confining pressure in the excavation of shallow tunnels. *Arab. J. Geosci.* **2021**, *14*, 1130. [CrossRef]
28. Kastner, H. *Statik des Tunnel-Und Stollenbaues auf Der Grundlagen Geomechanischer Erkenntnisse*; Springer: Berlin/Heidelberg, Germany, 2013.
29. Tu, H.; Zhou, H.; Lu, J.; Gao, Y.; Shi, L. Elastoplastic coupling analysis of high-strength concrete based on tests and the Mohr-Coulomb criterion. *Constr. Build. Mater.* **2020**, *255*, 119375. [CrossRef]
30. Zhang, J.; Wang, Z.; Song, Z. Numerical study on movement of dynamic strata in combined open-pit and underground mining based on similar material simulation experiment. *Arab. J. Geosci.* **2020**, *13*, 785. [CrossRef]
31. Liu, B.; Zhou, J.; Wen, X.; Guo, J.; Deng, Z.; Hu, X. Mechanical performance and failure criterion of coral concrete under combined compression-shear stresses. *Constr. Build. Mater.* **2021**, *288*, 123050. [CrossRef]
32. Wang, J.; Ren, L.; Xie, L.Z.; Xie, H.P.; Ai, T. Maximum mean principal stress criterion for three-dimensional brittle fracture. *Int. J. Solids Struct.* **2016**, *102–103*, 142–154. [CrossRef]

toxics

MDPI

Article

A Deep Learning Approach for Chromium Detection and Characterization from Soil Hyperspectral Data

Chundi Ma [1], Xinhang Xu [1], Min Zhou [1], Tao Hu [1] and Chongchong Qi [1,2,3,*]

[1] School of Resources and Safety Engineering, Central South University, Changsha 410083, China; machundi456@csu.edu.cn (C.M.); x578817460@163.com (X.X.); 225512080@csu.edu.cn (M.Z.); 225512073@csu.edu.cn (T.H.)
[2] School of Metallurgy and Environment, Central South University, Changsha 410083, China
[3] Fankou Lean-Zinc Mine, NONFEMET, Shaoguan 511100, China
* Correspondence: chongchong.qi@csu.edu.cn

Abstract: High levels of chromium (Cr) in soil pose a significant threat to both humans and the environment. Laboratory-based chemical analysis methods for Cr are time consuming and expensive; thus, there is an urgent need for a more efficient method for detecting Cr in soil. In this study, a deep neural network (DNN) approach was applied to the Land Use and Cover Area frame Survey (LUCAS) dataset to develop a hyperspectral soil Cr content prediction model with good generalizability and accuracy. The optimal DNN model was constructed by optimizing the spectral preprocessing methods and DNN hyperparameters, which achieved good predictive performance for Cr detection, with a correlation coefficient value of 0.79 on the testing set. Four important hyperspectral bands with strong Cr sensitivity (400–439, 1364–1422, 1862–1934, and 2158–2499 nm) were identified by permutation importance and local interpretable model-agnostic explanations. Soil iron oxide and clay mineral content were found to be important factors influencing soil Cr content. The findings of this study provide a feasible method for rapidly determining soil Cr content from hyperspectral data, which can be further refined and applied to large-scale Cr detection in the future.

Keywords: soil hyperspectral; deep learning; chromium; sensitive bands

1. Introduction

Chromium (Cr) contamination is a major global environmental problem [1]. With increasing urbanization, there has been a pronounced rise in industrial waste discharge from mining, metallurgy, and electronic equipment manufacturing industries, which has resulted in the continuous accumulation of Cr in the natural environment. For example, indiscriminate waste discharge from tanneries in India's Vellore district resulted in more than 65% of tested soil samples exceeding the maximum permissible Cr content limits established by environmental protection agencies [2–5]. Given the non-degradable and highly toxic nature of Cr, it poses a significant threat to the ecological environment [6,7].

Cr occurs stably in nature [8]. Cr can accumulate in living organisms by contaminating soil and water resources and then biomagnifying through the food chain [9,10]. In plants, high Cr content can damage cells and inhibit growth [11], while in humans, Cr can cause skin system, liver, and kidney dysfunction and even cancer [12,13]. Different countries have set different soil Cr thresholds (e.g., 100 mg/kg for Romania and 90 mg/kg for China), which shows that soil Cr content is a key environmental risk factor [14,15]. Therefore, detecting Cr content in soil is essential for evaluating soil Cr contamination and subsequent soil remediation.

Although it is one of the most widely used techniques for soil Cr detection, laboratory chemical analysis has some limitations, including slow speed, high cost, and environmental contamination by chemical reagents [16]. These issues would be worsened in the case of large-scale Cr studies, which are needed given the global scale of Cr pollution issues [17].

As an alternative to laboratory-based analytical methods, reflectance spectroscopy is increasingly being used to measure soil quality [18]. This method primarily relies on the principle that energy is absorbed or reflected by the vibrations of molecular bonds. This technique leverages the numerous advantages of spectroscopic measurements, including their rapid, non-destructive, cost-effective, and environmentally friendly nature, especially when processing numerous samples [19]. At present, most physical and chemical properties of soils can be predicted from soil spectra; however, there is still a significant knowledge gap in the detection of soil Cr content from soil spectra [19,20].

Soil spectra can be considered non-specific and high-dimensional data due to the overlapping reflectance signatures of many soil components [21–24]. In contrast to the commonly used principal component regression and partial least squares regression techniques, deep learning-based methods, which are directly trained on the data to predict the desired soil properties, are attracting increasing attention [25–28]. The nonlinear nature of deep learning makes it well suited to handling large volumes of high-dimensional data, and this approach has been successfully applied to similar environmental fields [29,30]. However, few studies to date have applied deep learning to Cr detection from soil spectra.

To address this research gap, this study employed a deep neural network (DNN) method to predict soil Cr content from soil spectra. As an important algorithm in the field of deep learning, the DNN approach can effectively handle large volumes of data with high dimensionality [31]; thus, this approach is suitable for large-scale Cr detection. The Land Use and Cover Area frame Survey (LUCAS) dataset was used to train the DNN model, whose performance was verified by several evaluation metrics. The trained DNN model was then interpreted to identify the spectral bands sensitive to Cr content.

2. Materials and Methods

As noted above, the LUCAS 2009 dataset and corresponding Cr content values were used to train and validate the DNN model. To determine the optimal spectral preprocessing methods for Cr detection, the first derivative, second derivative, convolution smoothing, and multivariate scattering correction approaches were compared. The optimal model's network structure and hyperparameters were determined using a grid search method. Model interpretation on the optimal DNN model was then performed to identify the spectral bands that were most sensitive to Cr content in soil.

2.1. Dataset

2.1.1. Data Collection

This study used the LUCAS 2009 dataset from the European Soil Data Center (ESDAC) with topsoil samples from 23 EU member states. A total of 18,675 topsoil samples were employed based on the LUCAS project (Figure 1a). In the original acquisition of this dataset, the samples from these points were collected using a standardized sampling procedure and sent to the same laboratory for physical and chemical analysis [32,33]. The topsoil samples were ground, dried, and scanned using an XDS™ Rapid Content Analyzer (Foss, Hillerød, Denmark) to obtain reflectance spectra data in the wavelength range of 400–2500 nm with a resolution of 0.5 nm; thus, the reflectance of 4200 wavelengths were measured in this region [34]. The hyperspectral curve is schematically illustrated in Figure 1b.

The LUCAS 2009 dataset has immense value for understanding soil characteristics and properties across the EU member states. By incorporating a large quantity of topsoil samples systematically collected from diverse locations, this dataset provides a comprehensive overview of soil variations and compositions within the EU. The standardized sampling procedure ensures data consistency and reliability across a large geographic area, making this dataset a valuable resource for soil-related studies.

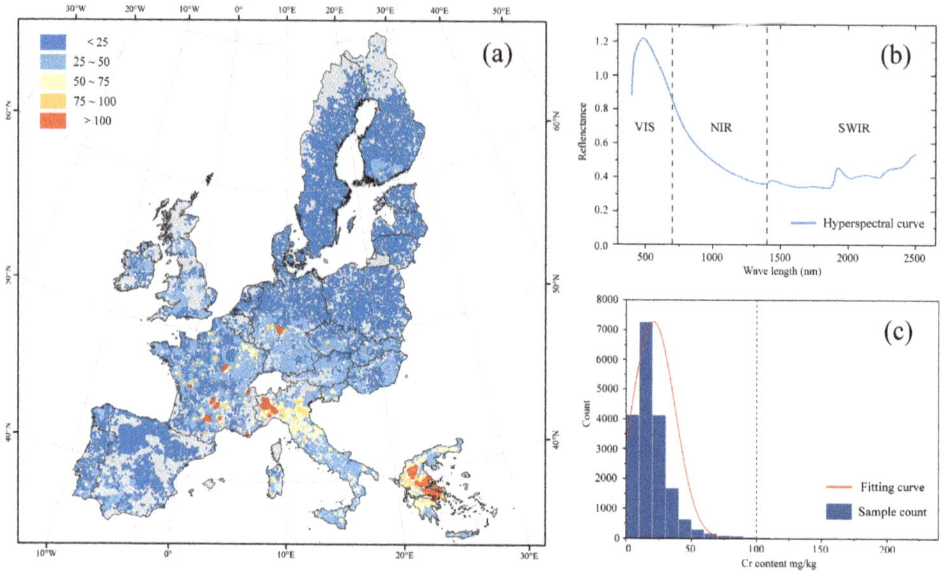

Figure 1. Descriptive diagram of the dataset: (**a**) Geological map of the sampling points; (**b**) Schematic diagram of the hyperspectral curve; (**c**) Distribution of Cr content.

The application of hyperspectral analysis enables detailed characterization of the reflectance spectra of the topsoil samples. This analysis allows for the identification of specific wavelengths that correspond to various soil properties and components. The reflectance spectra data, which span a broad range of wavelengths, provide important insights into the chemical, physical, and mineralogical properties of the topsoil samples [35]. Overall, the LUCAS 2009 dataset can be used to gain a deeper understanding of soil variability, nutrient content, organic matter composition, and other essential factors that influence soil health and fertility [36]. The rich and spatially extensive data provided by the LUCAS 2009 dataset can be used for in-depth soil studies, statistical modeling, and the development of predictive models to enhance soil management practices, land use planning, and environmental assessments.

The soil Cr content values were extracted from Cr maps [37]. Note that the Cr measurements were collected at the same sampling sites as the LUCAS 2009 study, ensuring consistency when integrating these two datasets. The distribution of Cr content indicates that most of the samples were not contaminated, when a Cr contamination threshold of 100 mg/kg was used (Figure 1c) [38]. The contaminated samples were primarily concentrated in countries located in South and South Central Europe, such as Italy and Greece.

2.1.2. Spectral Preprocessing Methods

Spectral preprocessing techniques can transform reflectance measurements using various mathematical methods that remove physical variability from light scattering and enhance features of interest [39]. Many studies have shown that applying appropriate spectral preprocessing can improve model accuracy [40–42]. In this study, five commonly used spectral preprocessing methods were applied and compared, including the first-order derivative (D1), second-order derivative (D2), Sazitzky–Golay (SG), multiplicative scattering correction (MSC), and standard normal variate normalization (SNV) (Figure 2) [43].

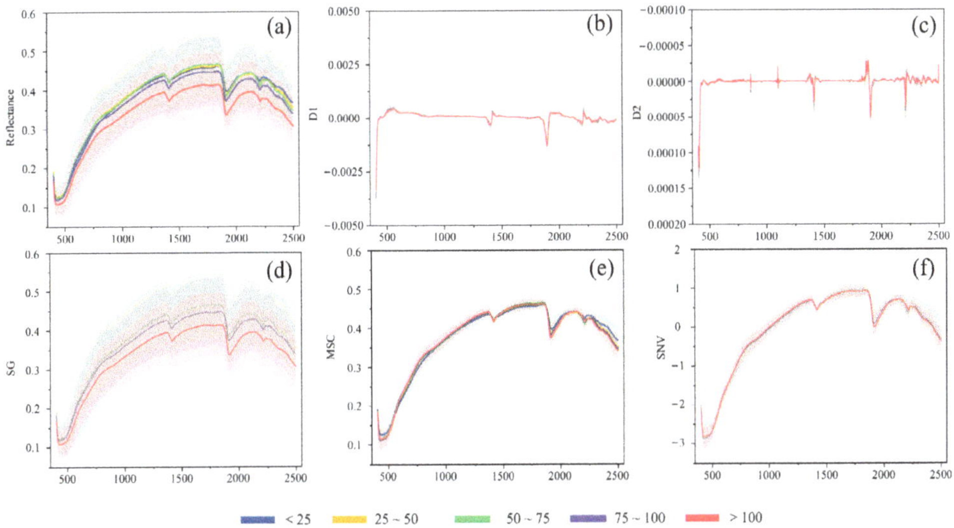

Figure 2. Hyperspectral curves with different spectral preprocessing: (**a**) Original spectral curve; (**b**) D1 preprocessed curve; (**c**) D2 preprocessed curve; (**d**) SG preprocessed curve; (**e**) MSC preprocessed curve; (**f**) SNV preprocessed curve. The curve represents the average value for each corresponding Cr group and the shadow represents the standard deviation.

All five hyperspectral preprocessing techniques can potentially enhance hyperspectral curve quality and analytical precision [44–48]. D1 and D2 effectively counteract baseline drift, refining peak recognition and quantification accuracy [49]. The advantage of SG smoothing is its ability to eliminate noise without altering peak attributes and improve signal-to-noise ratios and resolution; however, excessive smoothing can potentially worsen the peak definition [50]. The MSC method tackles scattering variations, amplifies spectrum-content correlations, and refines quantitative analysis, but its benefits may be limited in non-scattering-dominant scenarios [4]. SNV promotes spectrum comparability and aids feature identification; however, this technique may cause noise amplification, particularly in low-intensity zones, which can in turn influence subsequent interpretations [51]. These preprocessing methods can be applied to optimize specific data attributes and objectives and improve hyperspectral analysis precision; however, a comparison of these techniques is essential for optimal Cr detection.

2.2. Deep Learning

2.2.1. DNN Architecture

DNN models consist of multiple processing layers used to learn from and compute data [52]. Compared to other common regression models, the DNN approach is uniquely suited to compute large data with high dimensionality [53]. A DNN is composed of three main parts: an input layer, hidden layers, and an output layer, where each neuron within a layer is interconnected with all the neurons from the preceding layer, and nonlinearity is introduced by using activation functions on these connections (Figure 3) [54]. Each layer contains a specified number of neurons, and each neuron receives a set of inputs, which are weighted and aggregated then transformed via the activation function.

Activation functions are key components within a DNN, introducing non-linearity and enabling the network to capture complex patterns [55]. The selection of activation functions depends on the data characteristics and network architecture. Selecting suitable activation functions can significantly enhance the DNN's ability to capture intricate relationships in hyperspectral data, which can contribute to accurate Cr prediction [56].

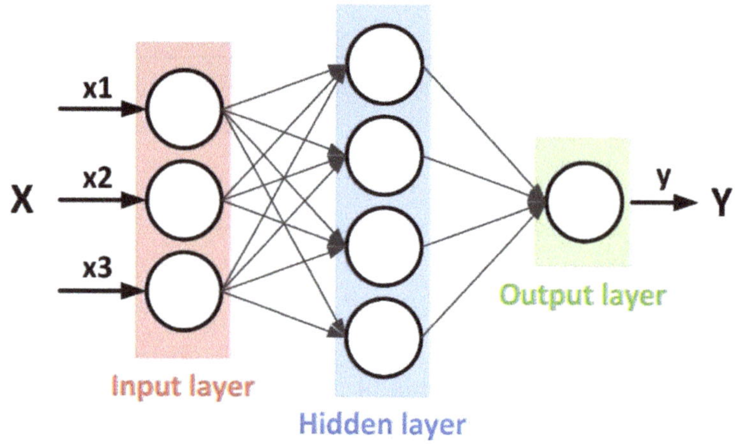

Figure 3. Schematic illustration of the DNN.

Layers represent the fundamental building blocks of a DNN and facilitate the extraction of hierarchical and abstract features from raw data. Each layer in a DNN serves as a specialized processing unit, transforming the input data through learned weights and activation functions [57]. The overall DNN structure typically consists of an input layer, multiple hidden layers, and an output layer. The data are received through the input layer, and feedforward processing is performed, during which the difference between the predicted and true values is calculated using a loss function. The parameter weights of each layer are then updated by an optimizer using backpropagation to minimize the misfit between the model predictions and true values. [58].

2.2.2. DNN Structure and Parameter Optimization

The construction of ML models must be optimized to achieve ideal performance on specific problems [59]. Both the DNN's structure and its hyperparameters will directly affect modeling performance. When a DNN is trained, the data are usually computed in batches, and the batch size will affect the model's training efficiency and generalization ability. Batch size is the number of training samples used in one iteration and plays a critical role in the training efficiency of a DNN. A larger batch size can accelerate the training process as more data are processed simultaneously, leading to faster convergence. However, this speed comes at the cost of accuracy. Larger batches provide a less accurate estimate of the gradient, potentially leading the training process to converge to suboptimal solutions. Conversely, a smaller batch size tends to provide a more accurate gradient estimate, enhancing the model's ability to generalize but slowing down the training process. Therefore, selecting an optimal batch size represents the balance between training speed and model accuracy. The larger the batch size, the faster the model's training speed but the lower its accuracy.

The dropout rate aims to reduce the risk of model overfitting by randomly dropping some neurons during the training process so that each neuron is more independent. During training, a randomly selected subset of neurons is ignored or 'dropped out'. This process prevents neurons from co-adapting too much, encouraging individual neurons to learn features independently, thereby reducing the model's reliance on any small set of neurons and thus mitigating overfitting. However, setting the dropout rate too high can lead to underfitting, where the model fails to learn the data's underlying pattern adequately. Thus, the dropout rate must be carefully calibrated to ensure the model learns sufficiently complex patterns without overfitting to the training data. Therefore, it is necessary to carefully optimize the batch size and dropout rate hyperparameters to enhance the model's performance.

In this study, the number of layers, the number of neurons, the activation function, batch size, dropout, and learning rate were optimized using the grid search approach

(Table 1) [60]. The performance of each combination was calculated as the average of 10 repetitions to avoid the effect of randomness during dataset splitting. The dataset was split in an 8:1:1 ratio of training, validation, and testing sets [61].

Table 1. Hyperparameter selection range.

Parameter	Search Range
Layer	[1, 2, 3, 4, 5, 6, 7, 8, 9]
Neurons	[100, 200, 400, 600, 800, 1000, 1500, 2000]
Activation function	[ReLU, Leakly_ReLU, Swish, Sigmoid]
Batch size	[25, 50, 75, 100, 500, 100]
Dropout rate	[0.1, 0.2, 0.3, 0.4, 0.5]
Learning rate	[0.01, 0.001, 0.0001]

2.2.3. Model Evaluation Metrics

Model evaluation is an important step in ML modeling and different evaluation metrics focus on different aspects of the trained models. In this study, the Pearson coefficient (R), root mean square error (RMSE), and mean absolute error (MAE) were chosen to evaluate the model's performance [62]. R measures the correlation between the model's predicted variables and the actual variables, thereby assessing the quality of the model's predictions. The RMSE is the square difference between the true and predicted values and represents the magnitude of the error generated in the model's predictions. The MAE is the average absolute difference between the true and predicted values. In general, lower RMSE and MAE values and higher R values indicate better model performance. The above metrics are calculated as follows:

$$R = \frac{Cov(X, Y)}{\sqrt{Var[X]Var[Y]}} \tag{1}$$

$$RMSE = \sqrt{\frac{1}{n}\sum_{i=1}^{n}(y_i - \hat{y}_i)^2} \tag{2}$$

$$MAE = \sum_{i=1}^{n}\frac{|x_i - y_i|}{n} \tag{3}$$

where n is the number of samples, y_i is the true Cr content of soil sample i, and $\hat{y}_i$ is the Cr content predicted by the model for sample i.

2.3. Model Interpretation

2.3.1. Overview of Model Interpretation

Deep learning models have proven to be highly effective in processing large-scale datasets, but their inherent "black box" nature often obscures the understanding of the internal mechanics influencing their results. This opacity can cast doubts on their reliability, especially in critical applications. To address this, the development of interpretable methods has become essential, providing insights into how these models arrive at their conclusions [63].

Interpretable methods in machine learning are broadly categorized into two types: global and local interpretable methods. Global interpretable methods aim to identify the overall internal working mechanisms of deep learning models. These methods are designed to provide a comprehensive view of how input data are transformed and processed through various layers and nodes of a neural network. By understanding these global mechanisms, it becomes possible to gain insights into the model's overall decision-making process, enhancing the transparency and trustworthiness of the model. In contrast, local interpretable methods focus on elucidating the causal relationships between specific inputs and their corresponding model predictions. They break down the prediction process for individual instances, enabling researchers to understand why a model made a particular decision for a specific input. This level of granular insight is invaluable for diagnosing and refining models, especially when dealing with complex datasets where the interactions between input variables can be intricate and non-intuitive [64]. In summary, interpretable models enhance the transparency and credibility of deep learning models, offering avenues for processing and optimizing the models.

2.3.2. Permutation Importance

Permutation importance is a valuable and insightful feature importance assessment technique that plays a crucial role in understanding and interpreting the outputs of machine learning (ML) models. This method, based on the predicted outputs of the ML model, offers a straightforward yet powerful way to determine the significance of different features in the model's predictions [65]. The process of permutation importance involves a systematic alteration of each feature in the dataset. To assess the importance of a particular feature, that feature's values are shuffled or 'permuted', while keeping the values of all other features unchanged. This shuffling disrupts the relationship between the feature and the target, essentially simulating a scenario where the feature does not provide any useful information to the model.

Once the data with the disrupted feature is prepared, it is fed back into the model for prediction. The key step in permutation importance is comparing the model's predictions on this perturbed data against its predictions on the original, unaltered data. The difference in performance, typically measured in terms of accuracy or error, indicates how much the model relies on the feature. A large degradation in the model's performance upon permuting a feature indicates its high importance. Conversely, if the model's performance remains relatively unchanged, the feature is likely less important or even redundant. This method of evaluating feature importance has several advantages. Firstly, it is model-agnostic, meaning it can be applied to any ML model regardless of its internal mechanics. This makes permutation importance particularly versatile and widely applicable across various types of models and algorithms. Secondly, it is computationally efficient, often requiring only a few additional rounds of prediction, making it suitable for large datasets and complex models [66,67]. Furthermore, permutation importance provides a more intuitive understanding of feature importance compared to other methods like coefficients in linear models or feature importance scores in tree-based models. It helps in identifying not just the highly influential features but also those which might be misleading or non-informative. This can guide the feature selection process, leading to simpler, more interpretable, and often more generalizable models.

2.3.3. Local Interpretable Model-Agnostic Explanations (LIME)

The main concept of the LIME method is to interpret a complex model by constructing a simple model. This method begins by perturbing the input data and creating a new, representative dataset that reflects the original data distribution and characteristics. The newly generated dataset is then used to train a simpler model, which is inherently more interpretable than the complex model. The key objective of LIME is to ensure that the predictions made by this simpler model on the new dataset closely align with those made by the complex model on the same data. By achieving this alignment, LIME effectively reveals how the complex model behaves locally around specific instances. It also verifies the local fidelity of the simple model to the complex global model. The goal of this method is to identify the features with a high degree of importance to the model [68]. The underlying principle of LIME is shown in Equation (4). To make the local model more accurately represent the global complex model, the error between the predicted values of the global complex model f and the new simple model g must be minimized. Subsequently, the locality-aware loss function is constructed as Equation (5).

$$Explanation(x) = \underset{g \in G}{\operatorname{argmin}} L(f, g, \pi_x) + \Omega(g) \quad (4)$$

$$L(f, g, \pi_z) = \sum_{z,z' \in Z} \pi_x(z)\left(f(z) - g(z')\right)^2 \quad (5)$$

where f denotes the global complex model, i.e., the model to be explained, g denotes the simple model, G is a collection of simple models, e.g., all possible linear models, π_x denotes the proximity measure of data z' in the new dataset to the original data z, and $\Omega(g)$ denotes the complexity of simple model g.

2.4. Implementation and Visualization

Python 3.8 was used as the programming language in the current study. Spectral preprocessing calculations were conducted using NumPy 1.25.2, Pandas, and SciPy 1.11.4. Keras 2.15.0 and scikit-learn 1.2.2 were utilized for the establishment, training, and performance assessment of the DNN models. The interpretability analysis of the models was conducted using Lime 0.2.0.1, SHAP 0.42.0, and eli5 0.13.0.

3. Result and Discussion

3.1. Model Optimization Results

In this study, the influence of five preprocessing methods (SNV, SG, D1, D2, SG, and MSC) on the modeling performance was compared based on the R value, as shown in Figure 4. D1 performed optimally on both the training and validation sets, with R values of 0.83 and 0.75, respectively. Compared to the original spectra, the DNN model trained on D1 preprocessed spectra exhibited a significant improvement, achieving the R increase of 0.30 in the training set and 0.247 in the testing set.

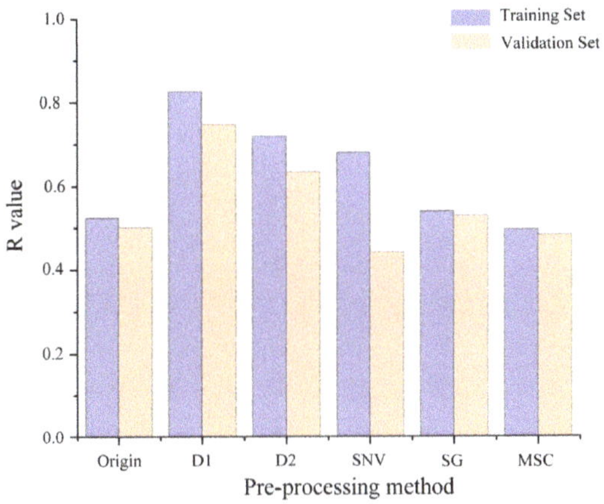

Figure 4. The influence of spectral preprocessing on DNN modeling performance.

To obtain the optimal DNN structure, the number of hidden layers and the number of neurons in the hidden layer were adjusted [59]. After the architecture optimization, the selected DNN structure comprised seven hidden layers. These layers contained 2000, 1500, 1000, 600, 400, 200, and 100 neurons, respectively, and were connected by the Leaky ReLU activation function. To prevent gradient explosion and overfitting, the dropout and early stopping mechanisms were also used during DNN model construction (Figure 5) [69,70].

As described above, different combinations of dropout rate, batch size, and learning rate will affect the modeling performance. After comparing the performance of learning rates of 0.01, 0.001, and 0.0001, a value of 0.001 was found to achieve better performance and was thus used for the subsequent optimization (Figure 6) [71]. The grid search of the dropout rate and batch size hyperparameters indicates that the model's performance was optimized when the dropout rate was 0.15 and the batch size value was 70. For the above optimal parameter combination, the DNN model achieved the R values of 0.85 and 0.79 on the training and validation sets, respectively.

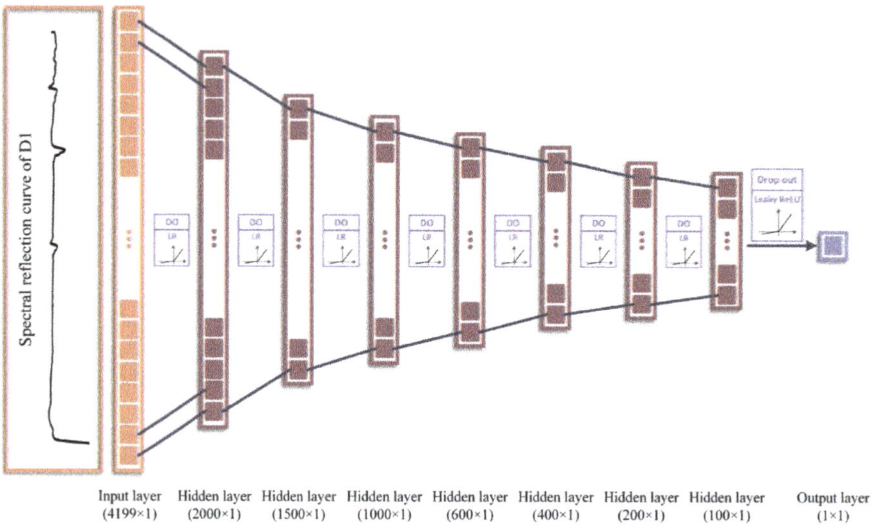

Figure 5. The optimized DNN structure.

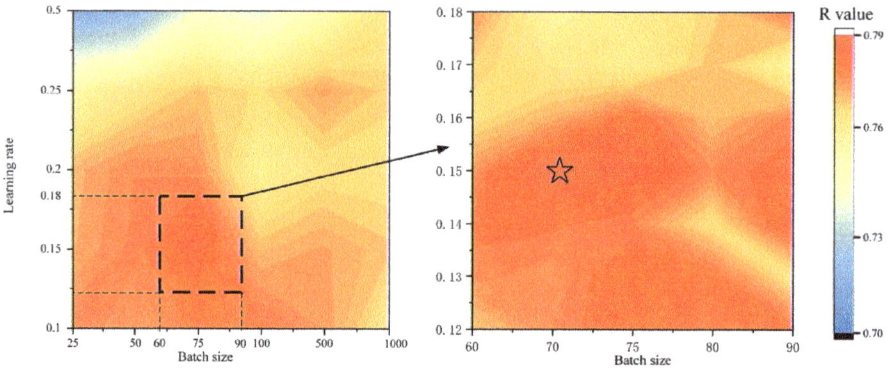

Figure 6. The optimized batch size and learning rate.

3.2. Model Evaluation

Figure 7a presents a comparative analysis of the DNN model's performance throughout the optimization process, ranging from the default model to the optimal model. A marked enhancement in the DNN's performance was observed during the whole model construction process. Specifically, the R of the model improved from 0.6 to 0.8 on the validation set, indicating a significant increase in the correlation between the model's predictions and the actual values (Figure 7b). In the meantime, RMSE was decreased from 168.6 to 68.4 and MSE was decreased from 8.16 to 5.84. These improvements in evaluation metrics demonstrate enhanced accuracy and reliability of the model in predicting chromium content in soil. The discrepancies between the actual and predicted values in the training, validation, and testing sets were predominantly concentrated within a range of 10, further underscoring the effectiveness of the DNN modeling (Figure 7c–e). On the testing set, the optimal DNN model exhibited robust predictive performance with an R value of 0.79, RMSE of 96.98, and MAE of 5.79. These results indicate that the model's capability to predict chromium content from soil hyperspectral data was enhanced after the optimization of DNN architecture and hyperparameters.

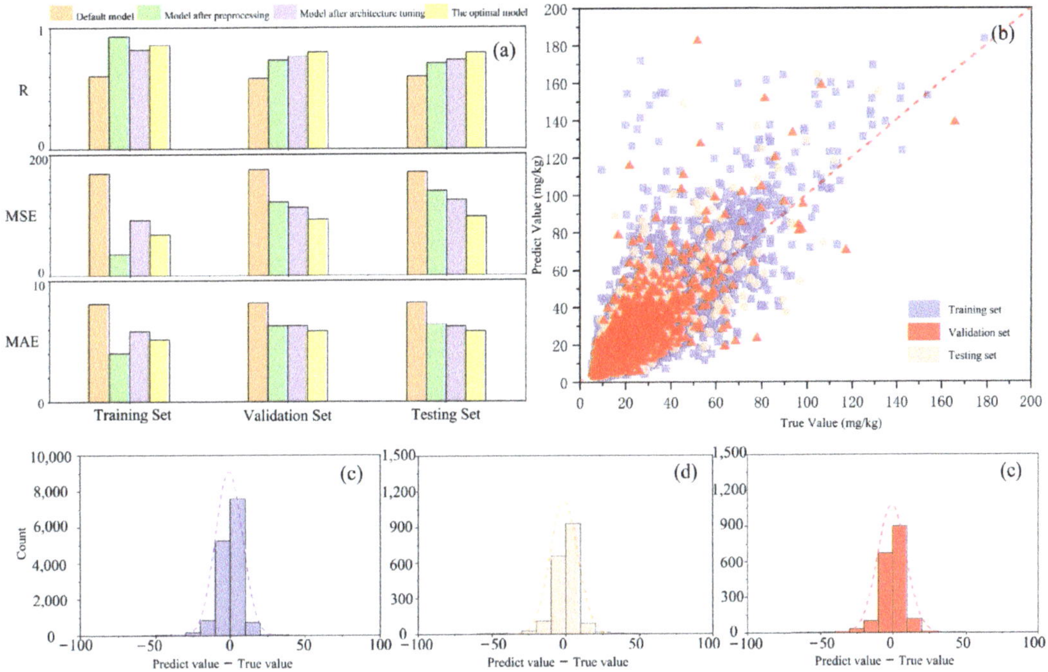

Figure 7. Evaluation of modeling performance: (**a**) Values of model evaluation metrics at each modeling stage; (**b**) Comparison of the actual and predicted Cr values using the optimal model; (**c**–**e**) Distribution of the difference between predicted and actual Cr values in the training, validation, and testing sets of the optimal model, respectively. The 'default model' refers to the initial DNN model, 'preprocess model' to the model post preprocessing, 'structure model' to the model after optimizing the neural network structure, and 'optimal model' to the model achieving the best performance.

3.3. Spatial Autocorrelation and Residual Analysis of the DNN Prediction

Considering that the soil chromium content might have some spatial patterns, Moran's index was calculated at 0.287 with a *p*-value of 0.001 using the residuals and geographical coordinates from the DNN model. This indicates a tendency for chromium to cluster within the spatial scope of the European Union, confirming the presence of significant spatial autocorrelation in the data.

Further analysis was performed on the residuals through kriging interpolation, as shown in Figure 8. This revealed that large residuals were predominantly distributed among Southern and Central European countries, while predictions for countries in Northern Europe were more reliable. In future research, more tailored DNN models could be established for different regions or countries within the European Union to enhance the robustness of the model's predictions.

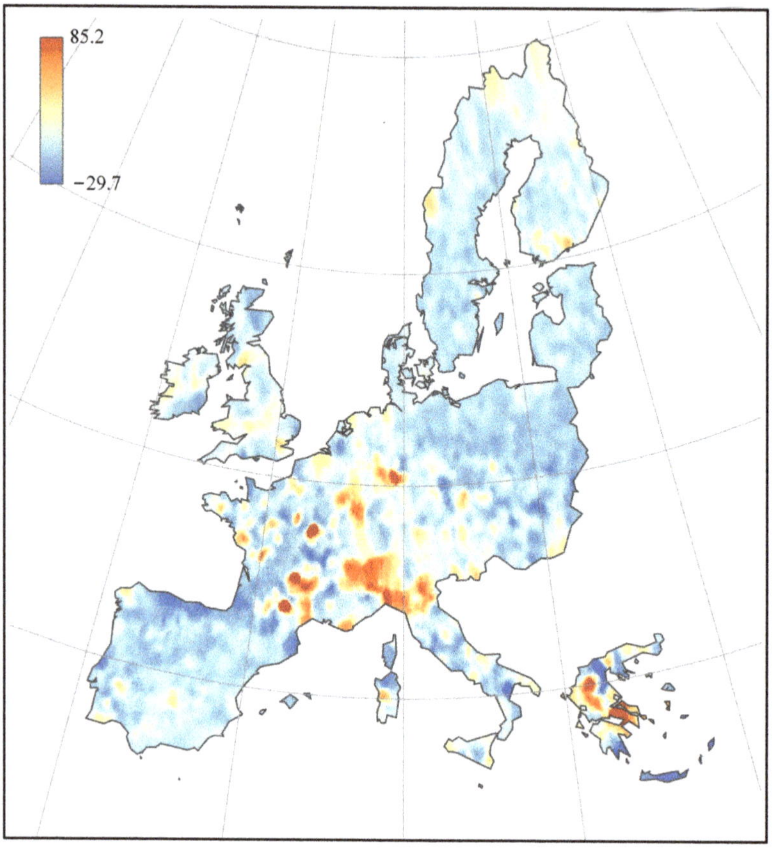

Figure 8. The distribution of the prediction residual across the EU.

3.4. Model Interpretation Analysis

As detailed above, the permutation importance can be calculated by randomly disrupting a single feature in all the samples. This can then be used to calculate the impact on the model's accuracy due to the change in that feature and, thus, the importance of each feature [72]. The calculated feature permutation importance using the optimized DNN model exhibited a relatively smooth trend (Figure 9); however, there were significant fluctuations in four specific band ranges: 400–439 nm (region I), 1364–1422 nm (region II), 1862–1934 nm (region III), and 2158–2499 nm (region IV). More pronounced absorption peaks were also observed in these ranges in the hyperspectral reflection curves processed using the first-order differential method.

Among these four band regions, region I is mainly attributed to the iron oxide content, while the peaks in regions II and III are primarily related to the presence of clay minerals and hydroxyl groups in water [73]. Region IV involves vibrations of metal–OH bonds and indicates a key absorption peak near 2200 nm; this peak is mainly influenced by Al–OH bonds, with the main contributing substances being kaolinite, montmorillonite, and illite [74].

The above permutation importance results indicate that soil Cr content is mainly correlated with clay minerals and iron oxides. Clay minerals directly influence soil texture and play a crucial role in the growth of plants and microorganisms. These factors, in turn, exert a significant influence on Cr content and flow within ecosystems [75]. For example, montmorillonite, a clay mineral characterized by a layered structure and active surface sites, can adsorb Cr through processes such as ion exchange with water within its lattice or complexation on its surface [75,76]. As an iron-loving element, Cr can be adsorbed with iron oxides to produce stable precipitates or surface complexes—these promote the binding of Cr to soil particles and thus reduce damage to the environment [77,78]. Therefore, the summarized high correlation between Cr content with clay minerals and iron oxides agrees well with the findings in the literature.

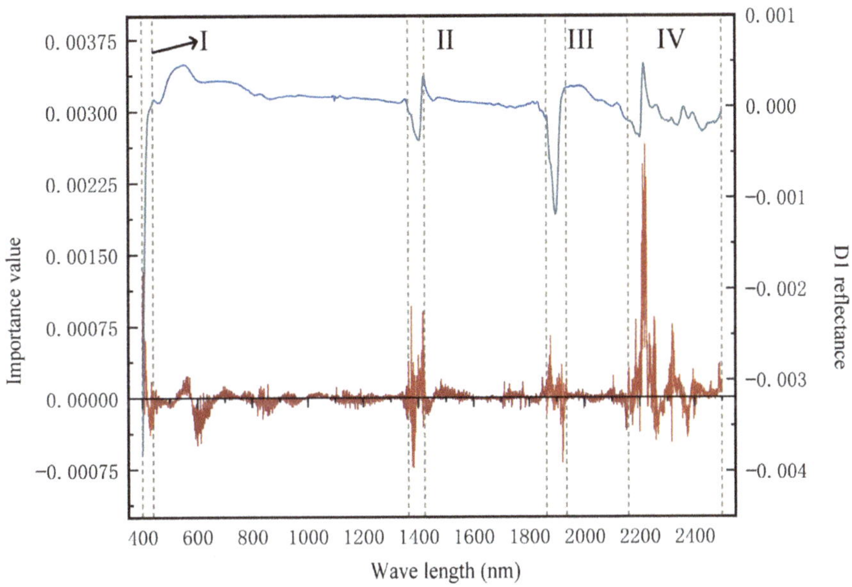

Figure 9. Permutation importance across the whole spectra. The blue curve is a representative hyperspectral curve after D1 preprocessing, the red curve represents the permutation importance value across the spectra, and the four regions are the sensitive band ranges.

In this study, 1000 samples were randomly selected for LIME analysis and their 10 most important features were classified based on the four important regions identified in Figure 8. The LIME results are summarized in Figure 10. As shown, 77% of the important features identified in the LIME analysis belonged to one of the four intervals, while the other 23% were irregularly distributed in the other regions. The above results indicate a good agreement between LIME and permutation importance results. Notably, the number of important features in region IV was much higher than those in regions I, II, and III, and the top 10 most important features identified from the permutation importance were also distributed in region IV. Therefore, clay minerals were observed to be the most important indicators for Cr content in soil.

The above importance analysis indicates that the chemical properties of soil significantly affect the accumulation of Cr in soil. To prevent excess Cr accumulation in the soil, heavy metal industries should not be distributed in soils with high clay minerals and iron oxide content. This measure can help minimize the extent and impact of Cr pollution resulting from industrial accidents and improper sewage treatment. For soils that have accumulated large amounts of Cr, remediation using iron oxide or clay minerals can be performed to absorbed Cr from soil.

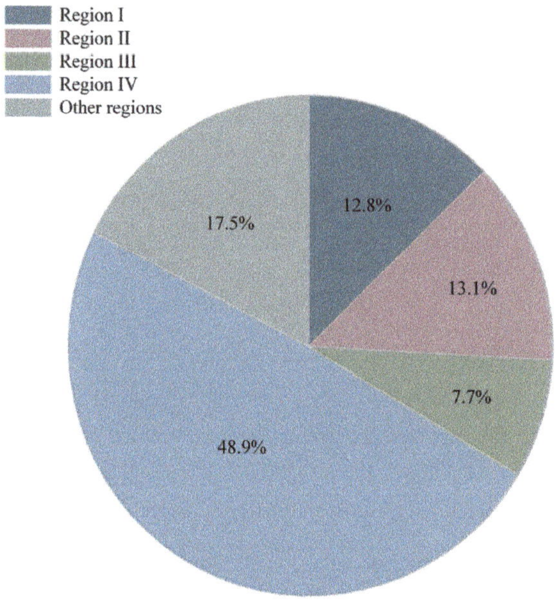

Figure 10. LIME importance analysis.

4. Conclusions

In this study, various analyses were performed based on the LUCAS dataset to establish a DNN model for Cr detection from soil spectra. As part of the optimization process, the optimal preprocessing method was determined, and the model's hyperparameters were tuned. The resulting optimum DNN model can accurately predict soil Cr content from soil spectra. Meanwhile, four Cr-sensitive bands were identified through the interpretation using the optimal DNN model. The main conclusions are as follows:

(1) D1 was identified as the optimal preprocessing method for the DNN model to predict soil Cr content. The R value of the DNN model increased from 0.50 to 0.75 on the testing set after spectral preprocessing.

(2) The adjustment of DNN architecture and hyperparameters resulted in the further improvements in the model performance. The R, RMSE, and MAE values of the optimal model on the testing set were 0.79, 96.98, and 5.79, respectively, which were significantly improved compared to the default model.

(3) Four important sensitive band regions of Cr content in soil were identified, namely, 400–439 nm (region I), 1364–1422 nm (region II), 1862–1934 nm (region III), and 2158–2499 nm (region IV). These bands correspond primarily to iron oxide and clay mineral content in the soil.

Author Contributions: C.M.: Conceptualization, Formal analysis, Methodology, Software, Visualization, Writing—Original draft preparation; X.X.: Formal analysis, Methodology, Software; M.Z.: Conceptualization, Software; T.H.: Conceptualization, Formal analysis; C.Q.: Data curation, Project administration, Writing—Original draft preparation, Funding. All authors have read and agreed to the published version of the manuscript.

Funding: This work was supported by the National Natural Science Foundation of China (No. 22376221, 52274151), Natural Science Foundation of Hunan Province, China (No. 2024JJ2074), Young Elite Scientists Sponsorship Program by CAST (No. 2023QNRC001), and Unveiling and Commanding Project from Fankou Lean-Zinc Mine. This work was also supported in part by the High Performance Computing Center of Central South University.

Institutional Review Board Statement: Not applicable.

Informed Consent Statement: Not applicable.

Data Availability Statement: The data that has been used is confidential.

Conflicts of Interest: Author Chongchong Qi was employed by the company NONFEMET. The remaining authors declare that the research was conducted in the absence of any commercial or financial relationships that could be construed as a potential conflict of interest.

References

1. Liu, Y.-R.; van der Heijden, M.G.A.; Riedo, J.; Sanz-Lazaro, C.; Eldridge, D.J.; Bastida, F.; Moreno-Jiménez, E.; Zhou, X.-Q.; Hu, H.-W.; He, J.-Z.; et al. Soil contamination in nearby natural areas mirrors that in urban greenspaces worldwide. *Nat. Commun.* **2023**, *14*, 1706. [CrossRef] [PubMed]
2. Coetzee, J.J.; Bansal, N.; Chirwa, E.M.N. Chromium in Environment, Its Toxic Effect from Chromite-Mining and Ferrochrome Industries, and Its Possible Bioremediation. *Expo. Health* **2020**, *12*, 51–62. [CrossRef]
3. Peng, H.; Guo, J. Removal of chromium from wastewater by membrane filtration, chemical precipitation, ion exchange, adsorption electrocoagulation, electrochemical reduction, electrodialysis, electrodeionization, photocatalysis and nanotechnology: A review. *Environ. Chem. Lett.* **2020**, *18*, 2055–2068. [CrossRef]
4. Buddenbaum, H.; Steffens, M. The Effects of Spectral Pretreatments on Chemometric Analyses of Soil Profiles Using Laboratory Imaging Spectroscopy. *Appl. Environ. Soil Sci.* **2012**, *2012*, 274903. [CrossRef]
5. Rangasamy, S.; Purushothaman, G.; Alagirisamy, B.; Santiago, M. Chromium contamination in soil and groundwater due to tannery wastes disposals at Vellore district of Tamil Nadu. *Int. J. Environ. Sci.* **2015**, *6*, 114–124.
6. Tumolo, M.; Ancona, V.; De Paola, D.; Losacco, D.; Campanale, C.; Massarelli, C.; Uricchio, V.F. Chromium Pollution in European Water, Sources, Health Risk, and Remediation Strategies: An Overview. *Int. J. Environ. Res. Public Health* **2020**, *17*, 5438. [CrossRef]
7. Prasad, S.; Yadav, K.K.; Kumar, S.; Gupta, N.; Cabral-Pinto, M.M.S.; Rezania, S.; Radwan, N.; Alam, J. Chromium contamination and effect on environmental health and its remediation: A sustainable approaches. *J. Environ. Manag.* **2021**, *285*, 112174. [CrossRef] [PubMed]
8. Jobby, R.; Jha, P.; Yadav, A.K.; Desai, N. Biosorption and biotransformation of hexavalent chromium [Cr(VI)]: A comprehensive review. *Chemosphere* **2018**, *207*, 255–266. [CrossRef]
9. Dhal, B.; Thatoi, H.N.; Das, N.N.; Pandey, B.D. Chemical and microbial remediation of hexavalent chromium from contaminated soil and mining/metallurgical solid waste: A review. *J. Hazard. Mater.* **2013**, *250*, 272–291. [CrossRef]
10. Zheng, C.J.; Yang, Z.H.; Si, M.Y.; Zhu, F.; Yang, W.C.; Zhao, F.P.; Shi, Y. Application of biochars in the remediation of chromium contamination: Fabrication, mechanisms, and interfering species. *J. Hazard. Mater.* **2021**, *407*, 124376. [CrossRef]
11. Xia, S.P.; Song, Z.L.; Jeyakumar, P.; Shaheen, S.M.; Rinklebe, J.; Ok, Y.S.; Bolan, N.; Wang, H.L. A critical review on bioremediation technologies for Cr(VI)-contaminated soils and wastewater. *Crit. Rev. Environ. Sci. Technol.* **2019**, *49*, 1027–1078. [CrossRef]
12. Kapoor, R.T.; Mfarrej, M.F.B.; Alam, P.; Rinklebe, J.; Ahmad, P. Accumulation of chromium in plants and its repercussion in animals and humans. *Environ. Pollut.* **2022**, *301*, 119044. [CrossRef] [PubMed]
13. Azeez, N.A.; Dash, S.S.; Gummadi, S.N.; Deepa, V.S. Nano-remediation of toxic heavy metal contamination: Hexavalent chromium [Cr(VI)]. *Chemosphere* **2021**, *266*, 129204. [CrossRef] [PubMed]
14. Jennings, A.A. Analysis of worldwide Regulatory Guidance Values for less frequently regulated elemental surface soil contaminants. *J. Environ. Manag.* **2013**, *128*, 561–585. [CrossRef] [PubMed]
15. Anca Maria, M. Assessment of Soil Pollution with Heavy Metals in Romania. In *Environmental Risk Assessment of Soil Contamination*; Maria, C.H.-S., Ed.; IntechOpen: London, UK, 2014; Chapter 9.
16. Kania, M.; Gruba, P.; Wiecheć, M. Applying the near-infrared spectroscopy to calculate the Soil Trophic Index. *Sylwan* **2017**, *161*, 935–939.
17. Hou, D.; Ok, Y.S. Soil pollution--speed up global mapping. *Nature* **2019**, *566*, 455–456. [CrossRef]
18. Jia, X.; O'Connor, D.; Shi, Z.; Hou, D. VIRS based detection in combination with machine learning for mapping soil pollution. *Environ. Pollut.* **2021**, *268*, 115845. [CrossRef]
19. Nocita, M.; Stevens, A.; van Wesemael, B.; Aitkenhead, M.; Bachmann, M.; Barthes, B.; Ben Dor, E.; Brown, D.J.; Clairotte, M.; Csorba, A.; et al. Soil Spectroscopy: An Alternative to Wet Chemistry for Soil Monitoring. *Adv. Agron.* **2015**, *132*, 139–159.
20. Nocita, M.; Stevens, A.; Noon, C.; van Wesemael, B. Prediction of soil organic carbon for different levels of soil moisture using Vis-NIR spectroscopy. *Geoderma* **2013**, *199*, 37–42. [CrossRef]
21. Qi, C.; Wu, M.; Liu, H.; Liang, Y.; Liu, X.; Lin, Z. Machine learning exploration of the mobility and environmental assessment of toxic elements in mining-associated solid wastes. *J. Clean. Prod.* **2023**, *401*, 136771. [CrossRef]
22. Wu, M.; Qi, C.; Derrible, S.; Choi, Y.; Fourie, A.; Ok, Y.S. Regional and global hotspots of arsenic contamination of topsoil identified by deep learning. *Commun. Earth Environ.* **2024**, *5*, 10. [CrossRef]
23. Zhou, M.; Hu, T.; Wu, M.; Ma, C.; Qi, C. Rapid estimation of soil Mn content by machine learning and soil spectra in large-scale. *Ecol. Inform.* **2024**, *81*, 102615. [CrossRef]

24. Xu, X.; Qi, C.; Aretxabaleta, X.M.; Ma, C.; Spagnoli, D.; Manzano, H. The initial stages of cement hydration at the molecular level. *Nat. Commun.* **2024**, *15*, 2731. [CrossRef]
25. Wold, S.; Ruhe, A.; Wold, H.; Dunn, W.J., III. The Collinearity Problem in Linear Regression. The Partial Least Squares (PLS) Approach to Generalized Inverses. *SIAM J. Sci. Stat. Comput.* **1984**, *5*, 735–743. [CrossRef]
26. Conforti, M.; Castrignanò, A.; Robustelli, G.; Scarciglia, F.; Stelluti, M.; Buttafuoco, G. Laboratory-based Vis–NIR spectroscopy and partial least square regression with spatially correlated errors for predicting spatial variation of soil organic matter content. *Catena* **2015**, *124*, 60–67. [CrossRef]
27. Morellos, A.; Pantazi, X.E.; Moshou, D.; Alexandridis, T.; Whetton, R.; Tziotzios, G.; Wiebensohn, J.; Bill, R.; Mouazen, A.M. Machine learning based prediction of soil total nitrogen, organic carbon and moisture content by using VIS-NIR spectroscopy. *Biosyst. Eng.* **2016**, *152*, 104–116. [CrossRef]
28. Wang, S.; Guan, K.; Zhang, C.; Lee, D.; Margenot, A.J.; Ge, Y.; Peng, J.; Zhou, W.; Zhou, Q.; Huang, Y. Using soil library hyperspectral reflectance and machine learning to predict soil organic carbon: Assessing potential of airborne and spaceborne optical soil sensing. *Remote Sens. Environ.* **2022**, *271*, 112914. [CrossRef]
29. Ahmed, A.N.; Othman, F.B.; Afan, H.A.; Ibrahim, R.K.; Fai, C.M.; Hossain, M.S.; Ehteram, M.; Elshafie, A. Machine learning methods for better water quality prediction. *J. Hydrol.* **2019**, *578*, 124084. [CrossRef]
30. Bellinger, C.; Mohomed Jabbar, M.S.; Zaïane, O.; Osornio-Vargas, A. A systematic review of data mining and machine learning for air pollution epidemiology. *BMC Public Health* **2017**, *17*, 907. [CrossRef]
31. Alom, M.Z.; Taha, T.M.; Yakopcic, C.; Westberg, S.; Sidike, P.; Nasrin, M.S.; Hasan, M.; Van Essen, B.C.; Awwal, A.A.S.; Asari, V.K. A State-of-the-Art Survey on Deep Learning Theory and Architectures. *Electronics* **2019**, *8*, 292. [CrossRef]
32. Orgiazzi, A.; Ballabio, C.; Panagos, P.; Jones, A.; Fernandez-Ugalde, O. LUCAS Soil, the largest expandable soil dataset for Europe: A review. *Eur. J. Soil Sci.* **2018**, *69*, 140–153. [CrossRef]
33. Gergely, T.; Arwyn, J.; Luca, M.; Christine, A.; Cristiano, B.; Florence, C.; Delphine, D.B.; Anna, G.R.; Ciro, G.; Tamás, H.; et al. *LUCAS Topoil Survey—Methodology, Data and Results*; Publications Office of the European Union: Copenhagen, Denmark, 2013.
34. Safanelli, J.L.; Chabrillat, S.; Ben-Dor, E.; Dematte, J.A.M. Multispectral Models from Bare Soil Composites for Mapping Topsoil Properties over Europe. *Remote Sens.* **2020**, *12*, 1369. [CrossRef]
35. Gao, Y.; Liu, L.; Zhang, X.; Chen, X.; Mi, J.; Xie, S. Consistency Analysis and Accuracy Assessment of Three Global 30-m Land-Cover Products over the European Union using the LUCAS Dataset. *Remote Sens.* **2020**, *12*, 3479. [CrossRef]
36. Panagos, P.; Meusburger, K.; Ballabio, C.; Borrelli, P.; Alewell, C. Soil erodibility in Europe: A high-resolution dataset based on LUCAS. *Sci. Total Environ.* **2014**, *479-480*, 189–200. [CrossRef] [PubMed]
37. Tóth, G.; Hermann, T.; Szatmári, G.; Pásztor, L. Maps of heavy metals in the soils of the European Union and proposed priority areas for detailed assessment. *Sci. Total Environ.* **2016**, *565*, 1054–1062. [CrossRef] [PubMed]
38. Tóth, G.; Hermann, T.; Da Silva, M.R.; Montanarella, L. Heavy metals in agricultural soils of the European Union with implications for food safety. *Environ. Int.* **2016**, *88*, 299–309. [CrossRef]
39. Dotto, A.C.; Dalmolin, R.S.D.; ten Caten, A.; Grunwald, S. A systematic study on the application of scatter-corrective and spectral-derivative preprocessing for multivariate prediction of soil organic carbon by Vis-NIR spectra. *Geoderma* **2018**, *314*, 262–274. [CrossRef]
40. Vasques, G.M.; Grunwald, S.; Sickman, J.O. Comparison of multivariate methods for inferential modeling of soil carbon using visible/near-infrared spectra. *Geoderma* **2008**, *146*, 14–25. [CrossRef]
41. Peng, X.; Shi, T.; Song, A.; Chen, Y.; Gao, W. Estimating Soil Organic Carbon Using VIS/NIR Spectroscopy with SVMR and SPA Methods. *Remote Sens.* **2014**, *6*, 2699–2717. [CrossRef]
42. Javadi, S.H.; Munnaf, M.A.; Mouazen, A.M. Fusion of Vis-NIR and XRF spectra for estimation of key soil attributes. *Geoderma* **2021**, *385*, 114851. [CrossRef]
43. Nawar, S.; Buddenbaum, H.; Hill, J.; Kozak, J.; Mouazen, A.M. Estimating the soil clay content and organic matter by means of different calibration methods of vis-NIR diffuse reflectance spectroscopy. *Soil Tillage Res.* **2016**, *155*, 510–522. [CrossRef]
44. Ben-Dor, E.; Inbar, Y.; Chen, Y. The reflectance spectra of organic matter in the visible near-infrared and short wave infrared region (400–2500 nm) during a controlled decomposition process. *Remote Sens. Environ.* **1997**, *61*, 1–15. [CrossRef]
45. Demetriades-Shah, T.H.; Steven, M.D.; Clark, J.A. High resolution derivative spectra in remote sensing. *Remote Sens. Environ.* **1990**, *33*, 55–64. [CrossRef]
46. Burger, J.; Geladi, P. Spectral pre-treatments of hyperspectral near infrared images: Analysis of diffuse reflectance scattering. *J. Near Infrared Spectrosc.* **2007**, *15*, 29–37. [CrossRef]
47. Savitzky, A.; Golay, M.J.E. Smoothing and Differentiation of Data by Simplified Least Squares Procedures. *Anal. Chem.* **1964**, *36*, 1627–1639. [CrossRef]
48. Barnes, R.J.; Dhanoa, M.S.; Lister, S.J. Standard Normal Variate Transformation and De-Trending of Near-Infrared Diffuse Reflectance Spectra. *Appl. Spectrosc.* **1989**, *43*, 772–777. [CrossRef]
49. Qiao, X.-X.; Wang, C.; Feng, M.-C.; Yang, W.-D.; Ding, G.-W.; Sun, H.; Liang, Z.-Y.; Shi, C.-C. Hyperspectral estimation of soil organic matter based on different spectral preprocessing techniques. *Spectrosc. Lett.* **2017**, *50*, 156–163. [CrossRef]
50. Stevens, A.; Udelhoven, T.; Denis, A.; Tychon, B.; Lioy, R.; Hoffmann, L.; van Wesemael, B. Measuring soil organic carbon in croplands at regional scale using airborne imaging spectroscopy. *Geoderma* **2010**, *158*, 32–45. [CrossRef]

51. Kobori, H.; Gorretta, N.; Rabatel, G.; Bellon-Maurel, V.; Chaix, G.; Roger, J.-M.; Tsuchikawa, S. Applicability of Vis-NIR hyperspectral imaging for monitoring wood moisture content (MC). *Holzforschung* **2013**, *67*, 307–314. [CrossRef]
52. Wang, L.; Yoon, K.J. Knowledge Distillation and Student-Teacher Learning for Visual Intelligence: A Review and New Outlooks. *Ieee Trans. Pattern Anal. Mach. Intell.* **2022**, *44*, 3048–3068. [CrossRef]
53. Maimaitijiang, M.; Sagan, V.; Sidike, P.; Hartling, S.; Esposito, F.; Fritschi, F.B. Soybean yield prediction from UAV using multimodal data fusion and deep learning. *Remote Sens. Environ.* **2020**, *237*, 111599. [CrossRef]
54. Khelifi, L.; Mignotte, M. Deep Learning for Change Detection in Remote Sensing Images: Comprehensive Review and Meta-Analysis. *IEEE Access* **2020**, *8*, 126385–126400. [CrossRef]
55. Sharma, S.; Sharma, S.; Athaiya, A. Activation functions in neural networks. *Towards Data Sci.* **2017**, *6*, 310–316. [CrossRef]
56. Karlik, B.; Olgac, A.V. Performance analysis of various activation functions in generalized MLP architectures of neural networks. *Int. J. Artif. Intell. Expert Syst.* **2011**, *1*, 111–122.
57. Albawi, S.; Mohammed, T.A.; Al-Zawi, S. Understanding of a convolutional neural network. In Proceedings of the 2017 International Conference on Engineering and Technology (ICET), Antalya, Turkey, 21–23 August 2017; IEEE: New York, NY, USA; pp. 1–6.
58. Guo, H.Q.; Zhou, J.; Koopialipoor, M.; Armaghani, D.J.; Tahir, M.M. Deep neural network and whale optimization algorithm to assess flyrock induced by blasting. *Eng. Comput.* **2021**, *37*, 173–186. [CrossRef]
59. Patel, A.K.; Ghosh, J.K.; Pande, S.; Sayyad, S.U. Deep-Learning-Based Approach for Estimation of Fractional Abundance of Nitrogen in Soil From Hyperspectral Data. *IEEE J. Sel. Top. Appl. Earth Obs. Remote Sens.* **2020**, *13*, 6495–6511. [CrossRef]
60. Padarian, J.; Minasny, B.; McBratney, A.B. Using deep learning to predict soil properties from regional spectral data. *Geoderma Reg.* **2019**, *16*, e00198. [CrossRef]
61. Geras, K.J.; Wolfson, S.; Shen, Y.; Wu, N.; Kim, S.; Kim, E.; Heacock, L.; Parikh, U.; Moy, L.; Cho, K. High-resolution breast cancer screening with multi-view deep convolutional neural networks. *arXiv* **2017**, arXiv:1703.07047.
62. Wang, F.; Yang, M.; Ma, L.; Zhang, T.; Qin, W.; Li, W.; Zhang, Y.; Sun, Z.; Wang, Z.; Li, F.; et al. Estimation of Above-Ground Biomass of Winter Wheat Based on Consumer-Grade Multi-Spectral UAV. *Remote Sens.* **2022**, *14*, 1251. [CrossRef]
63. Rudin, C. Stop explaining black box machine learning models for high stakes decisions and use interpretable models instead. *Nat. Mach. Intell.* **2019**, *1*, 206–215. [CrossRef]
64. Du, M.; Liu, N.; Hu, X. Techniques for interpretable machine learning. *Commun. ACM* **2019**, *63*, 68–77. [CrossRef]
65. Linardatos, P.; Papastefanopoulos, V.; Kotsiantis, S. Explainable ai: A review of machine learning interpretability methods. *Entropy* **2020**, *23*, 18. [CrossRef]
66. Huang, N.; Lu, G.; Xu, D. A permutation importance-based feature selection method for short-term electricity load forecasting using random forest. *Energies* **2016**, *9*, 767. [CrossRef]
67. Murdoch, W.J.; Singh, C.; Kumbier, K.; Abbasi-Asl, R.; Yu, B. Definitions, methods, and applications in interpretable machine learning. *Proc. Natl. Acad. Sci. USA* **2019**, *116*, 22071–22080. [CrossRef]
68. Ribeiro, M.T.; Singh, S.; Guestrin, C. "Why should i trust you?" Explaining the predictions of any classifier. In Proceedings of the 22nd ACM SIGKDD International Conference on Knowledge Discovery and Data Mining, San Francisco, CA, USA, 13–17 August 2016; pp. 1135–1144.
69. Srivastava, N.; Hinton, G.; Krizhevsky, A.; Sutskever, I.; Salakhutdinov, R. Dropout: A Simple Way to Prevent Neural Networks from Overfitting. *J. Mach. Learn. Res.* **2014**, *15*, 1929–1958.
70. Li, L.S.; Jamieson, K.; DeSalvo, G.; Rostamizadeh, A.; Talwalkar, A. Hyperband: A Novel Bandit-Based Approach to Hyperparameter Optimization. *J. Mach. Learn. Res.* **2018**, *18*, 1–52.
71. Bergstra, J.; Bengio, Y. Random search for hyper-parameter optimization. *J. Mach. Learn. Res.* **2012**, *13*, 281–305.
72. Altmann, A.; Toloşi, L.; Sander, O.; Lengauer, T. Permutation importance: A corrected feature importance measure. *Bioinformatics* **2010**, *26*, 1340–1347. [CrossRef]
73. Ge, Y.; Morgan, C.L.S.; Wijewardane, N.K. Visible and near-infrared reflectance spectroscopy analysis of soils. *Soil Sci. Soc. Am. J.* **2020**, *84*, 1495–1502. [CrossRef]
74. Stenberg, B.; Viscarra Rossel, R.A.; Mouazen, A.M.; Wetterlind, J. Chapter Five—Visible and Near Infrared Spectroscopy in Soil Science. *Adv. Agron.* **2010**, *107*, 163–215.
75. Griffin, R.A.; Au, A.K.; Frost, R.R. Effect of pH on adsorption of chromium from landfill-leachate by clay minerals. *J. Environ. Sci. Health Part A Environ. Sci. Eng.* **1977**, *12*, 431–449. [CrossRef]
76. de Pablo, L.; Chávez, M.L.; Abatal, M. Adsorption of heavy metals in acid to alkaline environments by montmorillonite and Ca-montmorillonite. *Chem. Eng. J.* **2011**, *171*, 1276–1286. [CrossRef]
77. Wu, Y.; Chen, J.; Ji, J.; Gong, P.; Liao, Q.; Tian, Q.; Ma, H. A Mechanism Study of Reflectance Spectroscopy for Investigating Heavy Metals in Soils. *Soil Sci. Soc. Am. J.* **2007**, *71*, 918–926. [CrossRef]
78. Buerge, I.J.; Hug, S.J. Influence of Mineral Surfaces on Chromium(VI) Reduction by Iron(II). *Environ. Sci. Technol.* **1999**, *33*, 4285–4291. [CrossRef]

toxics

MDPI

Article

Optimizing Heavy Metal Uptake in *Carpobrotus aequilaterus* Through Electrokinetic Treatment: A Comprehensive Study on Phytoremediation from Mine Tailings

Yasna Tapia [1], Osvaldo Salazar [1], Oscar Seguel [1], Jonathan Suazo-Hernández [2], Diego Urdiales-Flores [3], Humberto Aponte [4,5] and Cristian Urdiales [6,*]

[1] Departamento de Ingeniería y Suelos, Facultad de Ciencias Agronómicas, Universidad de Chile, Santiago 11315, Chile; yasnatapiafernandez@uchile.cl (Y.T.); osalazar@uchile.cl (O.S.); oseguel@uchile.cl (O.S.)
[2] Center of Plant, Soil Interaction and Natural Resources Biotechnology, Scientific and Biotechnological Bioresource Nucleus (BIOREN-UFRO), Universidad de La Frontera, Avenida Francisco Salazar, Temuco 01145, Chile; j.suazo06@ufromail.cl
[3] Institute of Earth Surface Dynamics, University of Lausanne, 1015 Lausanne, Switzerland; diego.urdialesflores@unil.ch
[4] Laboratory of Soil Microbial Ecology and Biogeochemistry, Institute of Agri-Food, Animal and Environmental Sciences (ICA3), Universidad de O'Higgins, San Fernando 3070000, Chile; humberto.aponte@uoh.cl
[5] Centre of Systems Biology for Crop Protection (BioSaV), Universidad de O'Higgins, San Fernando 3070000, Chile
[6] Sede Vallenar, Universidad de Atacama, Av. Costanera #105, Vallenar 1612178, Chile
* Correspondence: cristian.urdiales@uda.cl; Tel.: +56-998268437

Abstract: Copper mining drives economic growth, with the global demand expected to reach 120 million metric tons annually by 2050. However, mining produces tailings containing heavy metals (HMs), which poses environmental risks. This study investigated the efficacy of phytoremediation (Phy) combined with electrokinetic treatment (EKT) to increase metal uptake in *Carpobrotus aequilaterus* grown in tailings from the Metropolitan Region of Chile. The plants were exposed to varying voltages and treatment durations. In the control (no EKT), the root metal contents were Fe (1008.41 mg/kg) > Cu (176.38 mg/kg) > Mn (103.73 mg/kg) > Zn (30.26 mg/kg), whereas in the shoots, the order was Mn (48.69 mg/kg) > Cu (21.14 mg/kg) > Zn (17.67 mg/kg) > Fe (27.32 mg/kg). The optimal EKT (15 V for 8 h) significantly increased metal uptake, with roots accumulating Fe (5997.24 mg kg^{-1}) > Mn (672 mg kg^{-1}) > Cu (547.68 mg kg^{-1}) > Zn (90.99 mg kg^{-1}), whereas shoots contained Fe (1717.95 mg kg^{-1}) > Mn (930 mg kg^{-1}) > Cu (219.47 mg kg^{-1}) > Zn (58.48 mg kg^{-1}). Although EKT enhanced plant growth and biomass, higher voltages stressed the plants. Longer treatments were more effective, suggesting that EK–Phy is a promising method for remediating metal-contaminated tailings.

Keywords: environmental remediation; remediation efficiency; electrochemical treatment; hyperaccumulator plants; Mediterranean climate-types

1. Introduction

The evolution of mining has greatly contributed to the economic growth of many countries, underscoring the sector's global importance [1]. Copper (Cu) mining, in particular, plays a crucial role in the economic development of several countries. Currently, the global Cu production capacity is approximately 22 million metric tons (MMTs) annually, led by Chile, with 5.2 MMTs, followed by Peru and Congo (2.2 MMTs each), and China (1.9 MMTs). Projections indicate that by 2050, the global Cu demand is anticipated to reach between 86 and 102 MMTs annually [2,3]. However, the extraction and processing of copper produce significant amounts of mine tailings, which account for 97% of terrestrial minerals [4]. These tailings often contain high levels of heavy metals (HMs), metalloids, and other toxic substances, posing environmental risks to soil and groundwater [5,6]. The speciation of HMs within tailings affects their bioavailability and toxicity, allowing these metals to leach into groundwater or be taken up by plants, thereby entering the food chain [7].

In Chile, approximately 764 tailings deposits were reported by 2022, with over 14% being active and approximately 84% being inactive. Most deposits are located in arid northern regions, where dry Mediterranean climates lead to local pollution from wind erosion [8]. Tailings stability depends on the structure of the deposits, including sludge, fine-grained materials, and chemicals such as silicates and sulfides [9]. When sulfide minerals in tailings interact with rainwater, sulfuric acid forms, which in turn leads to leaching that contaminates surface waters with toxic HMs, including Cu, Zn, Fe, and Pb. In northern Chile, sulphate concentrations in mine tailings range from 1243 mg L^{-1} to 2912 mg L^{-1}. In comparison, natural water bodies contain 630 mg L^{-1}, 2 and 250 mg L^{-1} in lakes and 230 mg L^{-1} in groundwater [10–13]. These variations illustrate the substantial impact of mining on local water quality and availability for human consumption.

HMs go into crops through contaminated irrigation water, soil absorption, or leaf deposition and accumulate in edible parts via physiological processes [14,15]. The presence of HM in the food chain poses health risks to animals and humans [16] as even trace amounts of HM can be harmful [17]. Studies have detected HM levels in various food products, including beverages, juices, and wines [18–20]. In Chile's Central Zone, Cu levels in agricultural soil range from 307 mg kg^{-1} to 596 mg kg^{-1} [21]. In the Copiapó region, northern Chile, soils surrounding mine tailings have Cu concentrations between 68 mg kg^{-1} and 2116 mg kg^{-1} [22]. Currently, Chile lacks a defined soil quality standard for copper levels. In Brazil, guidelines establish an intervention value (IV) of 200 mg kg^{-1} (dry weight) [23] for Cu in agricultural soil. Copper levels in Copiapó agricultural soils often exceed this threshold. This raises concerns about potential health risks associated with elevated Cu concentrations in the region.

To reduce HM bioavailability, a wide range of remediation techniques, including chemical methods (e.g., mineral clays, nanoparticles, and biochar) and biological methods (plants, enzymes, and microorganisms), have been explored [24–27]. Phytoremediation (Phy) is a biological method in which plants are used to treat contaminated soil and water. It removes, degrades, or immobilizes harmful substances. This approach is cost-effective and sustainable. It also allows biodiversity restoration, improves aesthetics, and enhances carbon sequestration [26,27]. In northern Chile, native species such as *Atriplex nummularia*, *Prosopis tamarug*, and *Schinus molle* show promise for Cu phytostabilization [28]. In the Coquimbo Region, native/endemic plants such as *Carpobrotus aequilaterus* have demonstrated adaptability to nutrient- and water-scarce environments, whereas humic substances increase their Cu uptake [11,29]. Additionally, *C. aequilaterus* has proven effective in acid mine drainage treatment using the rhizofiltration technique [30].

Phytoremediation is a sustainable and cost-effective approach for remediating metal-contaminated soils, but its effectiveness is a standalone method. It is limited by challenges such as low metal bioavailability, phytotoxicity from high metal concentrations, and the slow pace of natural growth processes [31]. To overcome these limitations, various enhancement strategies have been developed. For instance, chelate-assisted uptake enhances metal bioavailability by forming soluble metal–chelate complexes, with synthetic chelating agents such as EDTA binding to various metal contaminants in the environment [32,33]. However, these agents pose significant environmental risks as they can mobilize heavy metals into deeper soil layers and groundwater, leading to contamination beyond the treatment zone [34,35]. Additionally, their slow degradation and high stability contribute to their persistence as environmental pollutants, sparking a growing interest in more eco-friendly, biodegradable alternatives. In response to these challenges, biodegradable chelating agents such as Ethylenediamine-N,N′-disuccinic acid (EDDS) and low molecular weight organic acids such as citric acid and glutamic diacetic acid (GLDA) have emerged as promising alternatives. These agents offer a dual benefit: they exhibit effective chelating properties while minimizing environmental impact [36]. Another technique to enhance the phytoremediation process is microbial-assisted phytoremediation, which leverages plant-microbe interactions to improve metal solubilization and absorption and offers additional potential [37,38]. These combined strategies aim to increase the efficiency and environmental sustainability of phytoremediation, particularly in addressing the challenges of highly contaminated environments such as mine tailings.

Considering that Phy is an effective process, it could struggle with highly contaminated tailings due to low nutrient levels, limited organic matter, and challenging soil conditions [39]. Electrokinetic treatment (EKT) offers a complementary approach in which direct electrical potential is applied through anodes and cathodes in tailings to mobilize contaminants toward electrodes for extraction. EKT enhances contaminant solubility and bioavailability, aiding HM uptake by plants when combined with Phy [40–42].

In EKT, electro-osmosis moves water toward the cathode, while electromigration directs ions toward oppositely charged electrodes, and both mechanisms improve contaminant mobility [42]. EKT induces physical and chemical changes in tailings that help to direct contaminants toward electrodes. Several studies have shown that electrodes placed in contaminated soil, combined with electrolytes to increase conductivity, improve treatment efficiency [43,44]. The tailings' pH also influences pollutant retention or leaching, while recycling cathodic solutions to anodic areas can mitigate acidic conditions. The use of buffer solutions such as citric or acetic acid further controls the pH through complexation or chelation [45,46]. Integrating EKT with Phy benefits mine tailing remediation by applying low-intensity electric fields that increase contaminant bioavailability for plants. Electric fields can increase pollutant uptake by plants. They also redistribute heavy metal concentrations in the soil. Higher levels accumulate near the anode; in contrast, lower levels are found near the cathode [47,48].

This study aimed to evaluate a synergistic approach by integrating the contaminant uptake capacity of *C. aequilaterus* with the enhanced mobility and bioavailability of contaminants facilitated through electrical stimulation. The effectiveness of electrokinetic treatments under varying durations and applied potentials was examined, alongside an assessment of the resulting micronutrient content in *C. aequilaterus* plant tissues following the treatments.

2. Materials and Methods

2.1. Mine Tailings Samples

Cu mine tailing composite samples were randomly collected from the operational Ovejería tailings dam (33.05° S; 70.79° W), which is located in the surroundings of the Huechún community in the Santiago Metropolitan Region, Chile (Figure 1). The tailings samples were homogenized, air-dried, and sieved to a particle size of <2 mm for later storage in sealed plastic bins at 20 °C.

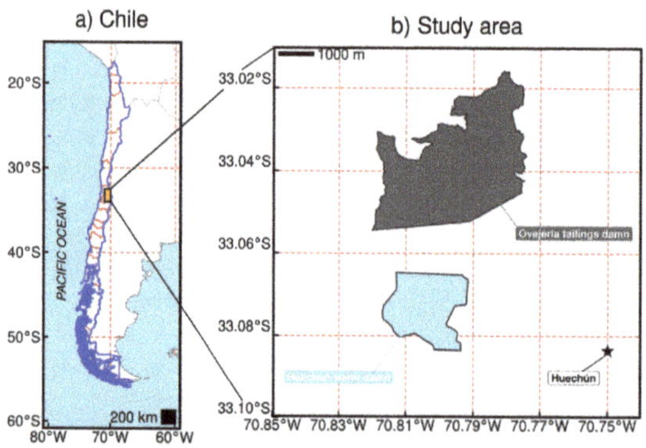

Figure 1. Geographical location of the study area according to (**a**) Chile reference and (**b**) the Ovejería tailings dam.

2.2. Determination of the Physicochemical Properties of Mine Tailings

The mine tailings were subjected to physical and chemical characterization procedures based on the methods described by Sadzawka et al. [49], including the determination of the pH of the aqueous suspension (1:2.5 w/v), electrical conductivity (EC) of the saturated extract, and OM content via the adapted Walkey–Black method. The available nitrogen (N) was measured via extraction with 2 M KCl, distillation, and titration with 0.001 M H_2SO_4. The Olsen method was used to determine the available phosphorus (P). For the available potassium (K) fraction, 1 M $NH_4CH_3CO_2$ was used as an extracting agent for subsequent atomic emission spectroscopy (AES) determination via a Perkin Elmer instrument (Waltham, MA, USA). Available sulfur (SO_4^{2-}) was extracted with 0.01 M $Ca(H_2PO_4)_2$ and measured later via turbidimetric methodology. The total contents of copper (Cu), iron (Fe), zinc (Zn), and manganese (Mn) in the tailing sample were determined via atomic absorption spectroscopy (AAS) (Perkin Elmer, Waltham, MA, USA) with previous microwave acid (HNO_3-HF) digestion [50]. An aqueous solution containing DTPA (0.005 M), $CaCl_2$ (0.01 M), and triethanolamine (0.1 M) adjusted to a pH of 7.3 was used for the extraction and assessment of available Cu, Fe, Mn, and Zn. The contents of these elements were subsequently determined using AAS. Each measurement was conducted in triplicate. The determination of soil texture was carried out using the Bouyoucos method [51]. In accordance with Stokes's law of sedimentation [52], the Bouyoucos method uses a 5% sodium hexametaphosphate solution and a high-speed mixer to disperse soil samples efficiently.

2.3. Experimental Columns Setup

Polyethylene columns with a height of 21 cm and an inner diameter of 9.56 cm were used (Figure 2). They were filled with mine tailings to a depth of 20 cm, ensuring uniform field bulk density. Drain holes were incorporated at the base of each column to collect the leachate. Two graphite electrodes were placed opposite each other on the column, 5 cm apart, and connected to an Ele-Tech HY3005E-2 Digital DC power supply (HYELEC, Hangzhou, China). For irrigation to provide essential macronutrients, a Hoagland-based nutrient solution was applied twice a week. Irrigation maintained the volumetric moisture content of the tailings above the experimentally determined field capacity (approximately 10% volumetric water content). A peristaltic pump connected to a reservoir facilitated the irrigation process.

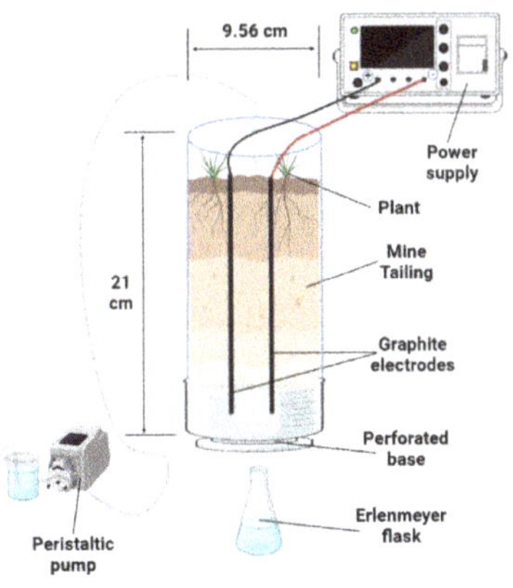

Figure 2. Experimental scheme of the assays.

Three *C. aequilaterus* plants were cultivated in each column of mine tailings, and these plants were set up through cuttings, which were directly inserted into the tailings for subsequent rooting. *C. aequilaterus* is a succulent native Chilean perennial plant used for ornamental purposes. The fruits and aerial parts of the plant are edible [29,53,54]. Additionally, *C. aequilaterus* is a recognized species capable of adapting and thriving spontaneously in mine tailings [29] because of its ability to absorb metals, metalloids, and sulphates.

This study evaluated the effectiveness of 15 V and 30 V voltage levels in enhancing HM extraction from mine tailings, which was selected on the basis of prior research highlighting their efficiency without hindering biological growth [55]. The optimization process involved preliminary trials that balanced effective HM mobilization with the ecological integrity of the medium, particularly preserving the growth of *C. chilensis*. The sustained application of voltage over seven days did not allow the development of *C. aequilaterus* environmental impacts. The treatments involved applying different voltages and durations to the mine tailings. In the first treatment (15 V-4 h), 15 volts were applied for 4 h twice a week. In the second treatment (15 V-8 h), 15 volts were applied for 8 h twice a week. In the third treatment (30 V-4 h), 30 volts were applied for 4 h on the same schedule. Finally, the fourth treatment (30 V-8 h) involved applying 30 volts for 8 h on the same schedule. Each treatment was replicated three times to ensure the reliability of the results and to optimize metal removal. For the control treatment, no voltage was applied, and the mine tailings were monitored under the same schedule without electrical intervention; the procedure was repeated three times to ensure the reliability of the results and to compare the metal removal efficiency with that of the treated samples.

The experiment lasted 45 days, during which the electrical conductivity of the leachate from each column was continuously monitored. Additionally, the soil water content was measured via a TDR probe from Campbell Scientific (Logan, UT, USA) to assess the dynamics of the water in the experimental setup.

2.4. Plant Growth and Biomass Accumulation

The initial area of the plant shoots (aerial part and stems) was determined before being planted in mine tailings. Throughout the growing season, we regularly measured the leaf area and root length to monitor the growth and accumulation of plant biomass in mine tailings. For leaf area determination, we used ImageJ version 1.54d image processing software.

The growth data from the plants cultivated in mine tailings were applied to a logistic growth model via Origin Pro 2019b software. The logistic growth model is defined by Equation (1):

$$Y = \frac{Y_{max}}{\left(1 + e^{-k(t-t_m)}\right)} \quad (1)$$

The symmetrical sigmoid model uses Y to represent the observed foliar area values.

Where Y_{max} signifies the maximum achievable foliar area (cm^2), k represents the growth rate (day^{-1}), and t_m is the time of the inflection point where growth slows down (day). The inflection point occurs at 50% of the horizontal asymptote ($0.5 \times Y_{max}$) in the logistic model.

2.5. Analytical Determinations in Plant Tissues

The EKT + Phy experiments were conducted at the laboratory of the Faculty of Agricultural Science, University of Chile (33.56° S; 70.63° W). The laboratory maintained an average temperature of 20 °C throughout the study period, which closely resembled the mean temperature of 18.1 °C reported by the weather station in the tailings pond area. The tests were conducted in the summer between January and December.

The *C. aequilaterus* plants were preserved for 45 days. Upon completion of the test, the plants were carefully washed with distilled water. The shoots and roots were subsequently separated, and their fresh weight (FW) was subsequently determined. The plant tissues were then subjected to drying in an oven at 65 °C until a constant weight was achieved, allowing for the determination of dry biomass (DW). The dried samples were ground separately in a mill for further analysis.

For the quantification of the Cu, Fe, Mn, and Zn contents in the plant tissues, the dried samples were digested in a microwave oven with HNO_3-HF [50] and analyzed via AAS.

2.6. Statistical Analysis

After evaluation of the statistical assumptions (normal distribution and homoscedasticity), one-way ANOVA with a Tukey HSD as a post hoc test was used to evaluate the individual influence of each treatment on the HM content in shoot and root tissues. The data were log10 transformed when the variables did not meet the statistical assumptions. All the statistical analyses were conducted with R statistic version 4.4.1 and the GraphPad Prism 8.0.2 software package.

3. Results and Discussion

3.1. Physicochemical Characterization of Mine Tailings

The physicochemical analysis of the Ovejería tailings sample revealed a moderately acidic pH, non-saline conditions, minimal OM content, and restricted access to essential nutrients such as N, P, and K (Table 1). The low macronutrient content (N, P, and K) creates significant challenges for reusing tailings as reforestation sites. It also hinders the establishment of self-sustaining ecosystems. This issue is widely reported in studies on Chilean tailings. Nutrient deficiencies and metal contamination further complicate efforts toward ecological restoration [11,29]. The mineral content processes resulted in significant increases in the sulphate content.

Table 1. Physicochemical characterization of the Ovejería mine tailings.

Parameter	Value
pH	5.80 ± 0.055
EC (mS cm^{-1})	2.74 ± 0.04
Organic matter (%)	0.55 ± 0.04
Available N (mg kg^{-1})	3.73 ± 0.47
Available P (mg kg^{-1})	1.64 ± 0.02
Available K (mg kg^{-1})	6.23 ± 0.17
Available SO_4^{2-} (mg kg^{-1})	410 ± 12
Bioavailable Cu (mg kg^{-1})	28.1 ± 1.04
Bioavailable Fe (mg kg^{-1})	307 ± 9.23
Bioavailable Mn (mg kg^{-1})	1.73 ± 0.23
Bioavailable Zn (mg kg^{-1})	4.37 ± 0.31
Total Cu (mg kg^{-1})	1924 ± 119
Total Fe (mg kg^{-1})	8769 ± 468
Total Mn (mg kg^{-1})	187 ± 3.58
Total Zn (mg kg^{-1})	92.6 ± 10.2
Bulk density (g cm^{-3})	1.67
Sand (%)	72.73
Silt (%)	7.22
Clay (%)	20.05

The results revealed a high total content, reflecting the mineral content in the tailings, but a relatively low bioavailable content. The physical analysis of the mine tailings revealed that sand accounted for 72.73% of the soil composition. The silt and clay fractions were low at 7.22% and 20.05%, respectively, indicating coarse-textured mine tailings with moderate water retention capacity compared with sandy topsoil [56].

3.2. Heavy Metal Contents in *Carpobrotus aequilaterus* Compartments

Figure 3 displays the HM contents (Cu, Fe, Mn, and Zn) in the shoots and roots of plants subjected to different electrical treatments. These treatments included 15 V for 4 h, 15 V for 8 h, 30 V for 4 h, and 30 V for 8 h, alongside a control group. In the shoots, the Cu content in the 15 V-8 h treatment group was the highest content, significantly greater than that in the other treatment groups, whereas that in the control group was the lowest. Similarly, the Fe content was highest in the 15 V-8 h treatment and much lower in the control and 30 V-4 h treatments. Mn followed the same trend, with the 15 V-8 h treatment yielding the highest concentration and the control group the lowest. Zn showed a slightly different pattern, with both the 15 V-8 h and 30 V-8 h treatments resulting in elevated concentrations that were significantly higher than those of the control and other treatments.

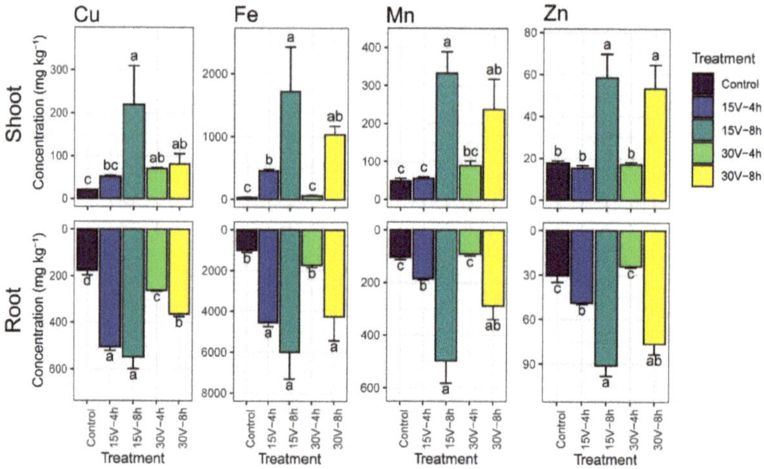

Figure 3. Heavy metal contents in plant tissues: Cu content; Fe content; Mn content; Zn content. Comparative analysis of the control and treatment groups: 15 V-4 h, 15 V-8 h, 30 V-4 h and 30 V-8 h. The error bar in the graph indicates the standard error of the mean, whereas different letters over the bars indicate statistically significant differences between means at $p < 0.05$.

In the roots, the HM content was significantly higher than that in the shoots, indicating an inverse relationship. The 15 V-8 h treatment produced the highest concentrations of Cu, Fe, and Mn, whereas significant reductions were observed in the 30 V-4 h treatment. Overall, the results suggest that electrical treatments strongly influence the distribution of metals between shoots and roots. Statistical analysis confirmed that, compared with the control, only the 15 V-8 h treatment significantly increased the Cu, Fe, Mn, and Zn contents in the roots. The Cu content was approximately three times higher than that of the control, the Fe content was six times higher, the Mn content was five times higher, and the Zn content was more than three times higher (Figure 3).

The optimized EKT at 15 V-8 h significantly increased Fe and Cu bioaccumulation within *C. aequilaterus*. The Fe content reached 5997.24 mg kg^{-1} in the roots and 1717.95 mg kg^{-1} in the shoots. These values were substantially higher than the levels recorded in prior studies with potassium humates over 120 days. In those studies, the Fe content was 2320 mg kg^{-1} in the roots and 546 mg kg^{-1} in the shoots. This study also surpassed the metal contents reported in earlier research by the same group using similar tailing samples and plant species [11].

Similarly, the optimized treatment showed superior efficacy in terms of Cu accumulation. The Cu content was 547.68 mg kg^{-1} in the roots and 219.4 mg kg^{-1} in the shoots, significantly exceeding earlier studies that reported a value of 29.3 mg kg^{-1} in the shoots [11]. These Cu levels were also comparable to those reported in *Achnatherum splendens Nevski*, which grows near mine tailings at the Ashel Cu–Zn and Kirk Tall Pb–Zn mines in Aletai, Xinjiang, China. In those areas, the Cu content was 554.2 mg kg^{-1} in the roots and 53.00 mg kg^{-1} in the shoots [57].

The Mn distribution pattern in this study differed from that reported in previous studies. The Mn content was 497.03 mg kg^{-1} in the root tissues and 332.82 mg kg^{-1} in the shoot tissues, which was significantly lower than the earlier results of 672 mg kg^{-1} in the roots and 930 mg kg^{-1} in the shoots. This variation may result from differences in soil composition, particularly the presence of potassium humates, which facilitate the formation of soluble metal complexes, thereby increasing metal bioavailability.

Zn accumulation was also slightly different. The Zn content was 90.99 mg kg^{-1} in the roots and 58.48 mg kg^{-1} in the shoots. These values were marginally lower than the earlier results in the roots (94 mg kg^{-1}) and shoots (57.4 mg kg^{-1}). Such differences may result from experimental conditions or genetic factors influencing Zn uptake and translocation. Zn plays essential roles in numerous enzymes and pathways, including carbohydrate and protein metabolism, auxin regulation, and membrane integrity. Zn plays essential roles as a structural component and a regulatory co-factor in more than three hundred enzymes, influencing critical pathways such as carbohydrate metabolism, which encompasses glycolysis and the citric acid cycle [58]. These pathways include the metabolism of protein and auxin, a key plant hormone responsible for growth regulation and response to light and gravity, pollen formation, maintenance of the integrity of biological membranes, and resistance to certain pathogens [59].

Several key insights emerged from the experiments. Compared with the 15 V treatment, the higher voltage (30 V) did not enhance the remediation process. Instead, stronger electric fields appeared to hinder contaminant removal by causing excessive migration toward the electrodes. This led to physiological stress in *C. aequilaterus*, reducing growth, biomass production, and nutrient extraction efficiency. Moreover, higher voltage levels may induce electrochemical reactions, degrading or transforming contaminants through oxidation or reduction.

The duration of voltage application is a critical factor influencing the effectiveness of the remediation process. Compared with the 4 h treatment, the 8 h treatment resulted in more substantial improvements in the metal content, mine tailing properties, and plant growth. Extended exposure to the electric field allowed for greater interactions between the electric field, plants, and mine tailings. This led to enhanced HM removal and improved conditions in the tailings. A longer treatment duration enabled more comprehensive redistribution and extraction of HMs. It also increased metal uptake and accumulation by plants. These processes are primarily driven by electromigration and electro-osmosis. In more detail, those mechanisms transport ionic species and fluids through the soil matrix under the influence of an electric field. This finding underscores that longer durations offer greater opportunities for achieving effective remediation outcomes.

The hierarchy of metal accumulation in plant tissues followed the order of Fe > Cu > Mn > Zn. This pattern contrasts with the bioavailability sequence in tailings, which is Fe > Cu > Zn > Mn (Table 1). This difference can be attributed to variations in metal mobility, plant uptake mechanisms, and the effects of electrokinetic treatments [40,47,48].

Electromigration and electro-osmosis enhance ion mobility within the soil matrix under an electric field. These processes preferentially mobilize Fe and Cu ions toward plant roots, aided by their high bioavailability and favorable physicochemical properties. The electric field also creates pH gradients in the soil, forming localized acidic zones that increase Fe and Cu solubility. Conversely, Mn and Zn have reduced solubilities under these conditions, limiting their mobility and plant uptake.

Plants have developed sophisticated mechanisms to acquire essential metals such as Fe and Cu. Root exudates chelate Fe ions, increasing their solubility and availability despite competition from other metals [60]. Copper uptake is supported by its vital physiological role and increased availability from electrokinetic treatments [61]. In contrast, competitive inhibition affects Mn and Zn uptake. Fe and Cu often share transport pathways with Mn and Zn, leading to preferential uptake of the former and limiting Mn and Zn accumulation in plant tissues.

Metal translocation within plant tissues follows distinct patterns. Owing to their high solubility and mobility in the vascular system, Fe and Cu are efficiently translocated from roots to shoots. In contrast, Mn and Zn exhibit lower translocation efficiencies, resulting in limited movement beyond the roots. This difference in translocation contributes to the observed hierarchy of metal accumulation in plant tissues.

In summary, metal accumulation in plant tissues is driven by electrokinetic enhancement and intrinsic plant uptake mechanisms. Owing to its high bioavailability and specialized uptake systems, Fe is dominant. Copper follows because of its physiological importance and increased availability. The limited bioavailability of Mn is due to its strong association with soil minerals, whereas Zn forms fewer mobile complexes and competitively inhibits Fe and Cu. Understanding these dynamics is essential for optimizing phytoremediation strategies and improving nutrient management in agricultural systems.

The experiments demonstrated the effectiveness of combining electrokinetic (EK) and phytoremediation (Phy) treatments for remediating mine tailings. These treatments effectively removed contaminants and improved environmental conditions. Key findings emphasized the importance of optimizing both voltage and duration to enhance treatment efficacy. The study revealed that 15 V was more effective than 30 V for contaminant removal without disrupting plant physiology. Prolonged exposure to an electric field significantly improved remediation outcomes. Longer treatment durations resulted in improved contaminant extraction and enhanced plant metal uptake. These findings highlight the potential of extended treatments to achieve more comprehensive and effective remediation results.

3.3. Plant Growth

The application of the logistic growth model provided valuable insights into plant growth dynamics. Figure 4 shows the logistic root growth curves and the experimental determination of the leaf area for the optimal EK treatment (15 V-8 h) and the control group. Table 2 shows the fitted parameters of the logistic growth model.

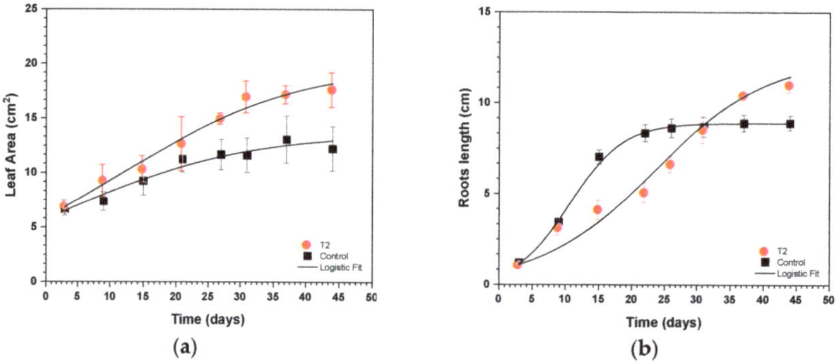

Figure 4. Variations in the leaf area of the designated *C. aequilaterus* shoots (**a**) and root length (**b**) during their expansion were noted for both the control group and the optimal treatment conditions (15 V-8 h). The research included four plants, with the vertical bars depicted on the graph indicating the standard error of the mean. The associations among the observed data points were delineated via a logistic growth model.

Table 2. Fitted parameters of the logistic growth model for the shoots and roots of *Carpobrotus aequilaterus* in the control and 15 V-8 h groups.

	Y_{max} (cm^2)	t_m (day)	K (day^{-1})	Reduced Chi-Sqr	R^2
			Shoots		
Control	13.63 ± 1.349	4.291 ± 2.737	0.07405 ± 0.02037	0.2466	0.9647
15 V-8 h	19.93 ± 1.176	11.86 ± 2.033	0.07304 ± 0.00834	0.2186	0.9939
			Roots		
Control	8.889 ± 0.1303	10.66 ± 0.2453	0.2495 ± 0.01121	0.3277	0.9985
15 V-8 h	12.54 ± 0.957	23.46 ± 2.497	0.1148 ± 0.01054	2.6585	0.9979

In the control group, the model parameters gradually increased. The maximum foliar area reached 13.63 cm^2, with an inflection point at 4.291 days. The goodness-of-fit indicators confirmed that the model effectively represented the growth dynamics of the control group. In contrast, the 15 V-8 h treatment resulted in a distinct growth trajectory. The parameters indicated a greater maximum foliar area of 19.93 cm^2 and a later inflection point at 11.86 days. Notably, the model's predictions closely matched the actual growth data, as evidenced by the high goodness-of-fit values.

The logistic growth model analysis of root growth also revealed significant findings. In the control group, root growth followed a well-fit logistic curve with parameters of Y_{max} = 8.89 cm, t_m = 10.66 days, and k = 0.2495 day^{-1}. This alignment demonstrated a strong correlation (R^2 = 0.9985) between the model and observed growth. Conversely, the 15 V-8 h treatment presented altered dynamics, with Y_{max} = 12.54 cm, t_m = 23.46 day, and k = 0.1148 day^{-1}. Although the model fit was slightly less accurate, as reflected by a greater reduction in the chi-square value of 2.6585, the R^2 value remained high at 0.9979, indicating a consistent growth trend.

The integration of EKT + Phy requires careful consideration of environmental risks. The introduction of an electric field can alter the soil pH, creating extremely acidic conditions near the anode and alkaline conditions near the cathode. These pH extremes negatively impact plant growth and microbial activity by causing nutrient deficiencies or toxicity. Additionally, electrical potential can significantly influence microbial and enzymatic activities [62]. Key enzymes such as urease, invertase, and phosphatase, which are critical for nutrient cycling and organic matter decomposition, may be adversely affected. Thus, the overall effectiveness of phytoremediation may be reduced.

The application of an electric field to mine tailings also generates heat due to the resistivity of the tailings [63,64], which can alter the physical properties of the tailings. This creates an unfavorable environment for both plants and soil. This research aims to optimize the use of EKT + Phy while minimizing environmental risk. Controlled electrical potentials between 15 V and 30 V were applied for 4 or 8 h, twice weekly, to avoid extreme acidic or alkaline conditions in the tailing soil profile. Additionally, this study explored succulent and shrubby plants that are tolerant to harsh conditions, aiding in contaminant degradation or stabilization. These strategies aim to increase the sustainability and effectiveness of phytoremediation technologies for managing mine tailings.

In summary, the results demonstrate that electrokinetic treatment significantly enhances metal uptake and accelerates plant growth. These findings establish it as a promising tool for optimizing phytoremediation in mine tailing reclamation. Careful calibration of electrokinetic parameters can improve remediation efficiency and plant health. These findings support the potential of this approach as a viable strategy for the ecological restoration of contaminated sites.

4. Conclusions

The combined use of electrokinetic treatment (EKT) and phytoremediation (Phy) represents a promising method for the remediation of metal-contaminated mine tailings, as evidenced by enhanced metal uptake and improved plant growth. This study demonstrated that applying a 15 V potential for 8 h significantly increased HM accumulation within plant tissues while simultaneously increasing plant growth rates, suggesting that this electrokinetic–phytoremediation approach could be an effective solution for ecological restoration in contaminated environments. However, our findings underscore that the effectiveness of this method is influenced by key factors, including the composition of tailings, pH levels, moisture content, and choice of plant species, each of which has a direct effect on the remediation process and growth dynamics.

Our results highlight the need for precise optimization of electrokinetic parameters, such as voltage application and duration, to maximize both metal uptake and plant health. The enhanced root dynamics and biomass accumulation observed in this study point to a potentially synergistic effect of EKT and Phy, which warrants further investigation to elucidate the underlying mechanistic interactions. For example, understanding how electric fields influence root architecture and growth kinetics could offer valuable insights into plant-based reclamation techniques and inform targeted strategies for other metal-polluted environments.

Future studies should also aim to evaluate the long-term ecological impacts of electrokinetic phytoremediation, including any possible effects on soil microbial communities, which play a crucial role in sustainable ecosystem recovery. Additionally, research into the scalability of this approach under field conditions is essential for assessing its feasibility in diverse mining environments, where soil properties and environmental conditions may vary widely. Further research will enable the development of more effective, site-specific strategies to enhance metal uptake and strengthen plant resilience, as well as to reduce the environmental footprint of mining activities. This knowledge will play a crucial role in designing sustainable, large-scale restoration techniques that prioritize ecological aspects.

Author Contributions: Conceptualization, C.U. and Y.T.; methodology, C.U. and Y.T.; formal analysis, J.S.-H. and D.U.-F.; investigation, C.U. and Y.T.; resources, Y.T. and O.S. (Osvaldo Salazar); data curation, O.S. (Osvaldo Salazar) and O.S. (Oscar Seguel); writing—original draft preparation, C.U. and J.S.-H.; writing—review and editing, D.U.-F., J.S.-H. and H.A.; visualization, H.A. and D.U.-F.; supervision, C.U.; project administration, C.U.; funding acquisition, C.U. All authors have read and agreed to the published version of the manuscript.

Funding: This research was funded by ANID FONDECYT Postdoctoral, grant number 3220201, 2022 of C. U. and FONDECYT postdoctoral project 3210752 of H.A.

Institutional Review Board Statement: Not applicable.

Informed Consent Statement: Not applicable.

Data Availability Statement: Data are contained within the article.

Acknowledgments: Cristian Urdiales thanks ANID Fondecyt Postdoctoral 2022, N° 3220201 and the Soil Chemistry Laboratory of the Agronomy Faculty of the University of Chile. J.S.-H. acknowledges ANID-FONDECYT/Post-Doctoral Grant N° 3230179.

Conflicts of Interest: The authors declare no conflicts of interest.

References

1. Candeias, C.; Ávila, P.; Coelho, P.; Teixeira, J.P. Mining Activities: Health Impacts. In *Encyclopedia of Environmental Health*; Elsevier: Amsterdam, The Netherlands, 2019; pp. 415–435. [CrossRef]
2. USGS Copper Statistics and Information. Available online: https://pubs.usgs.gov/periodicals/mcs2023/mcs2023-copper.pdf (accessed on 23 April 2023).
3. Seck, G.S.; Hache, E.; Bonnet, C.; Simoën, M.; Carcanague, S. Copper at the Crossroads: Assessment of the Interactions between Low-Carbon Energy Transition and Supply Limitations. *Resour. Conserv. Recycl.* **2020**, *163*, 105072. [CrossRef] [PubMed]
4. Edraki, M.; Baumgartl, T.; Manlapig, E.; Bradshaw, D.; Franks, D.M.; Moran, C.J. Designing Mine Tailings for Better Environmental, Social and Economic Outcomes: A Review of Alternative Approaches. *J. Clean. Prod.* **2014**, *84*, 411–420. [CrossRef]
5. Ramirez, M.; Massolo, S.; Frache, R.; Correa, J.A. Metal Speciation and Environmental Impact on Sandy Beaches Due to El Salvador Copper Mine, Chile. *Mar. Pollut. Bull.* **2005**, *50*, 62–72. [CrossRef]
6. Stauber, J.L.; Andrade, S.; Ramirez, M.; Adams, M.; Correa, J.A. Copper Bioavailability in a Coastal Environment of Northern Chile: Comparison of Bioassay and Analytical Speciation Approaches. *Mar. Pollut. Bull.* **2005**, *50*, 1363–1372. [CrossRef] [PubMed]
7. Zhang, X.; Yang, H.; Cui, Z. Migration and Speciation of Heavy Metal in Salinized Mine Tailings Affected by Iron Mining. *Water Sci. Technol.* **2017**, *76*, 1867–1874. [CrossRef]
8. Gerding, J.; Novoselov, A.A.; Morales, J. Climate and Pyrite: Two Factors to Control the Evolution of Abandoned Tailings in Northern Chile. *J. Geochem. Explor.* **2021**, *221*, 106686. [CrossRef]
9. Ritcey, G.M. Tailings Management in Gold Plants. *Hydrometallurgy* **2005**, *78*, 3–20. [CrossRef]
10. Santibañez, C.; De La Fuente, L.M.; Bustamante, E.; Silva, S.; León-Lobos, P.; Ginocchio, R. Potential Use of Organic- and Hard-Rock Mine Wastes on Aided Phytostabilization of Large-Scale Mine Tailings under Semiarid Mediterranean Climatic Conditions: Short-Term Field Study. *Appl. Environ. Soil Sci.* **2012**, *2012*, 895817. [CrossRef]

11. Tapia, Y.; Bustos, P.; Salazar, O.; Casanova, M.; Castillo, B.; Acuña, E.; Masaguer, A. Phytostabilization of Cu in Mine Tailings Using Native Plant Carpobrotus Aequilaterus and the Addition of Potassium Humates. *J. Geochem. Explor.* **2017**, *183*, 102–113. [CrossRef]
12. Nieva, N.E.; Borgnino, L.; García, M.G. Long Term Metal Release and Acid Generation in Abandoned Mine Wastes Containing Metal-Sulphides. *Environ. Pollut.* **2018**, *242*, 264–276. [CrossRef]
13. UNEP, G. *Water Data Summary 1985–1987*; Canada Centre for Inland: Burlington, VT, USA, 1990.
14. Kumar, S.; Prasad, S.; Yadav, K.K.; Shrivastava, M.; Gupta, N.; Nagar, S.; Bach, Q.V.; Kamyab, H.; Khan, S.A.; Yadav, S.; et al. Hazardous Heavy Metals Contamination of Vegetables and Food Chain: Role of Sustainable Remediation Approaches—A Review. *Environ. Res.* **2019**, *179*, 108792. [CrossRef] [PubMed]
15. Liu, X.; Song, Q.; Tang, Y.; Li, W.; Xu, J.; Wu, J.; Wang, F.; Brookes, P.C. Human Health Risk Assessment of Heavy Metals in Soil–Vegetable System: A Multi-Medium Analysis. *Sci. Total Environ.* **2013**, *463–464*, 530–540. [CrossRef] [PubMed]
16. Hosseinniaee, S.; Jafari, M.; Tavili, A.; Zare, S.; Cappai, G. Investigating Metal Pollution in the Food Chain Surrounding a Lead-Zinc Mine (Northwestern Iran); an Evaluation of Health Risks to Humans and Animals. *Environ. Monit. Assess.* **2023**, *195*, 946. [CrossRef] [PubMed]
17. EPA, U.S. *Integrated Risk Information System (IRIS) [Online Electronic Data File]*; US Environment Protection Agency, Office of Research and Development; Last updated December; National Center for Environment Assessment: Washington, DC, USA, 1996; Volume 1.
18. Wang, Z.; Jackson, L.S.; Jablonski, J.E. Factors Affecting the Levels of Heavy Metals in Juices Processed with Filter Aids. *J. Food Prot.* **2017**, *80*, 892–902. [CrossRef]
19. Ryan, R. Safety of Food and Beverages: Soft Drinks and Fruit Juices. *Encycl. Food Saf.* **2014**, *3*, 360–363. [CrossRef]
20. Asuku, A.O.; Ayinla, M.T.; Ajibare, A.J.; Adeyemo, M.B.; Adeyemo, R.O. Heavy Metals and Emerging Contaminants in Foods and Food Products Associated with Neurotoxicity. In *Emerging Contaminants in Food and Food Products*; CRC Press: Boca Raton, FL, USA, 2024; pp. 236–250. [CrossRef]
21. Neaman, A.; Reyes, L.; Trolard, F.; Bourrié, G.; Sauvé, S. Copper Mobility in Contaminated Soils of the Puchuncaví Valley, Central Chile. *Geoderma* **2009**, *150*, 359–366. [CrossRef]
22. Carkovic, A.B.; Calcagni, M.S.; Vega, A.S.; Coquery, M.; Moya, P.M.; Bonilla, C.A.; Pastén, P.A. Active and Legacy Mining in an Arid Urban Environment: Challenges and Perspectives for Copiapó, Northern Chile. *Environ. Geochem. Health* **2016**, *38*, 1001–1014. [CrossRef]
23. CETESB. *Companhia de Tecnologia de Saneamento Ambiental Decisão de Diretoria*; CETESB: Sao Paulo, Brazil, 2005.
24. Munir, M.A.M.; Irshad, S.; Yousaf, B.; Ali, M.U.; Dan, C.; Abbas, Q.; Liu, G.; Yang, X. Interactive Assessment of Lignite and Bamboo-Biochar for Geochemical Speciation, Modulation and Uptake of Cu and Other Heavy Metals in the Copper Mine Tailing. *Sci. Total Environ.* **2021**, *779*, 146536. [CrossRef] [PubMed]
25. Mujtaba Munir, M.A.; Liu, G.; Yousaf, B.; Ali, M.U.; Abbas, Q.; Ullah, H. Synergistic Effects of Biochar and Processed Fly Ash on Bioavailability, Transformation and Accumulation of Heavy Metals by Maize (*Zea mays* L.) in Coal-Mining Contaminated Soil. *Chemosphere* **2020**, *240*, 124845. [CrossRef]
26. Doku, E.T.; Sylverken, A.A.; Belford, J.D.E. Rhizosphere Microbiome of Plants Used in Phytoremediation of Mine Tailing Dams. *Int. J. Phytoremediat.* **2024**, *26*, 1212–1220. [CrossRef]
27. Yongpisanphop, J.; Babel, S.; Kruatrachue, M.; Pokethitiyook, P. Phytoremediation Potential of Plants Growing on the Pb-Contaminated Soil at the Song Tho Pb Mine, Thailand. *Soil Sediment Contam. Int. J.* **2017**, *26*, 426–437. [CrossRef]
28. Lam, E.J.; Cánovas, M.; Gálvez, M.E.; Montofré, Í.L.; Keith, B.F.; Faz, Á. Evaluation of the Phytoremediation Potential of Native Plants Growing on a Copper Mine Tailing in Northern Chile. *J. Geochem. Explor.* **2017**, *182*, 210–217. [CrossRef]
29. Orchard, C.; León-Lobos, P.; Ginocchio, R. Phytostabilization of Massive Mine Wastes with Native Phytogenetic Resources: Potential for Sustainable Use and Conservation of the Native Flora in North-Central Chile. *Cienc. Investig. Agrar.* **2009**, *36*, 329–352. [CrossRef]
30. Tapia, Y.; Salazar, O.; Joven, A.; Castillo, B.; Urdiales, C.; Garcia, A.; Ihle, C.; Acuña, E. Evaluation of Sulfate Rhizofiltration by *Carpobrotus chilensis* for Treating Mining Waters. *Int. J. Phytoremediat.* **2024**, *26*, 1556–1663. [CrossRef] [PubMed]
31. Adeoye, A.O.; Adebayo, I.A.; Afodun, A.M.; Ajijolakewu, K.A. Benefits and Limitations of Phytoremediation: Heavy Metal Remediation Review. In *Phytoremediation*; Biotechnological Strategies for Promoting Invigorating Environs; Academic Press: Cambridge, MA, USA, 2022; pp. 227–238. [CrossRef]
32. Shahid, M.; Austruy, A.; Echevarria, G.; Arshad, M.; Sanaullah, M.; Aslam, M.; Nadeem, M.; Nasim, W.; Dumat, C. EDTA-Enhanced Phytoremediation of Heavy Metals: A Review. *Soil Sediment Contam. Int. J.* **2014**, *23*, 389–416. [CrossRef]
33. Lim, J.M.; Salido, A.L.; Butcher, D.J. Phytoremediation of Lead Using Indian Mustard (*Brassica juncea*) with EDTA and Electrodics. *Microchem. J.* **2004**, *76*, 3–9. [CrossRef]
34. Guo, J.K.; Lv, X.; Jia, H.L.; Hua, L.; Ren, X.H.; Muhammad, H.; Wei, T.; Ding, Y. Effects of EDTA and Plant Growth-Promoting Rhizobacteria on Plant Growth and Heavy Metal Uptake of Hyperaccumulator Sedum Alfredii Hance. *J. Environ. Sci.* **2020**, *88*, 361–369. [CrossRef]

35. Beiyuan, J.; Tsang, D.C.W.; Bolan, N.S.; Baek, K.; Ok, Y.S.; Li, X.D. Interactions of Food Waste Compost with Metals and Metal-Chelant Complexes during Soil Remediation. *J. Clean. Prod.* **2018**, *192*, 199–206. [CrossRef]
36. Yin, F.; Li, J.; Wang, Y.; Yang, Z. Biodegradable Chelating Agents for Enhancing Phytoremediation: Mechanisms, Market Feasibility, and Future Studies. *Ecotoxicol. Environ. Saf.* **2024**, *272*, 116113. [CrossRef]
37. Aponte, H.; Sulbaran-Bracho, Y.; Mondaca, P.; Vidal, C.; Pérez, R.; Meier, S.; Cornejo, P.; Rojas, C. Biochemical, Catabolic, and PGP Activity of Microbial Communities and Bacterial Strains from the Root Zone of Baccharis Linearis in a Mediterranean Mine Tailing. *Microorganisms* **2023**, *11*, 2639. [CrossRef]
38. Novo, L.A.B.; Castro, P.M.L.; Alvarenga, P.; da Silva, E.F. Plant Growth–Promoting Rhizobacteria-Assisted Phytoremediation of Mine Soils. In *Bio-Geotechnologies for Mine Site Rehabilitation*; Elsevier: Amsterdam, The Netherlands, 2018; pp. 281–295. [CrossRef]
39. Wang, L.; Ji, B.; Hu, Y.; Liu, R.; Sun, W. A Review on in Situ Phytoremediation of Mine Tailings. *Chemosphere* **2017**, *184*, 594–600. [CrossRef] [PubMed]
40. Cang, L.; Wang, Q.Y.; Zhou, D.M.; Xu, H. Effects of Electrokinetic-Assisted Phytoremediation of a Multiple-Metal Contaminated Soil on Soil Metal Bioavailability and Uptake by Indian Mustard. *Sep. Purif. Technol.* **2011**, *79*, 246–253. [CrossRef]
41. Chen, Y.; Dong, M.; Lyu, P.; Wang, A.; Wang, H.; Li, J. Analysis of Metal(Loid) Pollution and Possibilities of Electrokinetic Phytoremediation of Abandoned Coking Plant Soil. *Sci. Total Environ.* **2023**, *870*, 161982. [CrossRef] [PubMed]
42. Cameselle, C.; Reddy, K.R. Development and Enhancement of Electro-Osmotic Flow for the Removal of Contaminants from Soils. *Electrochim. Acta* **2012**, *86*, 10–22. [CrossRef]
43. Stegmann, R.; Brunner, G.; Calmano, W.; Matz, G. *Treatment of Contaminated Soil*; Springer: Berlin/Heidelberg, Germany, 2001.
44. Robles, I.; Serrano, T.; Pérez, J.J.; Hernández, G.; Solís, S.; García, R.; Pi, T.; Bustos, E. Influence of EDTA on the Electrochemical Removal of Mercury (II) in Soil from San Joaquín, Querétaro, México. *J. Mex. Chem. Soc.* **2014**, *58*, 332–338. [CrossRef]
45. Popescu, M.; Rosales, E.; Sandu, C.; Meijide, J.; Pazos, M.; Lazar, G.; Sanromán, M.A. Soil Flushing and Simultaneous Degradation of Organic Pollutants in Soils by Electrokinetic-Fenton Treatment. *Process Saf. Environ. Prot.* **2017**, *108*, 99–107. [CrossRef]
46. Kim, G.N.; Jung, Y.H.; Lee, J.J.; Moon, J.K.; Jung, C.H. An Analysis of a Flushing Effect on the Electrokinetic-Flushing Removal of Cobalt and Cesium from a Soil around Decommissioning Site. *Sep. Purif. Technol.* **2008**, *63*, 116–121. [CrossRef]
47. Cameselle, C.; Chirakkara, R.A.; Reddy, K.R. Electrokinetic-Enhanced Phytoremediation of Soils: Status and Opportunities. *Chemosphere* **2013**, *93*, 626–636. [CrossRef]
48. Chirakkara, R.A.; Reddy, K.R.; Cameselle, C. Electrokinetic Amendment in Phytoremediation of Mixed Contaminated Soil. *Electrochim. Acta* **2015**, *181*, 179–191. [CrossRef]
49. Sadzawka, A.; Carrasco, M.A.; Grez, R.; Mora, M.L.; Flores, H.; Neaman, A. Métodos de Análisis de Suelos Recomendados Para Los Suelos de Chile: Revisión 2006. 2006, 34. Available online: https://biblioteca.inia.cl/items/9ec1aab6-f9aa-4e4b-b2c5-28d1fed798c3 (accessed on 25 November 2024).
50. United States Environmental Protection Agency. *Method 3052: Microwave Assisted Acid Digestion of Siliceous and Organically Based Matrices*; United States Environmental Protection Agency: Washington, DC, USA, 1996.
51. Bouyoucos, G.J. Hydrometer Method Improved for Making Particle Size Analyses of Soils1. *Agron. J.* **1962**, *54*, 464–465. [CrossRef]
52. Jury, W.A.; Horton, R. *Soil Physics*, 6th ed.; John Wiley & Sons: Hoboken, NJ, USA, 2004; ISBN 047105965X.
53. Villagrán, C.; Castro Rojas, M.V. *Ciencia Indígena de Los Andes Del Norte de Chile: Programa Interdisciplinario de Estudios En Biodiversidad (PIEB), Universidad de Chile*, 1st ed.; Biodiversidad; Editorial Universitaria: Santiago de Chile, Chile, 2004; ISBN 9561116979.
54. Rodriguez, R.; Marticorena, C.; Alarcón, D.; Baeza, C.; Cavieres, L.; Finot, V.L.; Fuentes, N.; Kiessling, A.; Mihoc, M.; Pauchard, A.; et al. Catálogo de Las Plantas Vasculares de Chile. *Gayana Bot.* **2018**, *75*, 1–430. [CrossRef]
55. Hansen, H.K.; Lamas, V.; Gutierrez, C.; Nuñez, P.; Rojo, A.; Cameselle, C.; Ottosen, L.M. Electro-Remediation of Copper Mine Tailings. Comparing Copper Removal Efficiencies for Two Tailings of Different Age. *Min. Eng.* **2013**, *41*, 1–8. [CrossRef]
56. Robson, T.; Golos, P.J.; Stevens, J.; Reid, N. Enhancing Tailings Revegetation Using Shallow Cover Systems in Arid Environments: Hydrogeochemical, Nutritional, and Ecophysiological Constraints. *Land. Degrad. Dev.* **2018**, *29*, 2785–2796. [CrossRef]
57. Liu, Z.; Hamuti, A.; Abdulla, H.; Zhang, F.; Mao, X. Accumulation of Metallic Elements by Native Species Thriving in Two Mine Tailings in Aletai, China. *Environ. Earth Sci.* **2016**, *75*, 781. [CrossRef]
58. Zeng, H.; Wu, H.; Yan, F.; Yi, K.; Zhu, Y. Molecular Regulation of Zinc Deficiency Responses in Plants. *J. Plant Physiol.* **2021**, *261*, 153419. [CrossRef]
59. Sadeghzadeh, B.; Rengel, Z. Zinc in Soils and Crop Nutrition. In *The Molecular and Physiological Basis of Nutrient Use Efficiency in Crops*; John Wiley & Sons, Inc.: Hoboken, NJ, USA, 2011; pp. 335–375. [CrossRef]
60. Clemens, S.; Weber, M. The Essential Role of Coumarin Secretion for Fe Acquisition from Alkaline Soil. *Plant Signal Behav.* **2016**, *11*, e1114197. [CrossRef] [PubMed]
61. Pandey, N. Role of Plant Nutrients in Plant Growth and Physiology. In *Plant Nutrients and Abiotic Stress Tolerance*; Springer: Singapore, 2018; pp. 51–93. [CrossRef]
62. Cang, L.; Zhou, D.M.; Wang, Q.Y.; Fan, G.P. Impact of Electrokinetic-Assisted Phytoremediation of Heavy Metal Contaminated Soil on Its Physicochemical Properties, Enzymatic and Microbial Activities. *Electrochim. Acta* **2012**, *86*, 41–48. [CrossRef]

63. Wu, C.; Fan, C.; Xie, Q. Study on Electrokinetic Remediation of PBDEs Contaminated Soil. *Adv. Mater. Res.* **2012**, *518–523*, 2829–2833. [CrossRef]
64. Fu, R.; Wen, D.; Xia, X.; Zhang, W.; Gu, Y. Electrokinetic Remediation of Chromium (Cr)-Contaminated Soil with Citric Acid (CA) and Polyaspartic Acid (PASP) as Electrolytes. *Chem. Eng. J.* **2017**, *316*, 601–608. [CrossRef]

minerals

MDPI

Review

Trends and Applications of Green Binder Materials for Cemented Paste Backfill Mining in China

Jiandong Wang [1,2], Bolin Xiao [1,3,*], Xiaohui Liu [4] and Zhuen Ruan [2,3]

1 State Key Laboratory of High-Efficient Mining and Safety of Metal Mine Ministry of Education, University of Science and Technology Beijing, Beijing 100083, China; wangjiandong@ustb.edu.cn

2 Key Laboratory of Safe and Green Mining of Metal Mines with Cemented Paste Backfill of the National Mine Safety Administration, University of Science and Technology Beijing, Beijing 100083, China; ustb_ruanzhuen@hotmail.com

3 Department of Mining Engineering, School of Civil and Resources Engineering, University of Science and Technology Beijing, Beijing 100083, China

4 School of Safety Engineering, North China Institute of Science and Technology, Langfang 065201, China; liuxiaohui28@ncist.edu.cn

* Correspondence: bxiao@ustb.edu.cn

Abstract: The backfill binder material is the key to the cost and performance of cemented paste backfill. This study aims to understand the current situation of metal ore backfill binders, identify industry challenges, inspire research ideas, and explore development directions. Current research investigates trends and developments of backfill binders through literature review, experience summary, field research, statistical analysis, and other methods. Firstly, the main backfill binder types are summarized, including cement, metallurgical slag, thermal slag, chemical slag, and tailings binders. Secondly, the research progress regarding reactivity activation, hydration mechanism, harmful ion solidification, energy conservation, and carbon reduction is summarized. Thirdly, three industrial applications of new backfill binders are introduced and summarized. Cement is still the most common, followed by slag powder binder. The cases of steel slag binder and semi-hydrated phosphogypsum backfill have shown significant effects. Solid waste-based backfill binder materials are gradually replacing cement, which is a trend. Finally, further research is discussed, including hydration modeling and simulation, material properties under extreme environments, hardening process control, and technical standards for backfill binders. This work provides a reference and basis for promoting green and efficient paste backfill and sustainable industry development.

Keywords: cemented paste backfill; binders; solid waste utilization; green material; slag

1. Introduction

The cemented paste backfill (CPB) technology creatively utilizes mining solid waste (waste tailings, rock, and water) to prepare a paste-like slurry backfilling for the excavated underground voids, eliminating tailings ponds and goaf; forming a high-recovery and low-dilution mining method; and providing a safe, green, and efficient solution for the industry [1]. CPB has become a prevalent practice for underground metal mining around the world.

When preparing the paste, a certain proportion of binder materials, which can account for 5 wt.% to 20 wt.%, needs to be added to meet the requirements of strength and safety. Binder materials are the main cost of CPB, accounting for 60% to 80% of the total [2]. For example, a mine with an annual production capacity of 5 million tons requires approximately

https://doi.org/10.3390/min15020097

200,000 to 300,000 tons of binder material, with a cost exceeding CNY 60 million [3]. CPB binder has become a limiting factor for the economic feasibility of mining base metals such as iron in China. Tailings are becoming finer and finer owing to the advancement of mineral processing technology. As a result, the bonding of ultra-fine tailings requires more cement to achieve the design strength [4]. Some mines that adopt the downward cut-and-fill mining method require high strength at a long-term scale and high early strength within 1–2 days in order to recover adjacent stopes and improve the efficiency of the mining–filling cycle [5]. The mining industry has increasingly stringent requirements for the cost and performance of CPB binders, where traditional Portland cement can no longer meet the demand. On the other hand, cement is one of the largest carbon-emitting industries, where approximately 0.9 tons of CO_2 is released in 1 ton of cement production. China produced over 50% of the world's total cement consumption, estimated at over 3 billion tons of CO_2 annually [6]. It is urgent to develop green cement materials to reduce greenhouse gases. Furthermore, millions of tons of solid waste are generated annually, occupying extensive lands, polluting water and solids, and threatening human health [7]. Solid industrial waste valorization in developing sustainable building materials exhibits encouraging prospects [8]. The research on developing green, low-carbon, low-cost, and efficient new backfill binder materials using bulk solid waste has become a hot topic in the current industry [9].

The development of backfill binder material began in 1969 when Australia first adopted cement for backfilling bottom pillars at the Mount Isa copper mine [10]. In order to reduce the binder costs, lead-zinc copper slag was used to partially replace cement [11]. In the early days of Canada, fly ash was used as a cement admixture in nonferrous and gold mines to reduce costs [12]. Motivated by occasional shortages and high costs of Portland cement, Glukhovsky first investigated the utilization of local raw materials, including soils and industrial wastes, as solid precursors for producing sodium silicate binders, which then were called "geopolymer" binder materials [13]. In recent decades, researchers have developed various new CPB binders using solid wastes such as slag, fly ash, and red mud, significantly reducing costs and promoting the development of CPB [14–17].

To help researchers and mining enterprises better understand the current situation of backfill binder materials in metal mines, this work reviews the current research and applications of green backfill binder material through a literature review, field research, statistical analysis, and case studies. The main content is divided into four parts. Firstly, CPB binder types are summarized and classified. Secondly, the research progress of CPB binders is reviewed. Thirdly, application cases of three binder types are introduced. Finally, the trends and further work in the research and application of CPB binders are discussed. The work aims to provide a reference and basis for promoting green and efficient paste fill and sustainable development.

2. Types of Backfill Binder Materials

The development of binder material has evolved from the earliest form of cement to various new materials. The main idea is to fully utilize multiple bulk solid wastes, which usually have potential hydraulic properties, also known as pozzolanic properties, and then prepare binders through various activation methods. According to the different originalities of pozzolanic solid wastes, CPB binders can be divided into the following types.

2.1. Portland Cement

Owing to the development of the construction industry, Portland cement is one of the earliest binder materials, and it has been widely used for hundreds of years. Jinchuan Nickel is the earliest mine in China to introduce the cemented backfill mining method [18].

It mainly uses Portland cement as the binder material, combined with coarse aggregates composed of rod-milled sand and waste rocks. The slurry concentration is 79–81 wt.%, with a cement content of 310 kg/m^3 to obtain the desired strength at 3, 7, and 28 days of 1.5, 2.5, and 5 MPa, respectively. The advantages of Portland cement are its stable and reliable quality and its wide range of sources. However, research shows that when ultra-fine tailings with particles below 74 μm exceed 75 wt.%, the 28-day CPB strength is only 1.6 MPa using Portland cement at a cement-to-sand mass ratio of 1:4 [19]. Portland cement can be considered unsuitable for cementing ultra-fine tailings. In addition, in some non-metallic mines such as potash mines, special cement, such as aluminum magnesium cement, should be used to bind the brine and salt [20]. Overall, Portland cement is still the most popular CPB binder, but it will gradually be replaced by newly developed green materials.

2.2. Metallurgical Slag Binders

According to the China Statistical Yearbook 2023, the general industrial solid waste production in China was 411.4 million tons, with a comprehensive utilization of 237 million tons, accounting for only 57%. General industrial waste usually includes blast furnace slag, steel slag, red mud, nonferrous metal slag, fly ash, coal slag, etc. Metallurgical solid wastes are widely used to develop backfill binder materials.

2.2.1. Iron and Steel Industry Slag

The production of pig iron and crude steel in China accounts for more than half of the world's total [3]. Blast furnace slag (BFS) and steel slag (SS), the most common and largest metallurgical solid wastes, have an annual output of 300 million tons. Their production process is shown in Figure 1. BFS has a glass content (CaO-SiO_2-Al_2O_3 solid solution) of over 80 wt.%, which is widely used as a supplementary cementitious material (SCM) in cement and concrete. BFS binder, which is composed of about 70 wt.% BFS powder and 30 wt.% activators (clinker, gypsum, and others) has been applied successfully. Studies show that under the same conditions, the CPB strength with BFS binder is 2–3 times that of cement CPB, while the manufacturing cost is only 40%–45% that of cement [21]. BFS can be activated by alkali and sulfate to generate large quantities of hydrated calcium silicate (aluminate) and ettringite hydration products, which have a good bonding effect on fine tailings [22]. The fineness of the slag and the curing temperature are crucial to CPB performance. It is usually necessary to grind it to a specific surface area above 400 m^2/kg. The finer the slag, the higher the activity. The curing temperature of the BFS binder should be above 10 °C, owing to the hydration rate, which is severely reduced in low-temperature environments. In contrast, a temperature around 30 °C can promote early strength without affecting later strength [23]. A case study in the Doyon mine (Quebec, QC, Canada) used a combination of 30 wt.% Portland cement and 70 wt.% BFS, representing 5 wt.% of the total dry tailings weight to improve its environmental behavior in the sulfide tailings CPB [24].

The BFS powder is used extensively in cement, and its price is close to that of cement in China. Many researchers have begun to use steel slag, developing binder materials to reduce costs and improve the utilization rate of solid waste. Steel slag is a by-product of steelmaking, which can be divided into basic oxygen furnace (BOF) slag, electric arc furnace (EAF) slag, refining furnace slag, etc. Most steel slag in China is BOF slag, and its main mineral composition is blite, which can be considered a "burnt clinker" that exhibits low hydrability. Therefore, it is usually necessary to combine it with other pozzolanic materials, such as slag and fly ash, to develop steel slag binders. There are several SS binder systems, including SS–cement, SS-BFS, SS-BFS-cement, SS-BFS-gypsum, SS-BFS-silica fume, and SS-gypsum-fly ash. The weight percentage of SS powders in binder materials can range from 30% to 60% [25]. The SS-BFS-gypsum system exhibits good performance, and its CPB strength is not lower than that of cement CPB under the same conditions, while its production cost is only about half that of cement [3]. Researchers from the US suggested that EAF slags should be aged for at least 6 months and BOF slags for at least 24 months in open air conditions before being used as backfill material, considering their mechanical performance, heavy metals leaching, and particle size distribution [26].

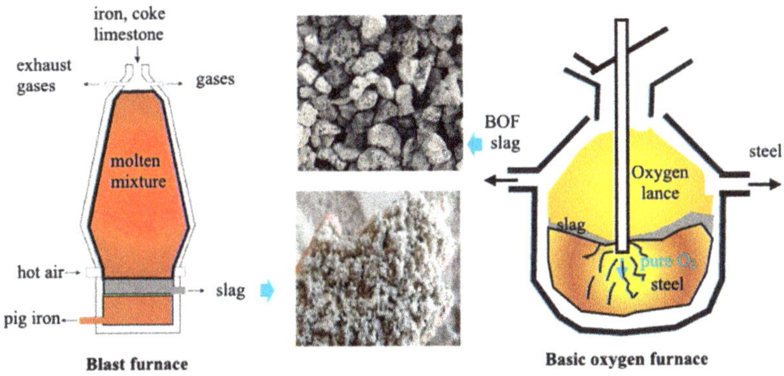

Figure 1. The production process and appearance of BFS and BOF slag.

In summary, BFS- and SS-based binders are the most likely to replace cement and have begun to be promoted in many mine practices.

2.2.2. Nonferrous Metal Metallurgical Slag

Copper, nickel, lead, zinc, and other nonferrous metallurgical slags, shown in Figure 2, are also important components of general industrial solid waste, with an annual yield of over 30 million tons. Compared with BFS, the content of calcium and aluminum oxides in nonferrous metallurgical slag is lower, and the content of iron oxides is higher, resulting in lower reactivity, higher specific area, sharp shape, and harmful ions content, which makes them challenging to utilize [27]. Many researchers have studied the utilization of nonferrous metallurgical slag in cement based on the experience of BFS. Zhu et al. [28] addressed the problem of low activity caused by high iron content and low calcium content in copper slag by using high-temperature reconstruction, mechanical activation, chemical excitation, and reduced quenching methods. The prepared binder CPB strength was still lower than that of cement. Liu et al. [29] developed a lead-zinc smelting slag-based ecological mine backfill by mechanical and alkali activation methods. The CPB strength met the requirements for high-strength grade self-weight backfill, the solidification ratio for each heavy metal exceeded 70%, and the cost of CPB was reduced. Wang et al. [30] produced a cemented backfill material with secondary smelting water-granulated nickel slag, composite activator (gypsum, carbide slag, Na_2SO_4, and clinker), and iron ore tailings. The 28-day flexural and compressive strength of CPB material reached 1.99 MPa and 3.38 MPa, which meet the requirements of mine backfill. Behera et al. [31] utilized lead-zinc mill tailings and lead-zinc smelter fuming furnace slag as paste backfilling for an underground metalliferous mine in India and found that the use of crushed fuming furnace slag as a cement replacement showed encouraging results of strength development, and the paste backfilling cost per ton was reduced significantly. Overall, due to properties of low reactivity, limited distribution, and hard grinding, utilization of nonferrous metallurgical slag in backfill binder materials is limited in individual cases. The properties of this type of slag determine that it is more suitable for use as backfill aggregate.

Figure 2. Appearance of some nonferrous metal metallurgical slags.

2.3. Thermal Slag

Thermal slag refers to residues generated after incineration in thermal power plants, incinerators, etc., mainly including fly ash and circulating fluidized bed (CFB) slag.

2.3.1. Fly Ash

Fly ash is a substance discharged from the flue of a power plant after coal combustion and collected by a dust collector. Its composition is similar to high-alumina clay, mainly in a glassy state, with SiO_2 and Al_2O_3 contents accounting for more than 80%. Under alkaline conditions, FA can undergo hydration reactions and produce gel-like products. According to China's national standard GB/T 1596 [32], FA can be classified into three grades based on fineness, water demand ratio, loss on ignition, moisture content, and sulfur trioxide content. FA grade I has been widely used as SCM in cement and concrete. In mine backfill, FA grades II and III are usually used to partially replace Portland cement, which can improve the flowability of the slurry and the later strength of the CPB but reduces the early strength [18]. FA is more used in coal mines because of its ability to improve the strength, impermeability, and chemical resistance of the backfill body [33]. In metal mines, Behera et al. [34] investigated the efficacy of FA as a partial replacement (25 wt.%) for Portland cement for paste backfill application, finding that UCS development is more sensitive toward FA replacement. Xiao et al. [35] proposed a new backfill binder made with 40 wt.% of low-quality Class F FA to totally replace Portland cement. The result showed that the new FA binder could meet the strength requirement of three different mines regarding different subsequent filling and cut-and-fill mining methods, and the binder costs were reduced by more than 30%. In Australia, backfilling with FA as a partial cement replacement and plasticizer enabled continuous longwall mining through old stub headings; however, proper drainage and prevention of erosion should be achieved to maximize its stabilization [36]. Due to the prevalence of FA in thermal power plants, utilizing FA to partially replace or prepare new binder materials has inspiring prospects.

2.3.2. Circulating Fluidized Bed Slag

The CFB slag is the combustion residue brown ash from the circulating fluidized bed, consisting of complex substances of fuel ash, unburned carbon particles, and other impurities. The main mineral composition includes quartz, calcite, hard gypsum, lime, and hematite. Zhang et al. [37] prepared clinker-free cemented backfill materials containing CFB slag (20–40 wt.%). The material exhibited excellent filling performance and promoted the formation of a greater polymerization degree of the C-A-S-H phase. Yang et al. [38] pointed out that CFB contains a large amount of calcium sulfite and free calcium oxide, which have risks of volumetric expansibility and instability. It can be combined with BFS slag to prepare binder materials, minimizing expansion. Liu et al. [39] reviewed the rheology, mechanics, microstructure, and durability of CFB binders and pointed out that adding nanomaterials or modifying CFB slag may become a large-scale utilization method. The utilization of CFB in backfill binders is still limited; further research should be conducted, since more than 280 million tons of CFB slag are produced every year in China [40]. Longo et al. [41] presented some case studies from South Africa and the US, demonstrating the recipes and process design that can be implemented to solve coal combustion residues by measuring the paint filter test, hydraulic conductivity, strength as defined by Unconfined Compressive Strength (UCS) tests, and acid generating potential.

2.4. Chemical Industry Slag

Chemical waste refers to solid waste generated during the production processes of the chemical industry. The pollution caused by chemical waste is widespread, making it difficult to control. The typical chemical slag in the mine backfill field is mainly phosphogypsum and red mud.

2.4.1. Phosphogypsum

Phosphogypsum (PG) is produced in tens of millions of tons annually from the production of phosphate fertilizers, phosphoric acid, and other phosphate chemicals. Most PG is stacked on the surface with extremely low utilization rates. Using PG as a raw material for filling underground voids has become an important disposal approach. The main component of PG is calcium sulfate dihydrate, which contains incompletely decomposed phosphate ore, residual phosphoric acid, fluoride, acid-insoluble substances, organic matter, etc. Min et al. [42] studied the strength performance of PC backfill through four binder types and found that the hydration process of CSA and CAC exhibited markedly fast hydration and UCS improvement. Nizevičienė et al. [43] investigated the neutralization process of the acid impurities of PG through mechanical activation together with the neutralizing zeolite additive. The hazardous phosphate impurities can be absorbed with higher compression strength. Wang et al. [44] proposed a hemihydrate PG backfill mixture consisting of 60 wt.% PG, 2.3 wt.% of quicklime, 5.7 wt.% of tailings, and 32 wt.% of water. Though the mechanical strength can meet requirements, interaction effects between temperature and pH on 28-day strength degradation were observed. The application of PG as backfill material is promising and of great significance in solving the problem of PG disposal in China. Hemihydrate PG can be prepared into binder materials for common CPB methods within the first 137 h of aging; thereafter, it can be used only as an inactive aggregate [45]. The application of PG still faces problems such as a fast reaction rate, easy pipe blockage, and emitting irritating odors after solidification and decay.

2.4.2. Red Mud

Red mud is an industrial waste generated during alumina production from bauxite, including sintering red mud and Bayer red mud (BRM). The BRM is the most common, producing 1.2–1.5 tons for every 1 ton of alumina. Wang et al. [46] investigated the durability, microstructure, permeability, and environmental properties of CPB with BRM, FA, and cement; the CPB strength reached 1.25 MPa and 2.37 MPa after curing for 3-day and 28-day, respectively, and the heavy metal concentration and pH value of the leachate were within the standard range. BRM is a strongly alkaline substance, which is prone to risks such as frosting, corrosion, and groundwater pollution. Strong alkalinity is the key factor restricting the utilization of BRM. However, the alkalinity can be used to promote the hydration of pozzolanic materials such as fly ash and slag. Li et al. [47] used BRM as an alternative activator to sodium hydroxide with BFS as the precursor; the result showed that CPB exhibited significantly better mechanical properties, with a 57%–94% increase in uniaxial compressive strength at 28 days compared with CPB made with Portland cement. Suchita et al. [48] pointed out that neutralization/treatment of red mud using different techniques such as using mineral acids, acidic waste (pickling liquor waste), coal dust, superphosphate, and gypsum as amenders; CO_2; or sintering with silicate material and seawater is the only alternative to make the bauxite residue environmentally benign. Further studies should focus on exploring the economic viability of these processes for better waste management and disposal of red mud.

2.5. Ultra-Fine Tailings

Tailings usually contain a large number of inert substances such as SiO_2 and are used as aggregates in backfill. However, converting these crystalline, non-reactive phases into more amorphous, reactive phases can activate the tailings' cementitious activity, which can then be used for cementitious material. Research showed that mechanical, thermal, chemical, and coupled activation technologies were usually used in tailings processing [49]. Grinding into ultra-fine particles is essential for activated tailings, which can increase the specific surface area, improve the surface energy and activity, increase the degree of participation in hydration reactions, promote the formation of a large number of acicular ettringite and C-S-H gel, improve the pore structure of hydration products, and lead to denser microstructure [50]. At present, mailing gold and iron tailings are used as mineral admixtures in cement and concrete with an incorporation rate of less than 10 wt.%.

For the above five main types of backfill binders, the active Ca-Si-Al chemical compositions, which are taken from the literature and field investigation, are shown in Figure 3. The diagram can provide a basis for activity evaluation, modification methods, and utilization of potential backfill binder material. Key properties and some examples of the green mine backfill binders are listed in Table 1.

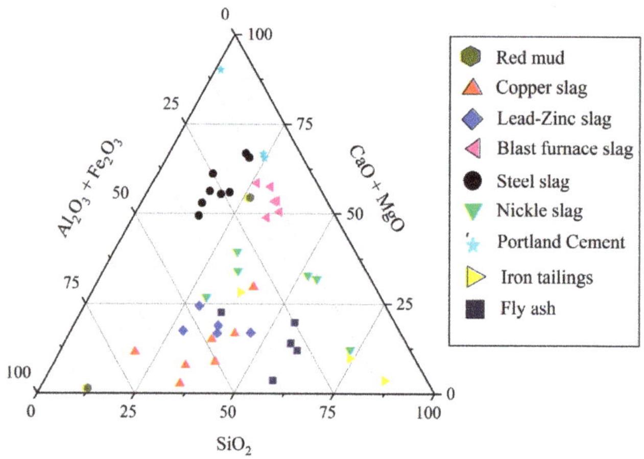

Figure 3. Phase diagram of Ca-Si-Al composition distribution of main solid waste materials.

Table 1. Key properties and examples of current green mine backfill binder types.

Binder Types	Examples	Key Features	References
Cement and blend cement	Ordinary Portland cement	Not applicable for ultra-fine tailings	[19]
	Slag, FA, etc., blend cement	SCM replacement ratio 20–80 wt.%	[51]
Metallurgical slag developed binders	Alkali + BFS + gypsum	70–90 wt.% BFS, excellent binding performance	[21,52,53]
	SS + BFS + gypsum	30–60 wt.% SS, good binding performance, low-cost	[3,54]
Nonferrous metallurgical slag binders	Cement partially replaced by copper, nickel, lead, zinc slag	Low reactivity of slag, slag replace ratio < 30 wt.%, heavy metal iron leaching risk	[29,30]
Thermal slag developed binders	FA + BFS + gypsum; FA + SS + gypsum	30–60 wt.% FA, good binding performance, low cost	[17,35,55]
Chemical industry slag-based binders	PG + BFS + alkali	Up to 70 wt.% of PG can be used, good binding performance, low cost	[45,56,57]
	BRM + BFS + gypsum	Good performance, low cost, alkaline leaching risk	[47,58]
Tailings-based binders	Tailings + alkali + BFS	Tailings ratio < 30 wt.%, limited tailing types	[34,59]

3. Research Development of Backfill Binder Materials

The previous section summarized five different types of backfill binder materials. Solid waste-based binders mostly require activation, have similar hydration mechanisms, and have advantages in performance, economy, and the environmental perspective. These properties are hotspots of current research.

3.1. Research on Slag Activation Methods

The primary raw materials for new backfill binders are bulk solid wastes with pozzolanic activity. Under normal circumstances, these materials do not undergo hydration reactions, except alkali and other activation methods are introduced.

The quality or reactivity of slag is critical for its utilization. From Ashfaque's study, there is a wide range of standardized and non-standardized methods used to evaluate reactivity, which can be divided into direct methods measuring physical properties or indirect methods measuring chemical properties [60]. Among them, the strength activity index, which is the compressive strength of concrete/mortar containing additives compared with reference concrete/mortar samples at different curing ages, is used in Chinese national standard GB/T 18046-2017 evaluating BFS's reactivity [61]. Various pre-treatment techniques and methods can be employed to enhance the reactivity, such as grinding, calcination, and modified catalysis. Grinding to finer particles can damage the structure of the glass body, causing cracks and distortions in the crystals, reducing crystallinity, and increasing the specific surface area of reaction, and thus, enhancing activity. Calcination can achieve the decomposition and recombination of minerals, obtaining more useful active components. Modified catalysis optimizes the functional groups on the particle surface by adding modifiers, achieving improved surface activity, solubility, and other properties.

The potential reactivity of slag needs to be activated to obtain hydraulic properties. Typical activation methods include mechanical, thermal, chemical, and combined techniques. Xie et al. [62] investigated the mechanical grinding activation of magnesium slag for backfill and showed that the optimal grinding time should consider energy consumption and technical benefits, and the corresponding specific surface area and 45 μm sieve residue are 503 m^2/kg and 3.83%, respectively. A study showed that the multi-phased cement clinker prepared at 1400 °C with the raw material containing 16.86% steel slag has comparable grindability and soundness to the normal Portland cement clinker [63]. Higher curing temperature can lead to a decrease in arsenic released from the slag–CPB, faster hydration rate and pozzolanic reaction, and lower volume of connective pores [64]. Alkali and sulfate, such as sodium hydroxide, calcium sulfate, and sodium sulfate, are commonly seen in chemical activation slag with excellent performance [3,47,49]. Industrial salt and alkali activators include desulfurization gypsum, fluorogypsum, water glass, clinker, carbide slag, steel slag, red mud, mirabilite, etc.

Figure 4 summarizes the current activation methods for CPB binders. Due to the differences in physicochemical properties, diverse slag types, and different application scenarios, it is difficult to determine the optimal activation scheme in practice, and combined activation experiments need to be conducted.

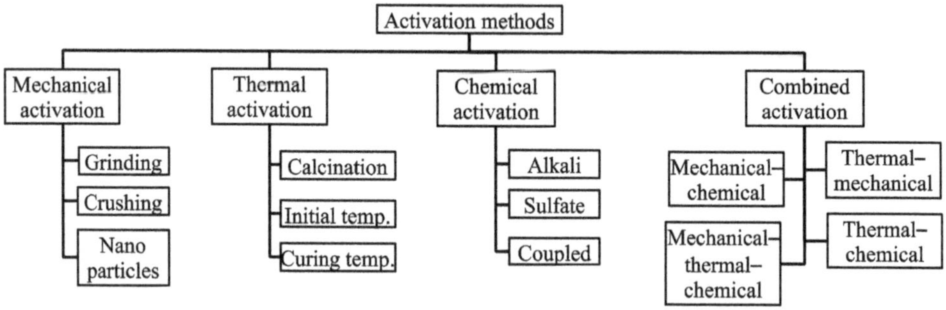

Figure 4. Current activation methods for solid waste based backfill binders.

3.2. Research on Hydration Mechanism

The study of the hydration mechanism of binder materials focuses on the feasibility of the hydration reaction, hydration product analysis, and reaction rate (hydration kinetics). The laws of thermodynamics and the principle of minimum Gibbs free energy are the theoretical bases. Although different types of raw solid waste are used, and the hydration mechanism varies for different binders, the hydration process has an outstanding common feature, which is the formation process of hydrated calcium silicate (aluminate) and ettringite (AFt), as shown in Equation (1):

$$\begin{cases} xCa^{2+} + OH^- + SiO_2(aq) + H_2O \rightarrow C-S-H + H^+ \\ Ca^{2+} + OH^- + Al_2O_3(aq) + H_2O \rightarrow C-A-H + H^+ \\ CaO{\cdot}Al_2O_3 + 2CaO + 3CaSO_4{\cdot}2H_2O + 30H_2O \rightarrow AFt \end{cases} \quad (1)$$

The hydration process of CPB binder materials can be divided into five stages. The first stage is the hydrolysis of the alkaline activator, forming an alkaline solution environment. The second stage is the equilibrium of precipitation dissolution of $Ca(OH)_2$ generated by OH^- and Ca^{2+}. In this stage, pozzolanic materials such as slag and fly ash are dissolved by OH^- to form $H_2SiO_4^{2-}$ and $H_2AlO_3^-$. In the third stage, Ca^{2+} reacts with $H_2SiO_4^{2-}$ and $H_2AlO_3^-$ to produce a quantity of gel products such as hydrated calcium silicate and hydrated calcium aluminates. The fourth stage involves the reaction of SiO_4^{2-} with $H_2AlO_3^-$ and Ca^{2+} to form microcrystalline AFm and AFt. The fifth stage is the gradual slag deposition and gel growth of hydration products on the surface of the particles, creating a complex structure of flocs. For example, the hydration process of the SS–BFS–gypsum binder system, modified from Hao's study [65], is shown in Figure 5. Helinski et al. [66] proposed a model showing that the pore-water pressure change depends on the amount of volume change associated with the cement hydration, the incremental stiffness change in the soil, and the porosity of the material. The rate of hydration and volumes of water consumed during hydration were unique for each cement–tailings combination regardless of the mix proportions.

The hydration heat of the metallurgical slag binders can be divided into the rapid reaction period, induction period, acceleration period, deceleration period, and slow hydration period. However, compared with Portland cement, the hydration heat release is significantly reduced, and the duration of the induction and acceleration periods is shortened considerably [67]. Low hydration heat is beneficial for reducing the risk of cracking or strength reduction caused by thermal stress in a massive filling body. The hydration kinetics model can be derived from the hydration heat monitoring curve. Then, a specific model for such materials can be established to describe the degree of hydration, hydration products, and other characteristics. Numerous hydration kinetics have been built from the literature, e.g., Byfors, Knudsen, Hansen and Pedersen, Basma et al., Nakamura et al., Cervera et al., Schindler and Folliard, and Bentz [68].

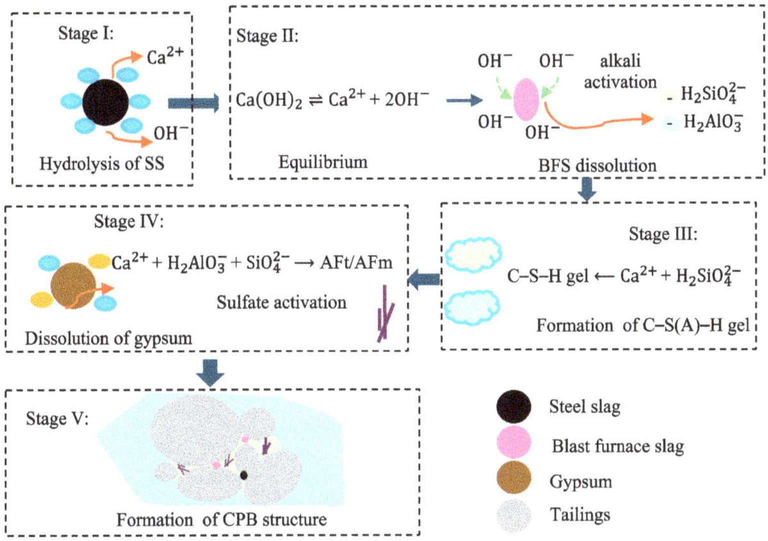

Figure 5. The hydration reaction process of the steel slag–slag–gypsum system.

3.3. *Research on Harmful Ion Solidification*

Metallurgical slag and tailings may contain harmful metal ions such as lead, arsenic, chromium, and cadmium, which are prone to migration. The issue of harmful ion leaching should be carefully examined before utilization. A study by Zhu et al. [69] showed that C–S–H exhibits various disordered layered structures at different Ca/Si ratios, exhibiting strong ion adsorption performance. The principle of heavy metal ions solidification in C–S–H gel can be divided into four ways: adsorption, ion substitution, insoluble matter generation, and encapsulation [69]. For example, in the arsenic solidification system of slag cement backfill, the calcium ions generated during the hydration process will react with the free arsenate or hydrogen arsenate ions in the pore fluid to achieve rapid arsenic fixation, that is, calcium–arsenic combination [70]. Zhang et al. [71] found that the metallurgical slag binder material has a higher lead solidification efficiency than Portland cement, mainly because of the adsorption of Pb^{2+} on the hydrated C–S–H gel and the synergistic effect of Pb^{2+} entering into the AFt and salt structure in the same phase. A suitable CPB mixture design to control the release of all heavy metals in CPB is of practical importance for groundwater quality [72]. The CPB prepared with metallurgical slag binders has smaller porosity, smaller pore diameter, and a large amount of gel-like hydration products, improving strength performance and facilitating the solidification of harmful ions [3].

3.4. *Research on Energy Conservation and Emission Reduction*

The cement industry, which produces 2.38 billion tons annually, contributes to more than 10% of the country's total carbon dioxide emissions. Each unit ton of clinker production emits 0.85–0.90 tons of carbon dioxide. Solid waste-based backfill binders are prepared by grinding industrial solid waste such as slag, steel slag, gypsum, and fly ash. This not only avoids environmental damage caused by the excavation of clay minerals required for clinker but also significantly reduces energy consumption and carbon emissions by requiring grinding without sintering. Traditional cement binders can be responsible for up to 70% of the greenhouse gas emissions in the mine backfill process [73]. Research has shown that the environmental impact of cementitious materials throughout their entire lifecycle can be divided into stages: raw material production, raw material transportation, cementitious material preparation, cementitious material transportation, and the construction stage. The carbon emissions of solid waste-based binders can be reduced by 18.62% to 38.35% compared with traditional Portland cement materials [74]. Liu et al. proposed the concept of carbon dioxide backfill, defined the calculation boundaries for carbon footprint and carbon consumption in CO_2 backfill, and calculated that the carbon consumption of 1 ton of CO_2 filling material throughout the entire lifecycle can reach 200–550 kg [75]. Therefore, solid waste-based backfill not only has advantages in the comprehensive utilization of solid waste but also in energy conservation and emissions reduction, forming zero carbon and negative carbon backfill.

4. Progress in the Application of New Backfill Binders

Owing to breakthroughs in theory and technology from a research perspective, various new backfill binder materials have been applied with good performance, low cost, and environmental friendliness. This section introduces three successful industrial cases.

4.1. *Jiaogu Powder*

Jiaogu powder (JGP), also known as BFS-based backfill binder material, is currently the most widely used binder material after Portland cement in China's metal mines. It usually comprises 70 wt.% BFS powder and 30 wt.% activator (clinker and gypsum), and it has achieved significant results in commercial applications. China's JGP originated in the Shandong province in 2004, where its first production line had an annual output of 20,000 tons for underground filling at Sanxin Company (Daye, China). Under the same cement dosage (dry contents 10 wt.%) and slurry concentration (74 wt.%), the strength of the slag-CPB was more than twice that of the cement-CPB, and the slurry flowability and workability were good [52]. After years of development and improvement, the JGP utilization has been shifted from combining the "C material" (activator) shipped from Shandong with local slag mixture to localized manufacture across the country. The raw materials are widely sourced, the grinding process is simple (industrial ball mills), the production cost is low, and the binding performance is good. According to incomplete statistics, there are currently over 50 JGP manufacturers in China. There are two utilization strategies, shown in Figure 6: (1) Finished JGP products manufactured in a grinding plant and then transported in tank cars to the backfill plant users. This method is convenient to apply, but the product price is high, and the users cannot control the quality. (2) Mine users buy the three raw materials separately and then mix them using a Raymond mill or place them in multiple powder silos in the backfill plant. This method has the lowest cost but requires accurate control of the mixing ratio to avoid activation failure accidents. If a mine has an annual usage of over 10,000 tons, it is also preferable to build a grinding plant next to the backfill plant with the benefits of quality and costs being controllable. The cost savings can enable the construction investment to be recovered within 3 years.

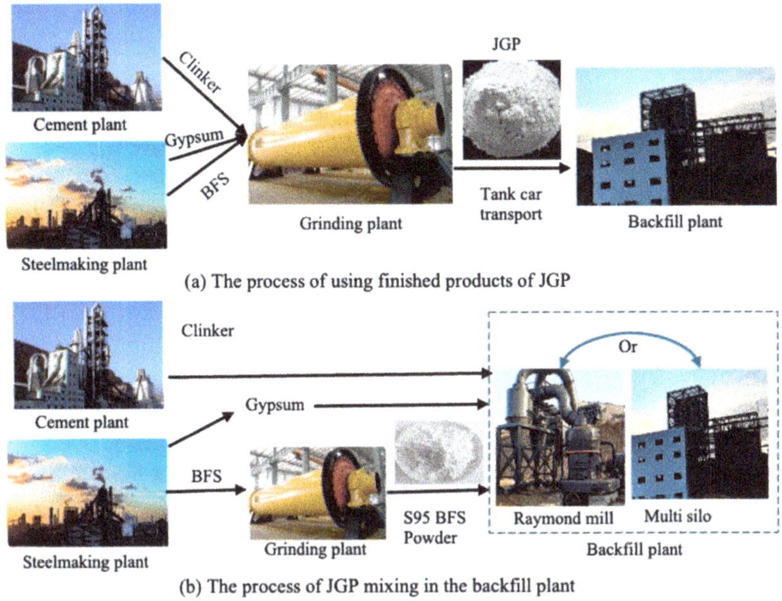

Figure 6. Two strategies for applying JGP utilization are (**a**) the process of using finished products and (**b**) the process of JGP mixing in the backfill plant.

4.2. Steel Slag Gujie Powder

Indoor experiments have shown that the backfill binder made with BFS-SS-gypsum has good binding properties and low production costs. In 2019, steel slag Gujie powder (SSGJP) achieved its first national large-scale industrial utilization in the Zhongguan Iron Mine, Hebei province, China. The SSGJP has SS content of 40 to 45 wt.%, showing excellent binding performance for ultra-fine tailings that have –74 μm fine particles accounting for 89.2 wt.%. The backfill plant built two powder silos and independently purchased raw material powders to be mixed in the backfill controlling system, as shown in Figure 7. The BFS powder and composite powder (SS and gypsum) are, respectively, transported to a double-axis horizontal mixer by a double screw feeder and then prepared into a paste slurry with a concentration of 58–62 wt.% through a high-speed activation mixing system. Finally, the paste is transported by gravity or pump to the underground voids. The underground strength sampling shows that the 28-day strength of SSGJP-CPB with a cement-sand mass ratio of 1:8 is greater than that of the cement-CPB with a cement-sand mass ratio of 1:4. Until the end of 2024, the SSGJP has been used continuously for over 350,000 tons in the Zhongguan Iron Mine (40,000 to 50,000 tons every year). The cost of the SSGJB is reduced by more than CNY 100/ton compared with cement. The material's performance and cost-effectiveness are significant, providing referencing examples for promoting SS binders in similar mines in China.

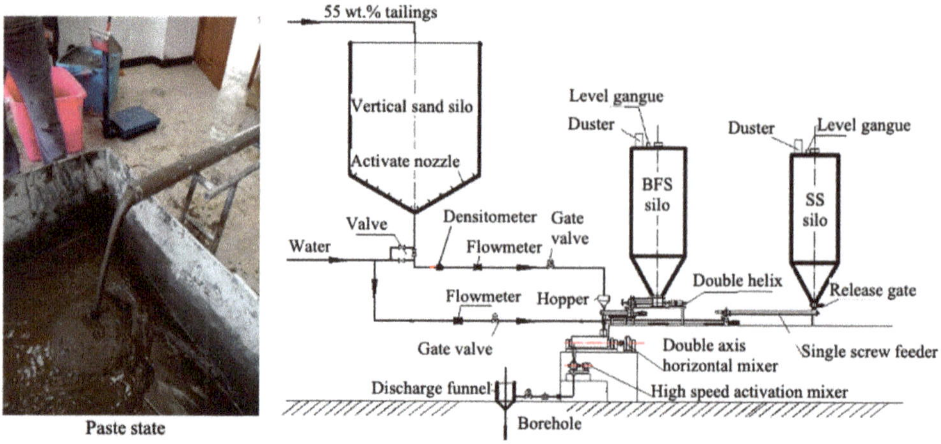

Figure 7. The schematic diagram of the SSGJP backfill process in Zhongguan Iron Mine.

4.3. Phosphogypsum Backfill

Utilization of PG in mine backfill may be the most effective technology for large-scale and green disposal of PG. As early as 2005, the Kaiyang Phosphate Mine (Guiyang, China) achieved industrial utilization of PG paste fill for the first time using cement, fly ash, and PG at a mass ratio of 1:1:(4–10). The backfill system has a capacity of 40 m^3/h and a CPB consolidation strength of 1.3–2 MPa. In recent years, ChanHen Chemical Co., Ltd. (Fuquan, China) developed a hemihydrate PG backfill technology (CHHPG), as shown in Figure 8. The CHHPG consists of hemihydrate PG accounting for over 95 wt.% and the rest alkaline modifiers. The paste concentration is 69 wt.%, and the strength of the CPB at all ages is greater than 3 MPa, with good fluidity, meeting the requirements of mines. ChanHen has built three backfill stations in Xiaoba, Xinqiao, and Gongjishan, forming a large-scale PG filling condition of 150,000 tons per month. The CHHPG backfill technology can solve the problem of PG and tailings storage in phosphate chemical enterprises, constructing a new circular economy model of the "integrated mining–chemical industry".

Figure 8. The industrial backfill of hemihydrate phosphogypsum backfill.

5. Discussion of Trends and Further Work of Backfill Binder Materials

Although significant progress has been made in the research and application of solid waste-based backfill binder materials, there are still many problems that urgently need to be solved.

5.1. Modeling and Simulation of Hydration and Hardening

Due to differences in material properties, equipment types, and production processes, the physicochemical properties of raw materials vary greatly. Substantial experiments usually need to be conducted to seek the optimal ratio, which can be time-consuming and cost-ineffective. Furthermore, in some cases, it is necessary to regulate the hydration reactions to meet the demands of early strength and slow setting. It is urgent to reveal the full hydration mechanism and to conduct hydration process modeling and simulating, and then, predict the hardening CPB properties such as hydration products, pore structure, and strength based on the physical and chemical properties of the materials and tailings.

5.2. Material Performance in Extreme Environments

With mining shifting to deep and high-altitude areas, the circumstances for backfill are becoming increasingly stringent, and the requirements for binder materials are becoming stricter in environments such as high temperature, low temperature, rich water, sulfur rich, and strong alkali. For example, research has shown that temperature significantly impacts the hydration performance of slag-based CPB. The hydration performance of slag powders is severely slowed down below 10 °C, while around 35 °C, it is beneficial for improving early strength without affecting later strength. At 50 °C, it is easy to cause internal cracking and a reduction in strength. In winter, the underground temperature of mines in Inner Mongolia, Xinjiang, and other places in China can be below zero degrees. It is worth further research and exploration on how to ensure material hydration and CPB performance, as well as to reduce the probability of freezing and pipe blockage, in such environments.

5.3. Technical Standards for Backfill Binder Materials

In the national standard GB175-2023, the strength grades of Portland cement are divided into six grades: 42.5, 42.5R, 52.5, 52.5R, 62.5, and 62.5R [76]. The strength grades of slag cement, fly ash cement, and volcanic cement are divided into 32.5, 32.5R, 42.5, 42.5R, 52.5, and 52.5R. The strength grades of composite Portland cement are divided into four levels: 42.5, 42.5R, 52.5, and 52.5R. The American Society of Testing Materials (ASTM) standard C150/C150M-24 specifies cement as types I, IA, II, IIA, III, IIIA, and V [77]. The ASTM standard C989/C989M-24 specifies slag cement for use in concrete and mortars [78]. These standards provide important guarantees for the application of construction cement. However, currently, there is no relevant grade standard or quality standard for backfill binder materials (aggregate tailings). In application, the binder material is verified based on a strength test sampled from each batch. If there are quality problems, there is no relevant evaluation standard, which may be detrimental to the industry's development.

There are significant drawbacks to using the same grade evaluation of cement standards. In more than ten years of application experience, a contradictory situation is often found: if the backfill binder grade is tested according to the cement standards, the grade may be extremely low; however, its bonding effect (strength) on the mine tailings is better than that of 32.5 or 42.5 cement CPB. Therefore, it is necessary to explore a new method for determining the grade and quality of backfill binder materials.

6. Conclusions

(1) New solid waste-based binders are gradually replacing traditional cement and are being applied to mine backfill through various activation methods. According to the types of solid waste raw materials, backfill binder materials can be divided into cement and blend cement, metallurgical slag, thermal slag, chemical slag, and tailings slag binders. Significant breakthroughs have been made in the research and application of binder materials.

(2) Concerning the research on new backfill binders, a lot of work has been carried out from the aspects of reactivity enhancement, reactivity activation, hydration mechanism, harmful ion solidification, energy savings, and carbon reduction. Generally, slag binders have a better binding effect, especially for ultra-fine tailings. The primary hydration process is the generation of gel-like and ettringite hydration products. Solid waste-based backfill binders often have less harmful ion leaching and advantages in energy savings and carbon emission reduction.
(3) In the industrial application of backfill binders, cement and blend cement are still the main types. However, blast furnace slag powder materials are rapidly becoming popular, exhibiting good performance and cost-effectiveness. Steel slag binders have achieved continuous industrial applications and have more cost advantages. Hemihydrate phosphogypsum backfill technology has achieved large-scale industrial utilization, providing a reference demonstration case for the comprehensive disposal of phosphogypsum.
(4) There are still some urgent problems to be solved in the research and application of backfill binder materials, including hydration modeling and simulation, properties prediction, performance regulation in extreme environments, and specific technical standards, which can build the foundation for promoting green and efficient cemented paste backfill and sustainable development.

Author Contributions: Conceptualization, B.X.; methodology, J.W.; software, X.L.; validation, Z.R. and J.W.; formal analysis, B.X.; investigation, X.L.; resources, J.W.; data curation, B.X.; writing—original draft preparation, B.X.; writing—review and editing, B.X.; visualization, J.W.; supervision, B.X.; project administration, B.X.; funding acquisition, B.X. and X.L. All authors have read and agreed to the published version of the manuscript.

Funding: This research was funded by Fundamental Research Funds for the Central Universities, grant number FRF-IDRY-23-015, and the National Natural Science Foundation of China, grant number 52074121.

Data Availability Statement: The original contributions presented in this study are included in the article. Further inquiries can be directed to the corresponding authors.

Acknowledgments: The authors are grateful to Zhongguan Iron Mine and ChanHen Chemical Co., Ltd. for providing the materials.

Conflicts of Interest: The authors declare no conflicts of interest.

References

1. Wu, A.; Wang, Y.; Ruan, Z.; Xiao, B.; Wang, J.; Wang, L. Key Theory and Technology of Cemented Paste Backfill for Green Mining of Metal Mines. *Green Smart Min. Eng.* **2024**, *1*, 27–39. [CrossRef]
2. Fang, K.; Zhang, J.; Cui, L.; Haruna, S.; Li, M. Cost Optimization of Cemented Paste Backfill: State-of-the-Art Review and Future Perspectives. *Miner. Eng.* **2023**, *204*, 108414. [CrossRef]
3. Xiao, B.; Wen, Z.; Miao, S.; Gao, Q. Utilization of Steel Slag for Cemented Tailings Backfill: Hydration, Strength, Pore Structure, and Cost Analysis. *Case Stud. Constr. Mater.* **2021**, *15*, e00621. [CrossRef]
4. Qiu, J.; Guo, Z.; Yang, L.; Jiang, H.; Zhao, Y. Effect of Tailings Fineness on Flow, Strength, Ultrasonic and Microstructure Characteristics of Cemented Paste Backfill. *Constr. Build. Mater.* **2020**, *263*, 120645. [CrossRef]
5. Behera, S.K.; Singh, P.; Mishra, D.P.; Mishra, K.; Kumar, A.; Mandal, S.K.; Mishra, A.K. Required Strength Design of Cemented Backfill for Underground Metalliferous Mine. *Int. J. Min. Reclam. Environ.* **2023**, *37*, 927–952. [CrossRef]
6. Naqi, A.; Jang, J.G. Recent Progress in Green Cement Technology Utilizing Low-Carbon Emission Fuels and Raw Materials: A Review. *Sustainability* **2019**, *11*, 537. [CrossRef]
7. Nanda, S.; Berruti, F. Municipal Solid Waste Management and Landfilling Technologies: A Review. *Environ. Chem. Lett.* **2021**, *19*, 1433–1456. [CrossRef]
8. Gehlot, M.R.; Shrivastava, S. Solid Industrial Waste Generation and Its Valorization in Developing Sustainable Building Materials—A State of the Art Review. *Mater. Today Proc.* 2023; *in press*. [CrossRef]

9. Tran, T.Q.; Kim, Y.; Dang, L.C.; Do, T.M. A State-of-the-Art Review on the Utilization of New Green Binders in the Production of Controlled Low-Strength Materials. *Constr. Build. Mater.* **2023**, *393*, 132078. [CrossRef]
10. Bloss, M.L. Evolution of Cemented Rock Fill at Mount Isa Mines Limited. *Miner. Resour. Eng.* **1996**, *05*, 23–42. [CrossRef]
11. Collins, R.J.; Miller, R.H. Utilization of Mining and Mineral Processing Wastes in the United States. *Miner. Environ.* **1979**, *1*, 8–19. [CrossRef]
12. Yu, T.R.; Counter, D.B. Use of Fly Ash in Backfill at Kidd Creek Mines. In *Canadian Mining and Metallurgical Bulletin*; Canadian Institute of Mining and Metallurgy location: Montreal, QC, Canada, 1988; Volume 81, p. 909.
13. Ponomar, V.; Luukkonen, T.; Yliniemi, J. Revisiting Alkali-Activated and Sodium Silicate-Based Materials in the Early Works of Glukhovsky. *Constr. Build. Mater.* **2023**, *398*, 132474. [CrossRef]
14. Zhang, M.; Li, K.; Ni, W.; Zhang, S.; Liu, Z.; Wang, K.; Wei, X.; Yu, Y. Preparation of Mine Backfilling from Steel Slag-Based Non-Clinker Combined with Ultra-Fine Tailing. *Constr. Build. Mater.* **2022**, *320*, 126248. [CrossRef]
15. Li, J.; Zhang, S.; Wang, Q.; Ni, W.; Li, K.; Fu, P.; Hu, W.; Li, Z. Feasibility of Using Fly Ash–Slag-Based Binder for Mine Backfilling and Its Associated Leaching Risks. *J. Hazard. Mater.* **2020**, *400*, 123191. [CrossRef]
16. Jiang, H.; Han, J.; Ren, L.; Guo, Z.; Yilmaz, E. Study of Early-Age Performance of Cementitious Backfills with Alkali Activated Slag under Internal Sulfate Attack. *Constr. Build. Mater.* **2023**, *371*, 130786. [CrossRef]
17. Behera, S.K.; Mishra, D.P.; Singh, P.; Mishra, K.; Mandal, S.K.; Ghosh, C.N.; Kumar, R.; Mandal, P.K. Utilization of Mill Tailings, Fly Ash and Slag as Mine Paste Backfill Material: Review and Future Perspective. *Constr. Build. Mater.* **2021**, *309*, 125120. [CrossRef]
18. Wei, H.; Xiao, B.; Gao, Q. Flow Properties Analysis and Identification of a Fly Ash-Waste Rock Mixed Backfilling Slurry. *Minerals* **2021**, *11*, 576. [CrossRef]
19. Qiu, J.; Yang, L.; Sun, X.; Xing, J.; Li, S. Strength Characteristics and Failure Mechanism of Cemented Super-Fine Unclassified Tailings Backfill. *Minerals* **2017**, *7*, 58. [CrossRef]
20. Ushakova, E.; Perevoshchikova, A.; Menshikova, E.; Khayrulina, E.; Perevoshchikov, R.; Belkin, P. Environmental Aspects of Potash Mining: A Case Study of the Verkhnekamskoe Potash Deposit. *Mining* **2023**, *3*, 176–204. [CrossRef]
21. Yang, X.; Xiao, B.; Gao, Q. Validating the Use of Slag Binder with 91 Percent Blast Furnace Slag for Mine Backfilling. *Adv. Mater. Sci. Eng.* **2020**, *2020*, 2525831. [CrossRef]
22. Zhao, Y.; Zhou, X.; Zhou, Q.; Zhu, H.; Cheng, F.; Chen, H. Development of Full-Solid Waste Environmentally Binder for Cemented Paste Backfill. *Constr. Build. Mater.* **2024**, *443*, 137689. [CrossRef]
23. Nasir, O.; Fall, M. Coupling Binder Hydration, Temperature and Compressive Strength Development of Nnderground Cemented Paste Backfill at Early Ages. *Tunn. Undergr. Space Technol.* **2010**, *25*, 9–20. [CrossRef]
24. Benzaazoua, M.; Bussière, B.; Demers, I.; Aubertin, M.; Fried, É.; Blier, A. Integrated Mine Tailings Management by Combining Environmental Desulphurization and Cemented Paste Backfill: Application to Mine Doyon, Quebec, Canada. *Miner. Eng.* **2008**, *21*, 330–340. [CrossRef]
25. Zhang, T.; Yu, Q.; Wei, J.; Li, J.; Zhang, P. Preparation of High Performance Blended Cements and Reclamation of Iron Concentrate from Basic Oxygen Furnace Steel Slag. *Resour. Conserv. Recycl.* **2011**, *56*, 48–55. [CrossRef]
26. Tozsin, G.; Yonar, F.; Yucel, O.; Dikbas, A. Utilization Possibilities of Steel Slag as Backfill Material in Coastal Structures. *Sci. Rep.* **2023**, *13*, 4318. [CrossRef] [PubMed]
27. Singh, J.; Singh, S.P. Geopolymerization of Solid Waste of Non-Ferrous Metallurgy—A Review. *J. Environ. Manag.* **2019**, *251*, 109571. [CrossRef] [PubMed]
28. Zhu, M.; Xiao, N.; Tan, L.; Zhong, S.; Lyu, X.; Chi, X. Preparation of New Cementitious Material by Reduction and Activation of Copper Slag and Its Application in Mine Filling. *Chin. J. Nonferrous Met.* **2020**, *30*, 2736–2745. [CrossRef]
29. Liu, W.; Du, R.; Zhao, Z.; Zhang, R.; Wan, Y.; Li, H. Development of Lead Smelting Slag-Based Ecological Mine Backfill Material: Performance Control and Heavy Metals Solidification. *Constr. Build. Mater.* **2023**, *367*, 130310. [CrossRef]
30. Wang, F.; Zheng, Q.; Zhang, G.; Wang, C.; Cheng, F.; Lin, G. Preparation and Hydration Mechanism of Mine Cemented Paste Backfill Material for Secondary Smelting Water-Granulated Nickel Slag. *J. New Mater. Electrochem. Syst.* **2020**, *23*, 51–59. [CrossRef]
31. Behera, S.K.; Ghosh, C.N.; Mishra, K.; Mishra, D.P.; Singh, P.; Mandal, P.K.; Buragohain, J.; Sethi, M.K. Utilisation of Lead–Zinc Mill Tailings and Slag as Paste Backfill Materials. *Environ. Earth Sci.* **2020**, *79*, 389. [CrossRef]
32. *GBT1596-2017*; Fly Ash Used for Cement and Concrete. National Standards of China: Beijing, China, 2017.
33. Feng, J.; Zhang, Z.; Guan, W.; Wang, W.; Xu, X.; Song, Y.; Liu, H.; Su, H.; Zhao, B.; Hou, D. Review of the Backfill Materials in Chinese Underground Coal Mining. *Minerals* **2023**, *13*, 473. [CrossRef]
34. Behera, S.K.; Ghosh, C.N.; Mishra, D.P.; Singh, P.; Mishra, K.; Buragohain, J.; Mandal, P.K. Strength Development and Microstructural Investigation of Lead-Zinc Mill Tailings Based Paste Backfill with Fly Ash as Alternative Binder. *Cem. Concr. Compos.* **2020**, *109*, 103553. [CrossRef]
35. Xiao, B.; Wang, J.; Wu, A.; Guo, R. Hydration and Hardening Properties of High Fly-Ash Content Gel Material for Cemented Paste Backfill Utilization. *Gels* **2024**, *10*, 623. [CrossRef]

36. Park, J.H.; Edraki, M.; Mulligan, D.; Jang, H.S. The Application of Coal Combustion By-Products in Mine Site Rehabilitation. *J. Clean. Prod.* **2014**, *84*, 761–772. [CrossRef]
37. Zhang, S.; Shi, T.; Ni, W.; Li, K.; Gao, W.; Wang, K.; Zhang, Y. The Mechanism of Hydrating and Solidifying Green Mine Fill Materials Using Circulating Fluidized Bed Fly Ash-Slag-Based Agent. *J. Hazard. Mater.* **2021**, *415*, 125625. [CrossRef] [PubMed]
38. Yang, Z.; Xiong, L.; Fang, L.; Gao, Q.; Tian, L. Preparation of New Filling Cementing Materials with Sintering Desulfurization Ash. *Nonferrous Met. Sci. Eng.* **2015**, *6*, 8–12. [CrossRef]
39. Liu, W.; Liu, X.; Zhang, L.; Wan, Y.; Li, H.; Jiao, X. Rheology, Mechanics, Microstructure and Durability of Low-Carbon Cementitious Materials Based on Circulating Fluidized Bed Fly Ash: A Comprehensive Review. *Constr. Build. Mater.* **2024**, *411*, 134688. [CrossRef]
40. He, P.; Zhang, X.; Chen, H.; Zhang, Y. Waste-to-Resource Strategies for the Use of Circulating Fluidized Bed Fly Ash in Construction Materials: A Mini Review. *Powder Technol.* **2021**, *393*, 773–785. [CrossRef]
41. Longo, S.; Pigeon, P.; Pretorius, C. Paste Technology–Not Just for Mining Anymore. In Proceedings of the Paste 2017: 20th International Seminar on Paste and Thickened Tailings, Beijing, China, 15–18 June 2017; Wu, A., Jewell, R., Eds.; University of Science and Technology: Beijing, China, 2017; pp. 235–242.
42. Min, C.; Liu, Z.; Shi, Y.; Lu, X. Improving the Strength Performance of Cemented Phosphogypsum Backfill with Sulfate-Resistant Binders. *Constr. Build. Mater.* **2023**, *409*, 133974. [CrossRef]
43. Nizevičienė, D.; Vaičiukynienė, D.; Kielė, A.; Vaičiukynas, V. Mechanical Activation on Phosphogypsum: Hydrosodalite System. *Waste Biomass Valorization* **2019**, *10*, 3485–3491. [CrossRef]
44. Wang, Z.; Wang, Y.; Dino, G.A.; Zhang, L.; Ruan, Z.; Zhang, M.; Li, J.; Wu, A. Degradation of Hemihydrate Phosphogypsum-Based Backfill in Underground Mining: Mechanical and Microstructural Insights on the Effects of PH and Temperature of Mine Water. *Process Saf. Environ. Prot.* **2025**, *193*, 272–285. [CrossRef]
45. Jiang, G.; Wu, A.; Wang, Y.; Wang, Y.; Li, J. Determination of Utilization Strategies for Hemihydrate Phosphogypsum in Cemented Paste Backfill: Used as Cementitious Material or Aggregate. *J. Environ. Manag.* **2022**, *308*, 114687. [CrossRef] [PubMed]
46. Wang, Z.; Wang, Y.; Wu, L.; Wu, A.; Ruan, Z.; Zhang, M.; Zhao, R. Effective Reuse of Red Mud as Supplementary Material in Cemented Paste Backfill: Durability and Environmental Impact. *Constr. Build. Mater.* **2022**, *328*, 127002. [CrossRef]
47. Li, X.; Wang, D.; Chen, Q.; Qi, C. Alkali Activation of Blast Furnace Slag Using Bayer Red Mud as an Alternative Activator to Prepare Cemented Paste Backfill. *Constr. Build. Mater.* **2024**, *453*, 139061. [CrossRef]
48. Rai, S.B.; Wasewar, K.L.; Agnihotri, A. Treatment of Alumina Refinery Waste (Red Mud) through Neutralization Techniques: A Review. *Waste Manag. Res.* **2017**, *35*, 563–580. [CrossRef]
49. Feng, W.; Yu, Z.; Bao, R.; Xiong, J.; Yan, K.; Liu, R.; Zhang, R.; Lu, X. Manufacture of Tailings-Based Cementitious Materials: Insights into Tailings Activation Strategies. *Constr. Build. Mater.* **2024**, *439*, 137194. [CrossRef]
50. Liu, J.; Ge, X.; Liu, P.; Song, G.; Hu, Z. Experimental Study on the Preparation of Cementitious Materials from Iron Ore Tailings by Activation. *Constr. Build. Mater.* **2023**, *385*, 131409. [CrossRef]
51. Sagade, A.; Fall, M. Study of Fresh Properties of Cemented Paste Backfill Material with Ternary Cement Blends. *Constr. Build. Mater.* **2024**, *411*, 134287. [CrossRef]
52. Jiang, H.; Qi, Z.; Yilmaz, E.; Han, J.; Qiu, J.; Dong, C. Effectiveness of Alkali-Activated Slag as Alternative Binder on Workability and Early Age Compressive Strength of Cemented Paste Backfills. *Constr. Build. Mater.* **2019**, *218*, 689–700. [CrossRef]
53. Krupnik, L.A.; Shaposhnik, Y.N.; Shaposhnik, S.N.; Nurshaiykova, G.T.; Tungushbaeva, Z.K. Technology of Backfill Preparation Based on Cement-and-Slag Binder in Orlov Mine. *J. Min. Sci.* **2017**, *53*, 77–83. [CrossRef]
54. Nunes, V.A.; Borges, P.H.R. Recent Advances in the Reuse of Steel Slags and Future Perspectives as Binder and Aggregate for Alkali-Activated Materials. *Constr. Build. Mater.* **2021**, *281*, 122605. [CrossRef]
55. Peyronnard, O.; Benzaazoua, M. Alternative By-Product Based Binders for Cemented Mine Backfill: Recipes Optimisation Using Taguchi Method. *Miner. Eng.* **2012**, *29*, 28–38. [CrossRef]
56. Zhou, S.; Li, X.; Zhou, Y.; Min, C.; Shi, Y. Effect of Phosphorus on the Properties of Phosphogypsum-based Cemented Backfill. *J. Hazard. Mater.* **2020**, *399*, 122993. [CrossRef]
57. Chen, Q.; Zhang, Q.; Fourie, A.; Xin, C. Utilization of Phosphogypsum and Phosphate Tailings for Cemented Paste Backfill. *J. Environ. Manag.* **2017**, *201*, 19–27. [CrossRef] [PubMed]
58. Sutar, H.; Chandra Mishra, S.; Kumar Sahoo, S.; Prasad Chakraverty, A.; Sekhar Maharana, H. Progress of Red Mud Utilization: An Overview. *Chem. Sci. Int. J.* **2014**, *4*, 255–279. [CrossRef]
59. Guner, N.U.; Yilmaz, E.; Sari, M.; Kasap, T. Cementitious Backfill with Partial Replacement of Cu-Rich Mine Tailings by Sand: Rheological, Mechanical and Microstructural Properties. *Minerals* **2023**, *13*, 437. [CrossRef]
60. Jhatial, A.A.; Nováková, I.; Gjerløw, E. A Review on Emerging Cementitious Materials, Reactivity Evaluation and Treatment Methods. *Buildings* **2023**, *13*, 526. [CrossRef]
61. *GB/T 18046-2017*; Granulated Blast Furnace Slag Powder Used in Cement, Mortar, and Concrete. National Standards of China: Beijing, China, 2017.

62. Xie, G.; Suo, Y.; Liu, L.; Zhu, M.; Xie, L.; Qu, H.; Sun, W. Mechanical Grinding Activation of Modified Magnesium Slag and Its Use as Backfilling Cementitious Material. *Case. Studies. Constr. Mater.* **2023**, *18*, e01778. [CrossRef]
63. Cao, L.; Shen, W.; Huang, J.; Yang, Y.; Zhang, D.; Huang, X.; Lv, Z.; Ji, X. Process to Utilize Crushed Steel Slag in Cement Industry Directly: Multi-Phased Clinker Sintering Technology. *J. Clean. Prod.* **2019**, *217*, 520–529. [CrossRef]
64. Bull, A.J.; Fall, M. Thermally Induced Changes in Metalloid Leachability of Cemented Paste Backfill That Contains Blast Furnace Slag. *Miner. Eng.* **2020**, *156*, 106520. [CrossRef]
65. Hao, J.; Zhou, Z.; Chen, Z.; Che, Z.; Wang, X. Review of Hydration and Hardening Properties of Steel Slag in Mine Filling Cementitious Materials. *J. Min. Sci. Technol.* **2024**, *9*, 573–585. [CrossRef]
66. Helinski, M.; Fourie, A.; Fahey, M.; Ismail, M. Assessment of the Self-Desiccation Process in Cemented Mine Backfills. *Can. Geotech. J.* **2007**, *44*, 1148–1156. [CrossRef]
67. Xin, J.; Liu, L.; Jiang, Q.; Yang, P.; Qu, H.; Xie, G. Early-Age Hydration Characteristics of Modified Coal Gasification Slag-Cement-Aeolian Sand Paste Backfill. *Constr. Build. Mater.* **2022**, *322*, 125936. [CrossRef]
68. Lin, F.; Meyer, C. Hydration Kinetics Modeling of Portland Cement Considering the Effects of Curing Temperature and Applied Pressure. *Cem. Concr. Res.* **2009**, *39*, 255–265. [CrossRef]
69. Zhu, K.; Wang, L.; Lv, G.; Zhang, D.; Liao, L.; Guo, L.; Hu, J.; Bai, Y. Research Progress of Heavy Metal Immobilization in Non-Ferrous Smelting Slag-Based Cementitious Materials. *Nonferrous Met. Eng.* **2024**, *14*, 185–193. [CrossRef]
70. Bothe, J.V.; Brown, P.W. Arsenic Immobilization by Calcium Arsenate Formation. *Environ. Sci. Technol.* **1999**, *33*, 3806–3811. [CrossRef]
71. Zhang, D.; Shi, S.; Wang, C.; Yang, X.; Guo, L.; Xue, S. Preparation of Cementitious Material Using Smelting Slag and Tailings and the Solidification and Leaching of Pb2+. *Adv. Mater. Sci. Eng.* **2015**, *2015*, 352567. [CrossRef]
72. Yılmaz, T.; Ercikdi, B.; Deveci, H. Evaluation of Geochemical Behaviour of Flooded Cemented Paste Backfill of Sulphide-Rich Tailings by Dynamic-Tank Leaching Test. *Int. J. Min. Reclam. Environ.* **2021**, *35*, 336–355. [CrossRef]
73. Romaniuk, N.A.; McFarlane, L.; Hariharan, N. Development of Slag Alternatives for Paste Backfill Operations. In Proceedings of the Paste 2024: 26th International Conference on Paste, Thickened and Filtered Tailings, Melbourne, Australia, 16–18 April 2024; Fourie, A.B., Reid, D., Eds.; Australian Centre for Geomechanics: Crawley, WA, USA, 2024; pp. 535–544.
74. Wang, C.; Qiao, H.; Li, Q.; Su, R.; An, B. Advances in Recycled Concrete Research – Improving Shrinkage for Sustainable Building Applications. *Archit. Eng. Des. Manag.* **2024**, *20*, 1589–1611. [CrossRef]
75. Liu, L.; Fang, Z.; Wang, S.; Gao, G.; Zhang, B.; Zhao, Y.; Zhu, M.; Liu, Z.; Wang, J.; Jia, Q.; et al. Theoretical Basis and Technical Conception of Backfill Carbon Fixation in Coal Mine. *Coal Sci. Technol.* **2024**, *52*, 292–308. [CrossRef]
76. *GB/T175-2023*; General Portland Cement. National Standards of China: Beijing, China, 2023.
77. *ASTM ASTM C150/C150M-21*; Standard Specification for Portland Cement. ASTM International: West Conshohocken, PA, USA, 2021; p. D3699.
78. *ASTM Committee C09.27 ASTM C989/C989M-14*; Standard Specification for Slag Cement for Use in Concrete and Mortars. ASTM International: West Conshohocken, PA, USA, 2014.

toxics

MDPI

Article

Pollution Risk Assessment of Potentially Toxic Elements in Soils Using Characterization and Microbiological Analysis: The Case of a Rare and Precious Metal Mining Site in Wuzhou, Guangxi

Yi Sun [1,2,*], Zixuan Yang [3], Kun Dong [3], Fujiang Hui [3], Dunqiu Wang [3] and Yecheng Huang [3]

1 Hubei Key Laboratory of Biologic Resources Protection and Utilization, Hubei Minzu University, Enshi 445000, China
2 College of Forestry and Horticulture, Hubei Minzu University, Enshi 445000, China
3 Guangxi Key Laboratory of Environmental Pollution Control Theory and Technology, Guilin University of Technology, Guilin 541004, China; 15293163375@163.com (Z.Y.); 2020005@glut.edu.cn (K.D.); h19507737192@139.com (F.H.); wangdunqiu@glut.edu.cn (D.W.); m18697920741@163.com (Y.H.)
* Correspondence: sunyi@hbmzu.edu.cn

Abstract: To understand the characteristics of the pollution risk of potentially toxic elements (PTEs) at a rare and precious metal mining site in Guangxi and to provide scientific evidence for the comprehensive evaluation and soil remediation of PTE pollution at the site, the Cd, As, Co, Cu, Cr, Ni, Pb, and Zn contents of five areas were determined. Laboratory testing was conducted on five soil plots in the selected five suspected contaminated areas (electroplating workshop, sewage treatment area, and boiler room). Correlation analysis, infrared spectroscopy (FTIR), X-ray diffraction (XRD), and X-ray photoelectron spectroscopy (XPS) were used to evaluate and analyze PTE pollution. The average contents of Cd, Co, As, Pb, Zn, and Cu at the site were higher than the background values in the Guangxi soil. The Probability Mass Function (PMF) model was used to perform a source apportionment of the PTEs and determine the main pollution sources and their contribution rates. The results of the single factor pollution of the PTEs showed that Cd, Ar, and Cr were heavy pollutants, and Co was a light pollutant. The Nemerow comprehensive pollution index analysis showed that the study area was heavily polluted. The Earth accumulation index results show that Cd exhibited a very serious accumulation, Cu and Zn exhibited mild to moderate accumulations, and As and Co exhibited moderate accumulations. The FTIR results showed that C=O in the soil was chelated with PTEs in some samples, which weakened the characteristic peaks of C=O in proteins and polypeptides. The XRD results showed that cadmium hydroxide, lead oxide, and zinc hydroxide were present in the soil samples. The XPS results showed that the production of O^{2-} in the O 1s high-resolution spectra mainly came from the metal oxides produced by the polluting metals. Meanwhile, the microbial results showed that the pollution risk of PTEs affected the soil microbial community structure and diversity to some extent.

Keywords: PTEs; soil pollution; characterization; risk assessment

1. Introduction

Soil is the base of agricultural production [1] and provides a living environment that promotes the growth and reproduction of plants and microorganisms [2]. Potentially toxic elements (PTEs) in agricultural soils, particularly cadmium (Cd) and lead (Pb), pose significant threats to human health and the environment [3–5]. They originate from various human activities. The topsoil near agricultural soils with PTEs acts as a direct sink, promoting the transfer of PTEs into the food chain. Even at ambient levels, PTEs migrate under acid conditions. Because they cannot be degraded, they may cause secondary pollution in water bodies and exacerbate ecological damage [6–9]. Under natural conditions, the transportation of weathered rock mineral products by rivers is the main source of PTEs in sediments. With societal progress in urban development, agricultural and industrial activities in lake basins, including urban transportation, fossil fuel combustion, mining, and metal smelting, and the use of fertilizers and pesticides, soil environmental pollution has expanded [10].

PTEs in soil solutions tend to accumulate on soil surfaces by adsorption, and some of them might be bacterially transformed into methyl compounds that are more toxic for us [11–13]. Studying soil pollution by PTEs and effectively controlling it, minimizing the pollution caused to the soil by the source, can improve the ecology of a soil and ensure the health of people. Research on soil PTE pollution mainly involves studying the basic physical and chemical properties of soil, analyzing the characteristics of PTE content, particularly different forms and distributions of PTEs in time and space, and combining different pollution assessment methods (for example, the Nemerow index method, single pollution index method, ratio of secondary phase and primary phase method, and risk assessment coding method) to analyze pollution level and ecological risk [14,15]. Several such studies have been published [14–17].

In the rare and precious metal mining area of Guangxi, there are metallurgical facilities where indium is extracted from zinc oxide through wet smelting and the resulting zinc sulfate solution is used to produce lithopone and solid zinc sulfate. The industrial solid waste produced is sent to the rotary kiln production line to recover valuable metals, and the resulting product, zinc oxide, is used as a supplementary raw material for the wet extraction of the indium production line. In this area, there is also a significant presence of solid waste (abandoned structures, construction waste, and residue waste) that could act as a secondary source of environmental pollution. In hot spots, such as the rare and precious metal mining area in Guangxi, where soil pollution levels exceed soil pollution risk control standards, relevant parties should conduct soil risk assessments for PTEs. These soil pollution risk assessment reports must aim to provide a reference for human health risk prevention and scientific basis for their risk control measures and restoration efforts in accordance with national regulations and requirements.

The difference between this area and others is that previous research has shown that for plots with soil pollution levels exceeding the soil pollution risk control standards, relevant parties should conduct soil pollution risk assessments and take corresponding risk control measures in accordance with national regulations and the requirements of soil pollution risk assessment reports. The present study considered the soils of a rare and precious metal site in Guangxi Province as the research object. By measuring the contents of Cd, As, Cu, Cr, Ni, Pb, and Zn in the soil at the site, the single pollution and Nemerow indices were calculated. Infrared spectroscopy (FTIR), X-ray diffraction (XRD), and X-ray photoelectron spectroscopy (XPS) were used to evaluate the soil pollution status and potential ecological risks. Correlation analysis was conducted to provide a reference for soil pollution prevention and risk control at the site. The single factor pollution method, Nemerow index analysis method, and geoaccumulation index method were used to evaluate the potential toxic element pollution source characteristics, distribution, pollutant types, and concentrations of the site's soils and conduct a risk assessment of soil pollution. The research results provide a reference for the treatment of contaminated soils in rare and precious metal mining areas.

2. Materials and Methods

2.1. Site Characteristics

The study area is a rare and precious metal site in Guangxi (Figure 1). It belongs to a subtropical monsoon climate with high temperatures (annual average temperature 21.6 °C) and is rainy and humid throughout the year. The site is generally high in the south and low in the north. Most surrounding areas contain industrial and residential buildings. The terrain of the area is relatively flat and there are no drinking water source protection areas, nature reserves, scenic spots, cultural relics, historic sites, or other environmentally sensitive objects that require special protection within 1 km of the site. The site primarily uses hydrometallurgy to produce secondary zinc oxide and refined iron ore from metallic zinc and iron slags.

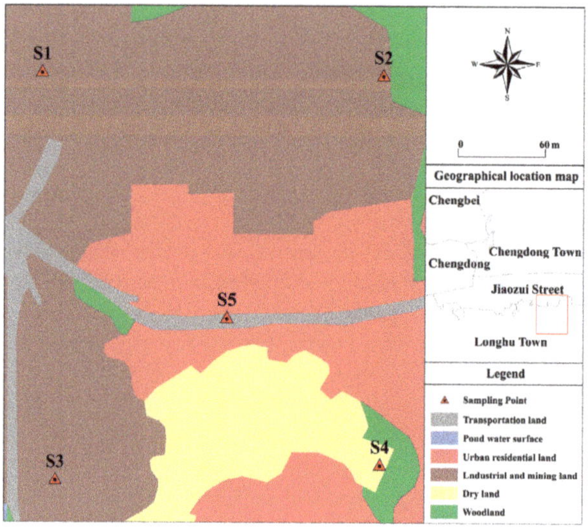

Figure 1. Overview of the study area. (S1–S5 are five sampling points).

This is a former mining site with metallurgical facilities. Excavation work is stopped, and the metallurgical production line is not in use. It awaits land restoration. Currently, the buildings and equipment in the plant area are dismantled, and the products and raw and auxiliary materials have been cleared and removed. The soil pollution status of the plots was investigated, and the area will be planned as a protective green space, storage, and logistics land, which will belong to the first class of land specified in the Standard for Soil Pollution Risk Management and Control of Construction Land for Soil Environmental Quality (GB36600-2018) [18].

The main process flow at this site was the extraction of indium from indium-containing zinc oxide via hydrometallurgy. The zinc sulfate solution produced was simultaneously used to produce lithopones and solid zinc sulfate. The industrial solid waste and metal waste produced in the process of hydrometallurgical indium extraction and lithopone production enter a rotary kiln production line to recover valuable metals, and the product, zinc oxide, was used as a supplementary raw material for hydrometallurgical indium extraction production. The production facilities in the plant area included a steam supply section, leaching slag yard, zinc barium white section, and steelmaking section.

2.2. Sample Collection and Processing

Based on preliminary and detailed investigations, monitoring points were planned to be set up in areas within the site that may be heavily polluted. Due to the use of a large amount of strong acidic materials in the production process, the tanks, valves, and other items of machinery are severely corroded, and the conveying pipelines and anti-corrosion lining are aging. Owing to leakage and seepage during the original production process, pollutants may have spread throughout the site, posing a significant threat to the safety of the soil, the groundwater, and the surrounding natural water environment. Due to years of outdated production activities, the soil at the site is polluted to varying degrees. The commonly used point sampling method is the five-point sampling method. First, the midpoint of the diagonal (in a square or rectangular sampling area, the diagonal refers to the straight line connecting the two diagonal vertices of the sampling area) was determined as the center sampling point, and then four points on the diagonal equidistant from the center sampling point were selected as sampling points. For S1–S5, we followed the five-point sampling method, where two samples of the same sample at each point are analyzed synchronously under identical conditions using the three-point sampling method. The soils at the rare and precious metal site were sampled from five plots, each with an area of 300 m^2, representing soils with different concentrations of PTEs.

We conducted sampling in the survey area using the five-point sampling method in May 2024. And three replicate samples were measured at each sampling point. The points in the production area, sewage treatment area, slag yard, storage area, and rotary kiln exhaust outlet of the factory were labeled S1–S5. A five-point sampling method was adopted for the soil sample collection from the quadrats, and three parallel samples were selected from each quadrat. The soil samples were placed in polyethylene bags, and debris (rocks, leaves, garbage, etc.) was removed, followed by air drying, grinding, and sieving through a 160 mesh. The total amount of PTEs in the soil was extracted using nitric acid, hydrochloric acid, hydrofluoric acid, and a microwave digestion system. The soil samples (0.2 g) were weighed and sieved into a digestion tube. Five regions with relatively high concentrations of PTEs were selected for characterization. Microwave digestion was carried out according to a "6 nitric acid, 6 hydrochloric acid, 3 hydrofluoric acid" acid mixture. A total of 2 mL of nitric acid and 1 mL of hydrofluoric acid were added to the digested samples and placed on an electric heating plate to reduce the acid volume to about 1–2 mL, and the mixture was then cooled to room temperature, made to a fixed volume in a 25 mL colorimetric tube, and filtered with a 0.45 μm water system filter membrane. Three blank samples and three soil reference materials (gss-8) were run through the same process with each sample digestion batch. An inductively coupled plasma mass spectrometer (Perkin Elmer, icp8000, Waltham, MA, USA) was used to detect the Cd, Co, As, Pb, Zn, and Cu in the digested samples. The recovery rate of the certified reference materials was 91–97%, and the test method was accurate and reliable. The functional groups of the samples were analyzed using a Fourier-transform infrared spectrometer (Bruker, ten-sor27, Billerica, MA, USA). The test conditions were as follows: scanning range was 400~4000 cm^{-1}, the scanning time was 16° min^{-1}, and the resolution was 4 cm^{-1}. An XRD diffractometer (Malvern Panalytical, X'Pert^3, Almelo, The Netherlands) was used to scan the crystal structure of the samples. The scanning angle was $2\theta = 10°\sim90°$, and the scanning speed was 5 (°) min^{-1}. The obtained XRD spectra were analyzed using Jade software 9.0. X-ray photoelectron spectroscopy (Thermo Fisher, escalab 250xi, Waltham, MA, USA) was used to analyze the elemental composition of the material surface, and the tested elements were charge-corrected with C 1s = 284.8 EV binding energy. The pH of the soil samples was determined using a 2.5:1 water–soil ratio (V:m) extraction potentiometric method.

2.3. Impact Assessment Method of PTE Pollution in Soil

In this study, the single pollution, Nemerow, geoaccumulation, and potential ecological risk indices were used to evaluate PTE pollution at the site. The background values of the Cd, As, Cr, Ni, Pb, Zn, and Cu in Guangxi were 0.267 mg/kg, 20.80 mg/kg, 82.10 mg/kg, 26.60 mg/kg, 24.00 mg/kg, 75.60 mg/kg, and 27.80 mg/kg, respectively [19].

2.3.1. Single Pollution Index Method

The single pollution index method evaluates the worst-performing indicator by comparing the measured data of each factor against the project's environmental quality standards, thereby determining their pollution category [20,21]. The single pollution index method can strengthen the objectivity of index weight determination. The single factor index method was used for the evaluation. The pollution index was calculated by comparing the measured values of the soil samples with their threshold limits. This study used the screening values from the GB36600-2018 standard. Since this standard does not specify a limit value for zinc, the soil pollution risk screening guideline value for construction land (500 mg/kg) was adopted as its reference to assess its contamination level.

The formula used was the following:

$$P_i = \frac{C_i}{S_i} \tag{1}$$

where P_i is the pollution index of each ith PTE pollutant, C_i is the concentration of each ith PTE pollutant, and S_i represents the evaluation criterion for the ith PTE pollutant.

2.3.2. Nemerow Index Method

The Nemerow index, which is a weighted multi-factor environmental quality index that considers both extreme values and prominent maximum values [22–24], highlights the evaluation factors with serious pollution. The classifications of the index are shown in Table 1.

Table 1. Classification of single factor pollution index and Nemerow comprehensive pollution index.

Single Factor Pollution Index	Level	Nemerow Comprehensive Pollution Index	Level
$P_i \leq 1$	Class I, without pollution	$P_{Comprehensive} \leq 0.7$	Class I, no pollution
$P_i \in (1, 2]$	Class II, slight pollution	$P_{Comprehensive} \in (0.7, 1.0]$	Class II, slight pollution
$P_i \in (2, 3]$	Class III, mild pollution	$P_{Comprehensive} \in (1.0, 2.0]$	Class III, mild pollution
$P_i \in (3, 5]$	Class IV, moderate pollution	$P_{Comprehensive} \in (2.0, 3.0]$	Class IV, moderate pollution
$P_i > 5$	Class V, severe pollution	$P_{Comprehensive} > 3.0$	Class V, severe pollution

The formula used was the following:

$$P_{Comprehensive} = \sqrt{\frac{P_i^2 + P_{max}^2}{2}} \tag{2}$$

where $P_{Comprehensive}$ represents the comprehensive pollution index of PTEs at the sampling point and P_{max} represents the maximum value of the single factor pollution index of each PTE pollutant at the sampling point.

2.3.3. Index of Geoaccumulation (Igeo) Method

The index of geoaccumulation (Igeo) method [25,26] for heavy metals is primarily used to study changes in the background value of PTE pollution caused by natural processes and the comprehensive impact of human activities on soil PTE pollution. The formula used was the following:

$$I_{geo} = \log_2\left(\frac{C_i}{LB_i}\right) \quad (3)$$

where I_{geo} represents the pollution index of geoaccumulation for PTEs, C_i represents the measured concentration of PTEs, B_i represents the background value of the PTEs used in this study (the soil background value of Guangxi Province was used in this study), and L is a constant. This is a coefficient that considers possible changes in background values caused by differences in rocks in different regions, and the value of L was 1.5 in this study. Classification of the index of geoaccumulation was shown in Table 2.

Table 2. Classification of the index of geoaccumulation (*Igeo*).

Igeo	$I_{geo} \leq 0$	$0 \leq I_{geo} < 1$	$1 \leq I_{geo} < 2$	$2 \leq I_{geo} < 3$	$3 \leq I_{geo} < 4$	$4 \leq I_{geo} < 5$	$5 \leq I_{geo} < 10$
Level	Without accumulation	Mild to moderate accumulation	Moderate accumulation	Medium to strong accumulation	Strong accumulation	Strong to extremely severe accumulation	Extremely severe accumulation

2.3.4. PMF Model Construction

The PMF model was used to apportion sources to the PTE data and determine the main pollution sources. The PMF model is a method first proposed by Paatero et al. in 1994 [27] and uses the weight coefficients of variables to determine the error of pollution components. It iteratively calculates the main pollution sources and their rates of contribution using the least-squares method. Owing to its advantages, including not requiring detailed source component spectra, being able to decompose factor loadings, and utilizing data uncertainty for optimization, the PMF model is widely used in apportioning heavy metal sources in soils. The basic principle of the PMF model is to decompose the original matrix X ($n \times m$) into a factor contribution matrix G ($n \times p$), factor composition matrix F ($p \times m$), and residual matrix E ($n \times m$).

When the concentration of the chemical components was less than or equal to the corresponding method detection limit (*MDL*), the uncertainty (*U*) was calculated using Equation (4):

$$U_{nc} = 5/6 \times MDL \quad (4)$$

When the concentration of the chemical components exceeded the corresponding *MDL*, *U* was calculated according to Equation (5):

$$U_{nc} = \sqrt{(c_{onc} \times EF)^2 + (0.05 \times MDL)^2} \quad (5)$$

where U_{nc} represents the uncertainty of the analyzed components (mg/kg); *MDL* represents the detection limit of the components (mg/kg); C_{onc} represents the mass concentration of the components (mg/kg); and *EF* (error fraction) represents the error coefficient of each component.

The indicator judgments Q (robust) and Q (true) were both 2.63102, with an error of less than 10%, and the model was interpretable. The BS inspection rate exceeded 80%; therefore, the results were usable. (Weak represents the component with high measurement uncertainty in Table 3).

Table 3. Parameter settings for PMF model construction.

Species			Category
Co			Weak
Cu			Weak
Zn			Weak
Pb			Weak
Ni			Weak
Cr			Weak
As			Weak
Cd			Weak
Number of base runs:			20
Base user-selected seed:			48
Number of factors:			3
Extra modeling uncertainty (%):			20

	Factor 1	Factor 2	Factor 3	Unmapped
Boot Factor 1	200	0	0	0
Boot Factor 2	1	199	0	0
Boot Factor 3	8	14	178	0

2.3.5. Extraction and High-Throughput Sequencing of Microbial DNA

The soil samples were pretreated, and the DNA was extracted using a DNA Extraction Kit (E.Z.A^{TM}mag-bindsoildnakit, Omega Bio-tek, Norcross, GA, USA). The target gene of the V3-V4 region was amplified by the 16S rRNA universal primers 341F (5′-CCTACGGGNGGCWGCAG-3′) and 805R (5′-GACTACHVGGGTATCTAATCC-3′). Illumina bridge PCR-compatible primers were used for the second round of amplification. The original data were spliced from MiSeq sequencing to distinguish the samples. Subsequently, the quality of the sequence was controlled and filtered, and the operational taxonomic unit (OTU) was used for the clustering and taxonomic analyses of species.

2.3.6. Data Analysis and Processing

Excel 2016 was used to statistically analyze the content of PTEs in the soil, and SPSS 26 and R studio 4.3.0 were used to analyze the correlation between the PTEs in the soil.

3. Results

3.1. PTE Content in Soil Samples

The basic physical and chemical properties of the soil samples from the five sites were determined. The soil type in this area belongs to the following solid waste pollution type: soil pollution caused by the stacking or disposal of industrial and mining waste, sludge, and urban garbage on the surface. The pH of the samples was between 6.81 and 7.33. The pH values of the five points S1–S5 were 6.81, 7.10, 6.89, 7.21, and 7.33, respectively. The PTE contents varied significantly. The average contents of Cd, Co, As, Pb, Zn, and Cu were higher than the background values for the soil in Guangxi (Table 4). The coefficient of variation (CV) was used to characterize the spatial dispersion of the soil samples. Compared to the screening value of the standard for the control of the soil pollution risk of construction land for soil environmental quality (GB36600-2018), the sample points exceeding the standard values of Cd, Co, As, Cr, and Pb were 80%, 60%, 60%, 40%, and 20%, respectively. The CVs of Cd, Co, As, Cr, Ni, Pb, Zn, and Cu were 7–142%, of which the coefficients of variation of Cd, As, Pb, and Cu were all greater than 1, namely 1.03, 1.42, 1.04, and 1.24, respectively, indicating that the elemental contents of the different samples in this study were quite different, with a high degree of dispersion, and were greatly affected by external factors. No points exceeded the standards for soil Ni, Zn, and Cu, that is, the content was lower than the standard for controlling the soil pollution risks of construction land, and the average content of Cd, Co, and As in the soil exceeded the screening value. Compared to the background values of the soil in Guangxi, the average contents of Cd, Co, As, Pb, Zn, and Cu were higher. From these results, the scope and degree to which the standard was exceeded were relatively large, which may be due to the fact that raw materials, including metal waste residue and other raw and auxiliary materials, entered the ground during production, processing, storage, transportation, and pollution treatment due to rainwater scouring and leakage through surface cracks, resulting in a degree of pollution in the soil.

Table 4. Characteristics of PTE content at the site.

PTE	Minimum (mg/kg)	Maximum (mg/kg)	Average (mg/kg)	Standard Deviation (mg/kg)	Coefficient of Variation	Screening Value (mg/kg)	Background Concentration
Cd	5.6	432.7	143.1	147.9	1.03	20.0	0.27
Co	12.4	93.3	47.6	31.6	0.66	20.0	10.40
As	21.5	425.2	110.2	157.2	1.42	20.0	20.50
Cr	80.6	92.5	82.8	6.5	0.07	90.0	82.10
Ni	32.5	38.6	30.3	4.4	0.14	150.0	26.60
Pb	5.1	728.8	264.3	274.2	1.04	400.0	24.00
Zn	130.9	318.6	187.9	72.6	0.39	500.0	75.60
Cu	32.5	286.9	83.5	104.0	1.24	2000.0	27.80

3.2. Assessment of PTE Pollution in Soil

3.2.1. Single Factor and Nemerow Index Methods

The pollution intensities of the PTEs ranged from strong to weak as follows: Cd (7.154) > As (5.312) > Co (2.378) > Pb (0.660) > Cr (0.625) > Zn (0.375) > Ni (0.166) > Cu (0.038). Ni, Pb, Zn, and Cu belonged to the non-pollution level (Table 5). The average content of Cd ranged from 0.279 to 20.284 mg/kg, which was much higher than the background value of the soil. As was between 0.0735 and 21.260 mg/kg, which is classified as severe pollution, while Co was between 0.618 and 4.484 mg/kg, which is classified as mild pollution, but its average content was more than two times higher than the background value of the soil.

Table 5. Single factor and comprehensive pollution index table of PTEs.

PTE	Single Factor Pollution Index Range	Average Value of Single Factor Pollution Index	Pollution Index	Comprehensive Pollution Index	Pollution Level
Cd	0.279~21.635	7.154	Severe pollution	10.42	Severe pollution
Co	0.618~4.484	2.378	Slightly pollution		
As	0.074~21.261	5.312	Severe pollution		
Cr	0.228~0.805	0.625	Without pollution		
Ni	0.083~0.257	0.166	Without pollution		
Pb	0.013~1.820	0.660	Without pollution		
Zn	0.261~0.637	0.375	Without pollution		
Cu	0.001~0.143	0.038	Without pollution		

3.2.2. Index of Geoaccumulation (*Igeo*)

The average accumulation of PTEs in the soil of the five sites was Cd (8.481) > Pb (2.876) >As (1.789) > Co (1.608) > Cu (0.895) > Zn (0.729) > Ni (−0.675) > Cr (−1.526)

(Table 6). Among these, the average value of the index of geoaccumulation of Cu and Zn was $0 \leq Igeo < 1$, reflecting mild to moderate accumulation, and the average value of the index of geoaccumulation of Ni and Cr was less than 0, indicating an absence of accumulation.

Table 6. Site's soil PTE indices of geoaccumulation (*Igeo*).

PTE	Index of Geoaccumulation	Pollution Level
Cu	0.895	Mild to moderate accumulation
Zn	0.729	Mild to moderate accumulation
Pb	2.876	Medium to strong accumulation
Ni	−0.675	Without accumulation
Cr	−1.526	Without accumulation
As	1.789	Moderate accumulation
Co	1.608	Moderate accumulation
Cd	8.481	Extremely severe accumulation

Sample points with extremely serious cumulative pollution levels of Cd and moderate to strong cumulative pollution levels of Pb in the soil were distributed in the replacement section, slag yard area, and extraction and reverse extraction workshop. Pollution sources were mainly related to the migration and transformation of geological background elements caused by high background values of Cd and Pb in the soil, as well as the artificial transportation of metal waste and other raw and auxiliary materials. This result is consistent with the conclusion of previous studies that the main pollutants in the soil near smelting activities are As, Cd, and Pb [28,29].

3.3. Correlation Analysis and Source Apportionment of PTEs in Soil

SPSS 26 correlation analysis of the PTEs in the soil of sites showed that a p value < 0.01 signified an extremely significant correlation, $0.01 < p < 0.05$ was a significant correlation, and $p > 0.05$ was not a correlation. Table 7 shows the correlation results of the PTEs in the soil at the five sites. Pb exhibited a very significant correlation with Ni, As, Co, and Cd, and a significant correlation with Cr. Ni exhibited a close relationship with Cr, As, Co, and Cd, showing a significant correlation. The main source of Ni in the soil was its geological form, and the pollution of Pb, As, and Co also reflected an anthropogenic impact caused by the construction waste and waste residue generated by the artificial metal waste residue and other raw and auxiliary materials in the production and leaching process areas. The pollution of soil Cd may have been caused by the discharge of industrial "three wastes" and the migration of toxic and harmful substances into the soil caused by a large number of metal warehouses. These correlations are consistent with the pollution of the soil at the site (see Figure 2).

Table 7. Correlation table of PTEs in soil.

Pearson	Cu	Zn	Pb	Ni	Cr	As	Co	Cd
Cu	1.000							
Zn	−0.322	1.000						
Pb	0.776	0.347	1.000					
Ni	0.776	0.347	1.00 **	1.000				
Cr	0.405	0.726	0.884 *	0.884 *	1.000			
As	0.681	0.473	0.990 **	0.990 *	0.942 *	1.000		
Co	0.643	0.516	0.981 **	0.981 *	0.959 *	0.999 **	1.000	
Cd	0.587	0.574	0.964 **	0.964 *	0.977 **	0.992 **	0.997 **	1.000

Note: ** indicates extremely significant correlation ($p < 0.01$), * indicates significant correlation ($0.01 < p < 0.05$), and $p > 0.05$ indicates no correlation.

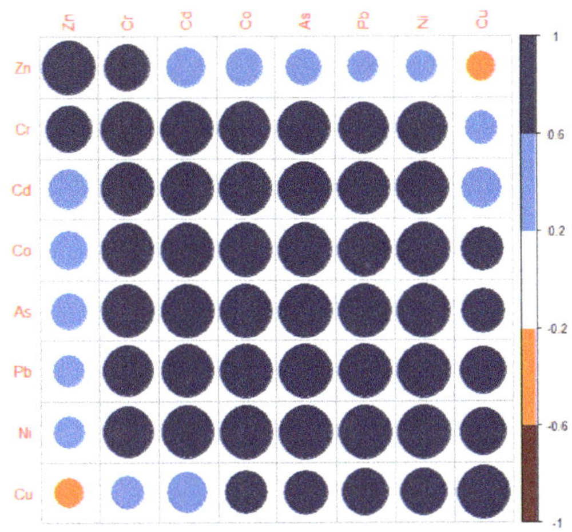

Figure 2. Correlation diagram of PTEs in soil.

To further analyze the pollution sources of the PTEs, a principal component analysis of eight types of PTEs was carried out. Three principal components were extracted, and the characteristic root values exceeded unity. Principal component 1 contributed 39.86%, principal component 2 contributed 29.15%, and principal component 3 contributed 22.23%, and their cumulative contribution was 91.24% of the total variation. The results of the principal components of the PTEs in the soil are shown in Table 8. As shown in Table 8 and Figure 3, the loads of Cu, Ni, and Pb in principal component 1 were relatively high. According to the distribution of pollution characteristics, these four PTEs were distributed near production and leaching process sites, replacement sections, and slag yard areas, and thus may be caused by the presence of solid wastes such as abandoned structures, construction waste, and slag in the plot and smoke, wastewater, and slag generated in the process of the preparation of raw materials and smelting, as well as by emissions, dripping, and leakage in the production process. The average content of the Cd, Co, and Cu PTEs exceeded the soil background value in Guangxi, and the range of their CVs was 0.66–1.24, indicating that these elements may be affected by human activities and that their non-uniformity may also be caused by hydrothermal and invasive directional processes. Their morphological changes were caused by changes in the solubility of each mineral. Although Co showed a relatively high loading in principal component 2, its depth concentration remained below the standard threshold, which may be due to the additional sources of Co possibly due to the inclusion of some production waste residue in the backfill or the soil outside the site that entered the plot to cause pollution.

Table 8. Soil PTE principal component analysis.

PTE	Component 1	Component 2	Component 3
Zn	−0.402	0.124	−0.860
Pb	−0.841	−0.005	0.464
Cr	−0.245	0.360	0.842
Co	0.632	0.739	0.181
Cd	0.570	−0.811	0.103
Cu	0.767	0.400	−0.081
Ni	0.899	0.172	0.054
As	0.374	−0.890	0.249

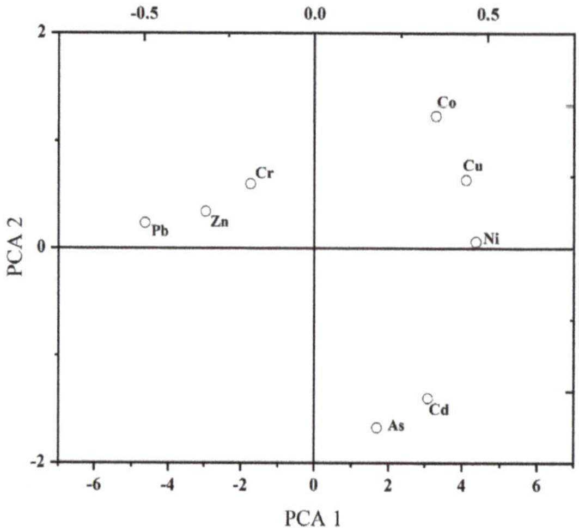

Figure 3. Soil PTE principal component analysis rotation space component diagram.

3.4. Source Apportionment of Potential Toxic Elements in PMF Model

The PMF model was used to identify and distinguish the potential sources of toxic elements in the soil. Table 3 lists the optimal number of factors. In this study, the number of input factors was set to three, and the indicator judgments Q (Robust) and Q (True) were both 2.63102, with an error of less than 10%. This model was interpretable, and the BS inspection rate exceeded 80%; therefore, the results are usable. Factor contribution rate of PMF source resolution was shown in Figure 4.

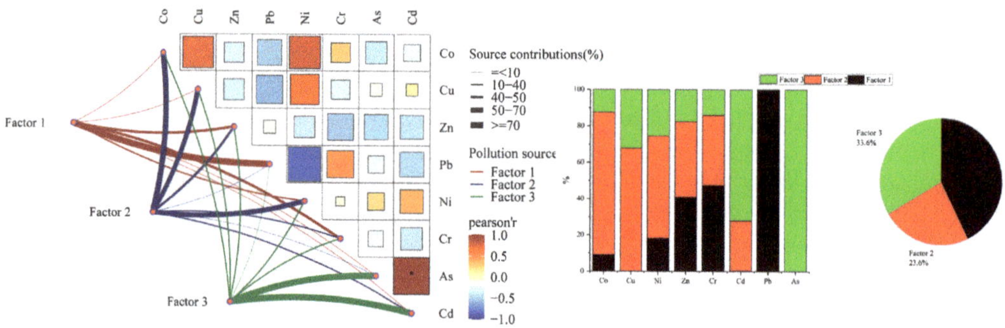

Figure 4. Factor contribution rate of PMF source resolution.

The contribution of Pb to factor 1 was as high as 99.9%, followed by Cr with a contribution of 47.3% and Zn with a contribution of 40.9%. The correlation between Pb and Cr was positive and equaled 0.884, whereas the contributions of Cr and Zn were negative and equaled 0.726. The high contribution of Pb reflects the presence of different forms of Pb at this site. Pb mainly comes from the combustion of fossil fuels, and due to human activities (such as the use of refined zinc ore to produce zinc oxide) and the input of agricultural chemicals, Pb accumulates in the soil. The study area has a history of using coal as a fuel during winter. Because of the stable air in winter, a temperature gradient is formed and the cold air at the bottom cannot rise to carry away pollutants, leading to their accumulation in the lower atmosphere and subsequent deposition onto the soil surface. The complex correlation of Pb-Cr-Zn reveals that this precious metal site is affected by numerous pollutants derived from traffic dust and metal-processing waste, whereas the negative correlation of Cr-Zn suggests the possibility of cross-contamination from different processes. Factor 1 was defined as a combination of transportation and industrial sources.

The contribution of Co to factor 2 was 78.6%, while the contributions of Cu, Ni, Zn, and Cr were 67.9%, 56.4%, 41.6%, and 38.6%, respectively. The correlations between Co and Cu and Co and Ni were positive, equaling 0.7 and 0.76, respectively. Consistent with the characteristics of nonferrous metal smelting sources, the Co–Ni–Cu combination is typical of the alloy manufacturing and electronics industries. This may have originated from the dismantling of waste electronic products, metal melting, and the process of industrial solid waste produced during the wet process of indium extraction and lithopone production entering the rotary kiln production line to recover valuable metals. Therefore, factor 2 reflects the influence of metal smelting and electronics manufacturing.

Factor 3 showed absolute dominance and a strong positive correlation (0.96) between As (100%) and Cd (72%), indicating agricultural non-point source pollution. The high coupling of As-Cd is consistent with the long-term application of arsenic-containing pesticides and phosphorus fertilizers. There was a certain amount of farmland around the plot, and because of the legacy of heavy metal raw materials, especially the arsenic and cadmium elements associated with phosphorus fertilizers, they accumulated in the surface soil through agricultural cultivation. Thus, factor 3 also reflects agricultural sources. The mixed sources of transportation and industry contributed 42.8%, indicating that emissions from transportation and various industrial production processes around precious metal companies are important factors leading to the heavy metal pollution of soils. The metal smelting and electronics manufacturing industries accounted for 23.6%. Owing to the characteristics of their precious metal companies' wet zinc-smelting production processes, these two industries often generate large amounts of heavy metal waste and harmful substances. Waste easily infiltrates the surrounding environment and becomes an important source of heavy metal pollution in the soil. Agricultural activities account for 33.6% of the total. Pesticides and fertilizers used in agricultural production, as well as manure produced from livestock and poultry farming, may contain heavy metals. The long-term and extensive use of these chemicals leads to their gradual accumulation in soil.

3.5. *Characterization of Soil*

3.5.1. FTIR Analysis

Figure 5 shows the infrared spectra of the soil at five different sampling sites. In the infrared spectra of these samples, there are characteristic peaks with a wavenumber of 3624 cm^{-1}, which are mainly caused by the stretching vibration of N-H in the amino acids and nucleic acids in the soil. The wide peak at a wave number of 3415 cm^{-1} belongs to the associated hydroxyl -OH stretching vibration peak in carbohydrates, hemicellulose, and carbohydrates. At 2987 cm^{-1} and 2874 cm^{-1}, the S3 sample shows the antisymmetric and symmetric stretching vibration peaks of methylene and methyl C-H in lipids, respectively. The 2510 cm^{-1} peak corresponds to the C≡N stretching vibration peak in unsaturated lipids. Meanwhile, the S1 and S3 samples show an obvious narrow peak at 1799 cm^{-1}, which corresponds to the C=O stretching vibration peaks of proteins, amino acids, and polypeptides. S2, S4, and S5 do not show the C=O characteristic peaks of protein and polypeptide substances near 1799 cm^{-1}, which may be due to the C=O in the soil chelating with PTEs, leading to the disappearance of this peak. Samples S1–S5 exhibit different-intensity peaks at 1632 cm^{-1} and 1432 cm^{-1}, which correspond to the C=O stretching vibration peak of the amide I band and the CO_3^{2-} characteristic peak of carbonate in the soil, respectively. The intensity of the peak at 1432 cm^{-1} is weak, which further indicates that the PTEs in the soil samples have a greater impact on the total carbohydrates in the soil. In the S3 sample, the C=O stretching vibration peak of the amide I band at 1632 cm^{-1} disappears, indicating that other PTE ions were combined with the C=O stretching vibration peak, resulting in a weakening of the peak's intensity. The peaks of the five samples at 1022 cm^{-1} are C-O stretching vibration peaks. In samples S1 and S3, the peak at 877 cm^{-1} is the characteristic peak of CrO_4^{2-}, indicating that the pollution of Cr in S1 and S3 is more serious than that in other samples. The characteristic peak of ZrO_4^{2-} appears near 799 cm^{-1} in S1–S5. Zr is widely distributed in nature. Zirconia (ZrO_2) is a naturally occurring raw material. Among the five soil samples, the characteristic peak intensity of ZrO_4^{2-} is the weakest. The characteristic peak of ZnO also appears at 467 cm^{-1}, indicating that there was a certain amount of ZnO pollution in the five soil samples.

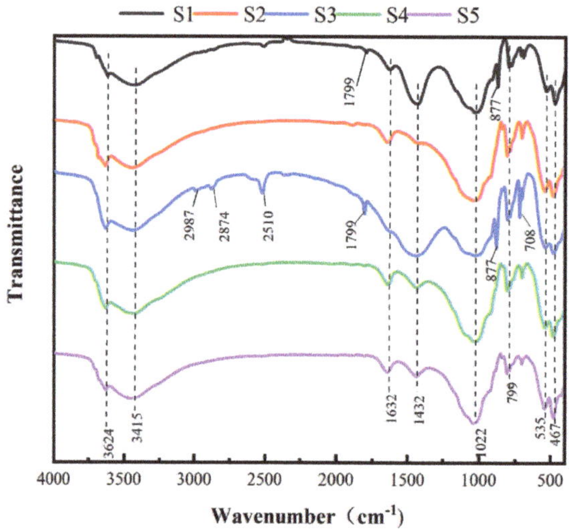

Figure 5. FTIR diagram of soil at five sample sites.

3.5.2. XPS Analysis

To further analyze the changes in the valence states and bond energies of the elements in the PTE-contaminated areas of the site, high-resolution XPS was applied to the samples taken from five different locations. The five groups of samples contained O, N, C, and other elements, and the positions of the diffraction peaks were consistent, indicating that the elements in each area were consistent and no additional new substances were generated (Figure 6). Figure 6b shows the high-resolution C 1s spectrum. The diffraction peaks at 283.84 eV, 284.80 eV, 286.23 eV, and 288.37 eV are attributed to metal carbides, C-C, C=O, and the CO_3^{2-} radical, respectively. It is speculated that the generation of metal carbides may have been caused by Pb and Cd pollution, and the binding energy at different positions did not shift, indicating that the chemical state of each area was consistent. Figure 6c shows the high-resolution spectrum of N 1s. The peaks at 398.75 eV, 399.76 eV, and 401.49 eV come from C-N=C, N-(C)3, and C-N-H, respectively. The binding energy and peak value at different positions show little change, consistent with element C. Figure 6d shows the high-resolution O 1s spectrum. The peaks at 530.6 eV, 531.56 eV, and 532.46 eV are mainly attributed to O^{2-}, -OH, and C=O, and it is speculated that the generation of O^{2-} is mainly from the metal oxides produced by the polluted metal elements.

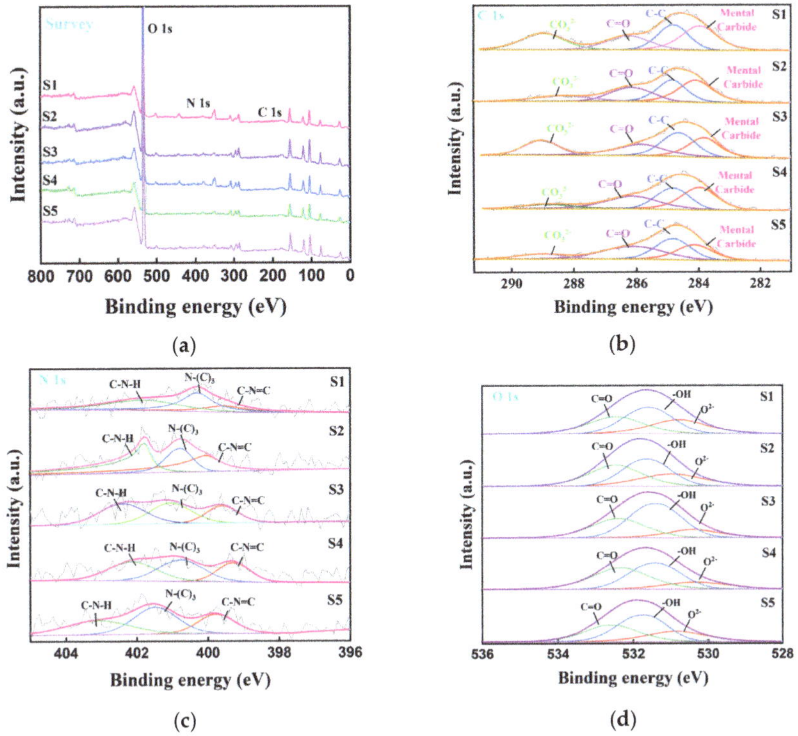

Figure 6. pXPS diagram of soil at five sample sites. (**a**) XPS full spectrum of five samples; (**b**) high-resolution spectrogram of C 1s; (**c**) high-resolution spectrogram of N 1s; (**d**) high-resolution spectrogram of O 1s.

3.5.3. XRD Analysis

The XRD patterns of the PTE-contaminated areas at the analysis site are shown in Figure 7. Within the diffraction ranges of the five samples, the distributions of the XRD characteristic peaks are relatively similar. The crystal peaks at the diffraction angles (2θ) of about 12.5°, 21°, and 45° are mostly consistent with the crystal plane spacing, crystal peak diffraction angle, and diffraction intensity of the Powder Diffraction File (PDF) standard card (PDF#01-079-6476). The phase indicated by the PDF standard card is kaolinite, and the corresponding crystal plane indices are (001), (-1-11), and (-203). The XRD pattern indicates kaolinite. Similarly, the weak diffraction peaks at the diffraction angle (2θ) of approximately 13.5° and 27.5° correspond to the phase potassium feldspar (PDF#01-073-9850), and the corresponding crystal plane index is (001) and (002). A small amount of quartz (PDF#01-086-2237) can be seen from the crystal peaks, with (100), (101), and (110) corresponding to the diffraction angles (2θ) of approximately 21°, 26.5°, and 36.5°. At the diffraction angles (2θ) of approximately 29°, 47.5°, and 48°, there were weak diffraction peaks. The crystal plane spacing, the crystal peak diffraction angle and diffraction intensity were consistent with those of calcite (PDF#01-072-4582), and the corresponding crystal plane indices were (104), (018), and (150). The crystal peaks at the diffraction angle (2θ) of approximately 19° and 35° are similar with the PDF standard card (PDF#01-073-0969) database in terms of crystal plane spacing, crystal peak diffraction angle, and diffraction intensity. It was determined that the XRD pattern contained cadmium hydroxide with corresponding crystal plane indices of (001) and (011), and the diffraction peak intensity was clearly apparent in samples S2, S3, and S5. At the same time, weak diffraction peaks appeared at the diffraction angles (2θ) of approximately 26°, 32°, 50°, 38°, and 51°, respectively. The corresponding substances were lead dioxide (PDF#04-002-2595), with (110), (101), and (211) facet indices, and zinc hydroxide (PDF#04-015-2996), with (011) and (102) facet indices, which appeared in the five samples. XRD pattern analysis showed that the soil in the PTE-contaminated area of the site contained quartz, kaolinite, muscovite, plagioclase, potassium feldspar, calcite, and other major components. In addition, cadmium hydroxide, lead oxide, and zinc hydroxide were found in the soil samples, indicating that the five sample sites contained multiple PTEs, including cadmium, lead, and zinc.

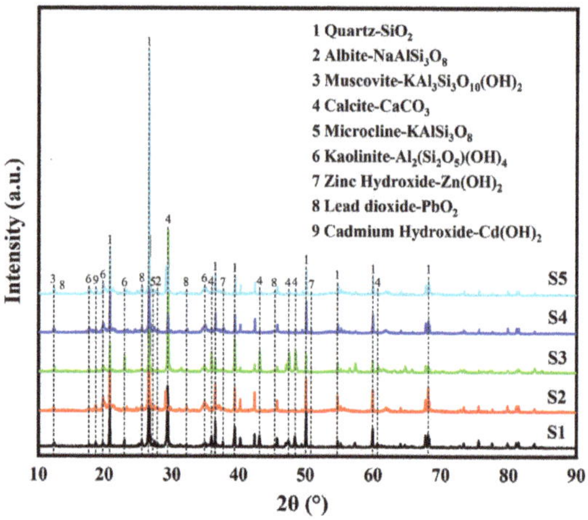

Figure 7. XRD diagram of soil at five sample sites.

3.6. Microbial Community Analysis

3.6.1. OTU Cluster Analysis

Significant differences were observed in microbial community structure. At the 97% nucleic acid similarity level, the five groups of samples obtained 701, 929, 951, 838, and 969 OTUs (Figure 8). The total number of OTUs in the five groups of samples was 303. The ranking of the OTU numbers is as follows: point 5 > point 3 > point 2 > point 4 > point 1. Among them, the point 5 sample contained the most OTUs, whereas points 4 and 1 contained the least. From the total number of OTUs in the samples at the four sites, the total number of OTUs at points 1234, 1235, 2345, 1245, and 2458 were 316, 347, 447, and 334, respectively, indicating that the microbial species at the four sites were the most similar.

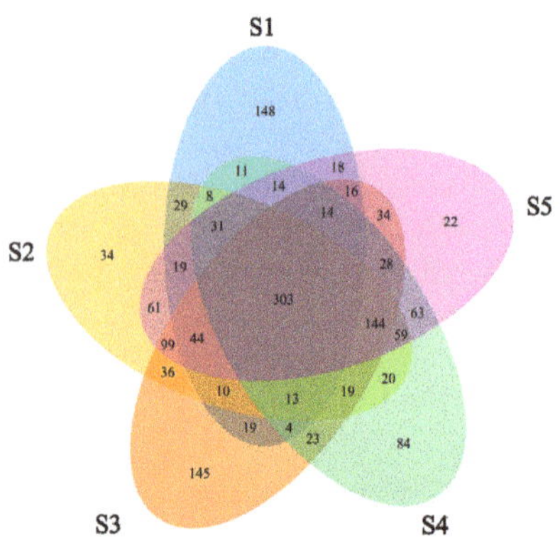

Figure 8. OTU clustering Venn diagram.

3.6.2. Diversity Index Analysis

The Shannon index of microorganisms is an index used to measure population diversity, which evaluates the diversity of microbial communities through specific calculation formulas. The Chao index is an index used to estimate the number of OTUs contained in a sample using the Chao1 algorithm. Chao1 is commonly used in ecology to estimate the total number of species. Abundance-based Coverage Estimator (ACE): The ACE index is an index used to estimate the number of operational taxonomic units (OTUs) present in a community and is one of the commonly used indices in ecology for estimating the total number of species. The ACE index is similar to the Chao1 index, both based on abundance coverage estimation methods, but their algorithms are different. The ACE index takes into account the influence of low-abundance groups. The Simpson index was proposed by Edward Hugh Simpson in 1949 and is commonly used in ecology to quantitatively describe the biodiversity of a region. The Shannon index is an indicator used to evaluate the evenness of species distribution in ecosystems. It is commonly used in conjunction with the Shannon–Wiener diversity index to comprehensively describe the biodiversity of an ecosystem [30]. Coverage refers to the microbial coverage rate, and the higher its value, the lower the probability of new species not being detected in the sample. The index actually reflects whether the sequencing results represent the true situation of the sample.

The alpha diversity index was calculated at a 97% similarity level, and results are summarized in Table 9. The Chao1 and Shannon indices reflect the richness and diversity of the microbial community. The larger the value, the higher the richness and diversity of the microbial community. The Simpson index reflects the uniformity of the microbial community. The smaller the value, the more uniform the community distribution. The Chao1 and Shannon indices of samples S1 and S5 were relatively large (Table 9), indicating that the richness and diversity of the microbial communities were high at these points, whereas the Simpson index of sample S1 was the lowest. No plant or leaching residue sites were established at site S1. The degree of PTE pollution was low, and the microbial community uniformity of the sample at this point was higher than that of the other samples with a high degree of pollution. The intake of PTE pollutants reduced soil microbial richness, resulting in low diversity and an uneven community distribution.

Table 9. APLHA diversity index.

Sample Number	Chao1	Shannon	ACE	Simpson	Shannoneven	Coverage
1	746.69	4.26	750.33	0.05	0.65	1.00
2	998.07	4.62	986.71	0.03	0.78	1.00
3	965.35	4.81	965.30	0.03	0.70	1.00
4	872.17	4.66	861.87	0.03	0.69	1.00
5	1094.20	4.72	1088.85	0.04	0.69	1.00

3.6.3. Microbial Community Structure at the Phylum Level in the Samples Collected at Different Sites

The sequencing results were annotated to obtain information on the different sequences at different classification levels (phylum, class, order, family, and genus). Based on the annotation results, the taxon results for phyla and genera were selected for statistical analysis. The relative abundances of the dominant bacterial phyla in the soil samples from each sampling point were analyzed. *Actinobacteria* and *Proteobacteria* had the highest bacterial content at the phylum level, followed by *Firmicutes*, *Bacteroidetes*, *Chloroflexi*, *Acidobacteria*, and *Cyanobacteria_chloroplast* (Figure 9). These seven phyla accounted for more than 80% of the average total bacterial count. *Actinobacteria* had the highest relative abundance, indicating its strong adaptability to PTE-contaminated soil environments (Figure 8). *Proteobacteria* also showed an abundance advantage in the five samples. In the S1–S5 samples, the S4 samples were significantly different from the other samples at the phylum level, and the microbial compositions of the other samples were slightly different from each other. Compared with the S1 sample without plant or leaching residue sites and with low pollution, the relative abundances of *Actinobacteria* and *Proteobacteria* in S2, with a high PTE content, increased by 4.72% and 4.25%, respectively. The relative abundance of *Firmicutes* decreased by 7.65% due to an increase in the concentration of PTEs. The results showed that *Actinobacteria* and *Proteobacteria* increased and *Firmicutes* decreased in PTE-contaminated soil.

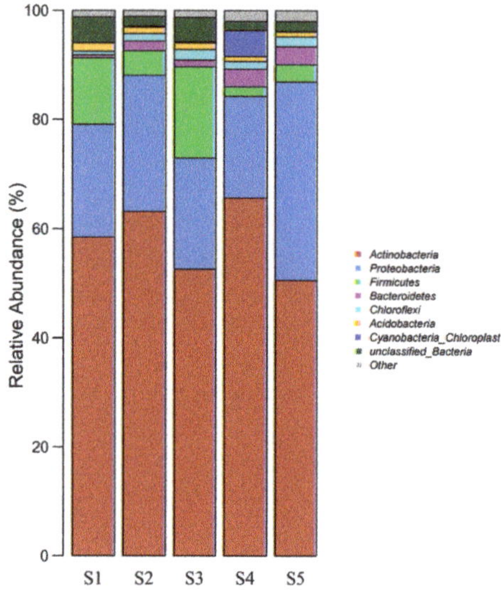

Figure 9. Microbial community structure at the phylum level in the samples collected at different sites.

3.6.4. Microbial Community Structure at the Genus Level in the Samples Collected at Different Sites

The relative abundances of the top 40 dominant bacterial species in the samples were analyzed. *Nocardioides*, *Janibacter*, *Dietzia*, *Micromonospora*, *Saccharopolypora*, and *Pseudoarthrobacter* were the dominant species (Figure 10). The abundances of these five groups of bacteria in the five samples (S1–S5) were 47.16%, 42.36%, 35.6%, 47.27%, and 36.37%, respectively. Pan et al. found that *Nocardioides* had a strong tolerance to one or more PTEs and could be used as a biological indicator of PTE pollution [31]. From S1 to S5, the proportions of *Nocardioides* were 17.15, 8.26, 10.49, 14.55, and 8.01%, respectively. The results showed that actinomycetes may be enriched in soils with a high PTE content. With an increase in metal content, the abundance of *Janibacter* groups also increased, similar to the results of the present study [32]. The content of *Janibacter* in the S1 sample without plant or leaching residue sites and with low pollution was 1.6%, and that in S2–S5 was 8.29%, 10.19%, 5.46%, and 16.60%, respectively. As the dominant species, *Dietzia*, *Micromonospora*, *Saccharopolypora*, and *Pseudoarthrobacter* all had the effect of removing high concentrations of PTEs [33–35].

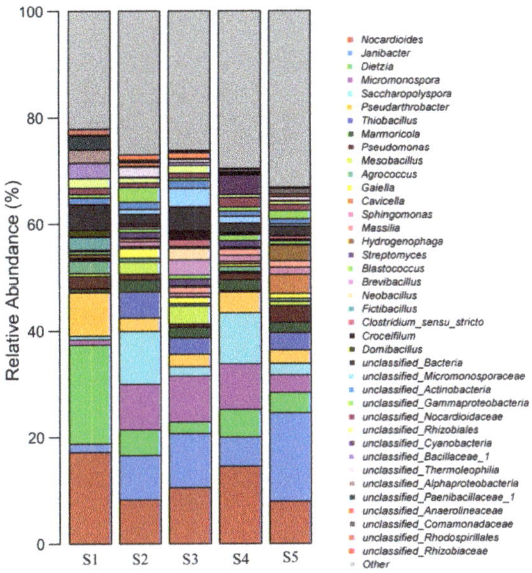

Figure 10. Microbial community structure at the genus level in the samples collected at different sites.

4. Conclusions

(1) The average contents of Cd, Co, As, Pb, Zn, and Cu in the soil at the site were higher than the background levels in the soils of Guangxi Province. The CV of Cd, As, Pb, and Cu exceeded 1, with a high degree of dispersion, which was greatly affected by external factors.

(2) The single factor pollution index showed that Cd and As were heavily polluting, Co was slightly polluting, and the other PTEs were non-polluting. The geoaccumulation index showed that Cd was highly accumulated; Pb, As, and Co were moderately accumulated; Cu and Zn were mildly accumulated; and Ni and Cr were not accumulated. The load of Co in principal component 2 was relatively high, and there was a correlation with most other PTEs, which may be due to pollution caused by the inclusion of some of the production waste residue and exogenous soil in backfill of the site.

(3) The results of infrared spectroscopy showed that C=O in the soil was chelated with PTEs, leading to the disappearance of C=O's characteristic peaks in protein and polypeptide substances. The PTEs in the soil samples exerted greater damage to the total carbohydrates in the soil, and there was a variable amount of PTE pollution. XPS analysis showed that metal carbides appeared in the high-resolution C 1s spectrum, O^{2-} appeared in the high-resolution O 1s spectrum, mainly from the metal oxides produced by the PTEs, and the soil was polluted with Pb, Zn, and Cd to a certain extent. The XRD results showed that cadmium hydroxide, lead oxide, and zinc hydroxide were present in all five sampling sites (S1–S5), indicating that the five samples were contaminated by PTEs such as cadmium, lead, and zinc.

(4) The risks of Cd, Zn, Cu, and Pb in the soil of the study area were high, mainly due to the joint effects of human activities and natural conditions. Measures should be taken to improve the quality of soil and the surrounding environment and reduce pollution by PTEs.

(5) *Actinobacteria* and *Proteobacteria* were the dominant microbial phyla in soil in the study area. The dominant bacterial genera were *Nocardioides*, *Janibacter*, *Dietzia*, *Micromonospora*, Saccharopolyspora, and Pseudarthrobacter.
(6) According to the PMF model, the soil properties in the study area are jointly influenced by agricultural activity sources, transportation and industry, and the metal smelting and electronic manufacturing industries.

Author Contributions: Conceptualization, Y.S.; methodology, Y.S.; software, K.D.; validation, Y.S. and K.D.; formal analysis, D.W.; investigation, Y.H.; resources, F.H.; data curation, K.D.; writing—original draft preparation, Y.S.; writing—review and editing, Y.S.; visualization, Y.S.; supervision, Y.S.; project administration, Z.Y.; funding acquisition, K.D. All authors have read and agreed to the published version of the manuscript.

Funding: This research was funded by the National Natural Science Foundation of China (42367070) and the Open Fund of the Hubei Key Laboratory of Biological Resources Protection and Utilization (Hubei Minzu University) (KYPT012405).

Institutional Review Board Statement: Not applicable.

Informed Consent Statement: Not applicable.

Data Availability Statement: The original contributions presented in this study are included in the article, and further inquiries can be directed to the corresponding authors.

Conflicts of Interest: The authors declare no conflicts of interest.

References

1. Gao, H.Z.; Lu, Q.P. Near Infrared Spectral Analysis and Measuring System for Primary Nutrient of Soil. *Spectrosc. Spect. Anal.* **2011**, *31*, 1245–1249. [CrossRef]
2. Dangi, S.R.; Bañuelos, G.; Buyer, J.S.; Hanson, B.; Gerik, J. Microbial community biomass and structure in saline and non-saline soils associated with salt- and boron-tolerant poplar clones grown for the phytoremediation of selenium. *Int. J. Phytoremediat.* **2018**, *20*, 129–137. [CrossRef]
3. Liu, X.; Yang, Z.; Li, B.; Wu, Z.; Wang, L.; Yu, T.; Li, C.; He, Z.; Xie, M.; Deng, C.; et al. Source Apportionment and Analysis of Potentially Toxic Element Sources in Agricultural Soils Based on the Positive Matrix Factorization and Geo-Detector Models. *Land* **2025**, *14*, 146. [CrossRef]
4. Aoshima, K. *Itai-itai* disease: Renal tubular osteomalacia induced by environmental exposure to cadmiumhistorical review and perspectives. *Soil Sci. Plant Nutr.* **2016**, *62*, 319–326. [CrossRef]
5. Qin, S.; Liu, H.; Nie, Z.; Rengel, Z.; Gao, W.; Li, C.; Zhao, P. Toxicity of cadmium and its competition with mineral nutrients for uptake by plants: A review. *Pedosphere* **2020**, *30*, 168–180. [CrossRef]
6. Wang, X.; Han, Z.; Wang, W.; Zhang, B.; Wu, H.; Nie, L.; Zhou, J.; Chi, Q.; Xu, S.; Liu, H.; et al. Continental-scale geochemical survey of lead (Pb) in mainland China's pedosphere: Concentration, spatial distribution and influences. *Appl. Geochem.* **2019**, *100*, 55–63. [CrossRef]
7. Shi, P.; Zhang, Y.; Li, Z.; Li, P.; Xu, G. Influence of land use and land cover patterns on seasonal water quality at multi-spatial scales. *Catena* **2017**, *151*, 182–190. [CrossRef]
8. Ma, L.; Abuduwaili, J.; Liu, W. Spatial Distribution and Ecological Risks of the Potentially-Toxic Elements in the Surface Sediments of Lake Bosten, China. *Toxics* **2020**, *8*, 77. [CrossRef]
9. Akindele, E.O.; Omisakin, O.D.; Oni, O.A.; Aliu, O.O.; Omoniyi, G.E.; Akinpelu, O.T. Heavy metal toxicity in the water column and benthic sediments of a degraded tropical stream. *Ecotoxicol. Environ. Saf.* **2020**, *190*, 110153. [CrossRef]
10. Chen, Z.W.; Xu, J.; Yang, F.; Hou, Z.L.; Ren, K.; Yu, L.S.; Yang, S.Y.; Li, Z.; Zhang, X.B. Comprehensive Monitoring and Ecological Risk Assessment of Heavy Metals in Soil and Surface Water of Chishui River Basin in Upper Reaches of the Yangtze River. *Water* **2023**, *15*, 2069. [CrossRef]
11. Guo, G.H.; Wu, F.C.; Xie, F.Z.; Zhang, R.Q. Spatial distribution and pollution assessment of heavy metals in urban soils from southwest China. *J. Environ. Sci.* **2012**, *24*, 410–418. [CrossRef]
12. Yu, Y.; Lu, Y.-N.; Wang, W.-J.; Qu, X.-D.; Liu, J.-T.; Wen, C.-Y. Spatio-Temporal Distribution and Risk Assessment of Heavy Metals in Middle and Lower Reaches of Le'an River. *Environ. Sci.* **2020**, *41*, 691–701. [CrossRef]

13. Alengebawy, A.; Abdelkhalek, S.T.; Qureshi, S.R.; Wang, M.-Q. Heavy Metals and Pesticides Toxicity in Agricultural Soil and Plants: Ecological Risks and Human Health Implications. *Toxics* **2021**, *9*, 42. [CrossRef] [PubMed]
14. Su, K.; Wang, Q.; Li, L.; Cao, R.; Xi, Y. Water quality assessment of Lugu Lake based on Nemerow pollution index method. *Sci. Rep.* **2022**, *12*, 13613. [CrossRef]
15. Su, K.; Wang, Q.; Li, L.; Cao, R.; Xi, Y.; Li, G. Water quality assessment based on Nemerow pollution index method: A case study of Heilongtan reservoir in central Sichuan province, China. *PLoS ONE* **2022**, *17*, e0273305. [CrossRef]
16. Zhao, D.; Sang, Y.M.; Zhang, Q.; Man, Y.; Han, Y.H.; Wu, B.; Guo, C.Y.; Gu, Q.B. Spatial Distribution, Risk Assessment, Source Apportionment and Stabilization Performance of Potentially Toxic Elements in Sediments of Dongdagou River, Northwest China. *Water Air Soil Pollut.* **2024**, *235*, 352. [CrossRef]
17. Shi, L.Y.; Ma, Z.B.; Zhang, D.E.; Xue, S.G.; He, X.S.; Huo, K.; Wang, G.H.; Yi, J. Pollution evaluation and source analysis of heavy metals in copper smelting site in Yellow River Basin. *Chin. J. Nonferrous Met.* **2023**, *33*, 3462–3475. [CrossRef]
18. Guo, W.; Zhang, H.; Wang, M.; Yin, X.; Wang, Z. Arsenic Removal from Contaminated Soil Inside Non-Ferrous Metal Smelter by Washing. *Soil Sediment Contam.* **2020**, *29*, 151–164. [CrossRef]
19. Tang, X.X. Study on the Interaction of Cadmium, Zinc, and Selenium in Soil of Typical High Cadmium Geological Background Areas in Guangxi. Master's Thesis, Chongqing Three Gorges University, Chongqing, China, 2024. [CrossRef]
20. Hou, X.J.; Zhang, Y.L.; Liu, X.; Zhou, C.S.; Dang, X.J.; Hu, J.N. An evaluation and analysis method of heavy metal pollution in soil of agricultural land around mining areas based on double index method. *Int. J. Environ. Pollut.* **2024**, *74*, 35–49. [CrossRef]
21. Li, W.X.; Zhang, X.X.; Wu, B.; Sun, S.L.; Chen, Y.S.; Pan, W.Y.; Zhao, D.Y.; Cheng, S.P. A comparative analysis of environmental quality assessment methods for heavy metal-contaminated soils. *Pedosphere* **2008**, *18*, 344–352. [CrossRef]
22. Wei, J.; Zheng, X.; Liu, J. Modeling Analysis of Heavy Metal Evaluation in Complex Geological Soil Based on Nemerow Index Method. *Metals* **2023**, *13*, 439. [CrossRef]
23. Wang, Y.; Sun, W.; Zhao, Y.L.; He, P.Y.; Wang, L.D.; Nguyen, L.T.T. Assessment of Heavy Metal Pollution Characteristics andEcological Risk in Soils around a Rare Earth Mine in Gannan. *Sci. Program.* **2022**, *2022*, 5873919. [CrossRef]
24. Yang, Y.L.; Zhang, D.W.; Quan, J.; Wang, P.W.; Xu, Y. Water quality assessment of Middle Route of South-North Water Diversion Project based on modified Nemerow index method. *Water Supply* **2021**, *21*, 1005–1015. [CrossRef]
25. Odewande, A.A.; Abimbola, A.F. Contamination indices and heavy metal concentrations in urban soil of Ibadan metropolis, southwestern Nigeria. *Environ. Geochem. Health* **2008**, *30*, 243–254. [CrossRef]
26. Mookan, V.P.; Machakalai, R.K.; Srinivasan, S.; Sigamani, S.; Kolandhasamy, P.; Gnanamoorthy, P.; Moovendhan, M.; Srinivasan, R.; Hatamleh, A.A.; Ai-Dosary, M.A. Assessment of metal contaminants along the Bay of Bengal-Multivariate pollution indices. *Mar. Pollut. Bull.* **2023**, *192*, 115008. [CrossRef]
27. Paatero, P.; Tapper, U. POSITIVE MATRIX FACTORIZATION—A Nonnegative Factor Model with Optimal Utilization of Error-Estimates of Date Values. *Environmetrics* **1994**, *5*, 111–126. [CrossRef]
28. Deng, J.Y.; Yu, D.; Wu, H.; Du, J.Y.; Pu, S.Y. Soil contamination characteristics and source analysis of a copper smelting site in Guangxi. *Acta Sci. Circumstantiae* **2023**, *43*, 290–299. [CrossRef]
29. Shao, H.Q.; Zhang, Y.W.; Qu, C.; Li, W.H.; Zhao, Y.J.; Liu, N.; Cai, H.M.; Wu, C.D.; Liu, J.M. Analysis of heavy metal contamination in the soil and enrichment capabilities ofterrestrial plants around a typical vanadium smelter area. *Chin. J. Eng.* **2020**, *42*, 302–312. [CrossRef]
30. Zhao, J.; Gao, Y.; Wang, Y.; Zhao, Y.J. The effect of different afforestation tree species on plant diversity after 50 years on Mount Tai, China. *Appl. Ecol. Environ. Res.* **2022**, *19*, 4515–4526. [CrossRef]
31. Pan, X.; Zhang, S.; Zhong, Q.; Gong, G.; Wang, G.; Guo, X.; Xu, X. Effects of soil chemical properties and fractions of Pb, Cd, and Zn on bacterial and fungal communities. *Sci. Total Environ.* **2020**, *715*, 136904. [CrossRef]
32. Větrovsky, T.; Baldrian, P. An in-depth analysis of actinobacterial communities shows their high diversity in grassland soils along a gradient of mixed heavy metal contamination. *Biol. Fertil. Soils* **2015**, *51*, 827–837. [CrossRef]
33. Ausuri, J.; Dell'Anno, F.; Vitale, G.A.; Esposito, F.P.; Funari, V.; Franci, G.; Galdiero, M.; Della Sala, G.; Tedesco, P.; Coppola, D.; et al. Bioremediation of Multiple Heavy Metals Mediated by Antarctic Marine Isolated *Dietzia psychralcaliphila* JI1D. *J. Mar. Sci. Eng.* **2022**, *10*, 1669. [CrossRef]
34. Bofill, R.; Orihuela, R.; Romagosa, M.; Domenech, J.; Atrian, S.; Capdevila, M. *Caenorhabditis elegans* metallothionein isoform specificity—Metal binding abilities and the role of histidine in CeMT1 and CeMT2. *FEBS J.* **2009**, *276*, 7040–7069. [CrossRef] [PubMed]
35. Ortuzar, M.; Trujillo, M.E.; Román-Ponce, B.; Carro, L. Micromonospora metallophores: A plant growth promotion trait useful for bacterial-assisted phytoremediation? *Sci. Total Environ.* **2020**, *739*, 139850. [CrossRef]

applied sciences

MDPI

Article

Effects of Coal Mining Subsidence on Loess Slope Morphology and Soil Erosion in the Middle Reaches of the Yellow River

Shijie Song [1,2,3,*], Ruilin Niu [1,2,3], Shuai Yang [4], Xing Cheng [1,2,3], Hao Ruan [1,2,3], Baodeng Chen [1,2,3], Yuanhong Li [1,2,3] and Lijun Tang [5]

[1] College of Geology and Environment, Xi'an University of Science and Technology, Xi'an 710054, China; 22209226109@stu.xust.edu.cn (R.N.); 22209226107@stu.xust.edu.cn (X.C.); 22209226108@stu.xust.edu.cn (H.R.); 23209226129@stu.xust.edu.cn (B.C.); 23209226101@stu.xust.edu.cn (Y.L.)
[2] Research Institute of Coal Green Mining Geology, Xi'an University of Science and Technology, Xi'an 710054, China
[3] Key Laboratory of Geological Guarantee for Coal Green Development of Shaanxi Province, Xi'an 710054, China
[4] Shanxi Satellite Application Technology Center for Natural Resources, Xi'an 710002, China; shuai.yang.cug@hotmail.com
[5] Ningxia Hui Autonomous Region Land and Resources Investigation and Monitoring Institute, Yinchuan 750002, China; tang_lijun2024@163.com
* Correspondence: songshijie@xust.edu.cn

Abstract: How to solve the contradiction between coal mining and soil and water conservation is a key scientific issue in the achievement of high-quality development in the middle reaches of the Yellow River. In this paper, the northern Shaanxi mining area in the middle reaches of the Yellow River is taken as the research area, and the surface loess micro-topography is taken as the entry point. The numerical simulation test and soil loss model calculation are used to reveal the different types of loess natural slope morphology (straight slope, concave slope, convex slope, and composite slope) and the natural slopes (5°, 15°, 25°, 35°, 45°). The influence characteristics and laws of the same mining on the surface loess slope morphology in the coal mining subsidence area are analyzed, and the soil erosion effect on the slope scale is analyzed. The results show that: (1) Coal mining subsidence will lead to an increase in the slope of the loess slope, and the smaller the natural slope, the greater the increase in slope. Among them, the influence of coal mining subsidence on the 'concave loess slope with natural slope of 15°' is the most significant, and the natural slope of 15° is the key dividing point for the transformation of the sensitive slope shape of the loess slope in the coal mining subsidence area of northern Shaanxi. (2) Coal mining subsidence will lead to the decrease in slope length of a loess natural slope, and the smaller the natural slope, the greater the decrease in slope length. Among them, coal mining subsidence has the most significant impact on the 'concave loess slope with a natural slope of 25°'. The natural slope of 25° is the key point of the sudden change rate of the slope length of the loess slope in the coal mining subsidence area of northern Shaanxi. (3) Coal mining subsidence will lead to the increase in the soil erosion modulus on the surface loess slope under the scale of 'annual erosion rainfall' and 'typical field erosion rainfall', and the smaller the natural slope, the greater the increase in the soil erosion modulus. The natural slopes of 15° and 25° are the key points of the abrupt change in soil erosion intensity on the loess slope in the coal mining subsidence area of northern Shaanxi under the scales of 'annual erosion rainfall' and 'typical erosion rainfall', respectively. Under the scale of annual erosion rainfall, the increment of the 15° slope was 1.65 times, 1.12 times, 1.11 times, and 1.02 times that of the 5°, 25°, 35°, and 45° slopes, respectively. Under the typical erosion rainfall scale, the increment of the 25° slope was 4.22 times, 1.32 times, 1.04 times, and 1.15 times that of the 5°, 15°, 35°, and 45° slopes, respectively.

 https://doi.org/10.3390/app15105684

(4) For the loess subsidence slope with any slope shape, the increase in slope gradient is the main factor for the increase in the soil erosion modulus. Under the annual erosion rainfall scale, the contribution of slope increase was 92.9%. Under the typical erosion rainfall scale, the contribution of slope increase was 79.1%. The research results can provide scientific guidance for soil erosion and control in the northern Shaanxi mining area in the middle reaches of the Yellow River Basin.

Keywords: soil erosion effect; the morphology of loess slope; coal mining subsidence; numerical simulation; middle Yellow River

1. Introduction

As an important basic energy in China, coal plays the role of ballast and stabilizer in ensuring national energy security and supporting economic development [1]. In 2024, China's total raw coal production reached 4.78 billion tons, an increase of 1.2% [2]. Driven by the strategy of "carbon peak, carbon neutrality", the proportion of coal in China's primary energy consumption has entered a downward channel. It is predicted that the proportion of coal consumption will be reduced to 42.51% in 2030 [3]. The status of coal will change from being the dominant energy source to one of the major energy sources, but it will still provide a guarantee for national energy security [4]. As the main producing area of coal resources in China, the Yellow River Basin is densely distributed with large coal bases planned and constructed by nine countries [5]. As of 2023, the coal production in the Yellow River Basin has reached 3.76 billion tons, accounting for 79.1% of the total coal production in the country. Among them, the coal production in the middle reaches of the Yellow River has exceeded 1.89 billion tons, accounting for about 40% of the total coal production in the country [6]. The coal mining area in northern Shaanxi has large resource reserves, good coal quality, and superior mining conditions. It plays an important role in coal production in the middle reaches of the Yellow River and in national energy supply [7]. However, due to the constraints of geological occurrence conditions and the ecological environment quality, large-scale coal mining has caused serious coal mining subsidence and derivative soil erosion problems in the coal mining area of northern Shaanxi. More importantly, the coal mining area in northern Shaanxi is spatially highly overlapped with the Yellow River sediment-laden coarse sand national soil and water loss key control area [8]. This phenomenon aggravates the conflict between coal mining and soil and water conservation needs, especially in the loess gully landform area [9]. Therefore, how to scientifically prevent and control the soil erosion effect of coal mining subsidence in the loess mining area has become one of the key points of ecological environment protection and high-quality development in the northern Shaanxi coal mining area and even the middle reaches of the Yellow River [10].

As the direct embodiment of the complex terrain in the loess gully region, the natural form (slope) of the loess layer on the surface is not only an important geological factor affecting the coal mining subsidence, but also an important topographic basis for shaping the final shape of the surface after the subsidence, so that there is an obvious mutual feedback effect between the 'shape and deformation of the loess slope' and the 'characteristics and laws of coal mining subsidence'. That is, the loess layer affects the occurrence and development of coal mining subsidence. At the same time, coal mining subsidence will also significantly change the shape of the loess layer on the surface. At present, the relevant research results at home and abroad mainly focus on the aspect of 'loess layer affecting coal mining subsidence'. Shijie Song [11] and Wang X [12] studied the influence of the thickness of the loess layer on the subsidence coefficient; Lian X [13] analyzed the influence of the mechanical properties of loess joints on the development law of coal mining subsidence. Marek Drewnik [14] studied the effect of chemical properties of loess soil on the surface morphology. Tang Fuquan [15], Yao Y [16], Wang L [17], Vanapalli S [18], and Smalley [19] studied the effect of the collapsibility of the loess layer on the surface movement and damage effect. However, there are few reports on the research results of 'coal mining subsidence affecting loess layer' [20,21]. The essence of the influence of coal mining subsidence on the shape of the loess slope is that with the high-intensity coal mining, after a large area of goaf is formed underground, the upper rock layer loses its support, and the original stress balance is destroyed. This stress change is transmitted to the loess layer on the surface through the overlying rock layer, which significantly changes the original terrain (the shape of the surface slope) [22]. Through numerical simulation, He L [23] found that the stress change in the goaf caused the roof to break and transmit to the surface, resulting in the movement and deformation of the surface. AN C [24] showed that the vertical extrusion and horizontal stretching caused by coal mining have a significant effect on slope deformation. As we all know, coal mining subsidence often has the distinct characteristics of short duration from occurrence to stability, and soil erosion factors such as precipitation [25], soil [26], vegetation [27], and water conservation measures [28] generally do not change significantly in such a short period of time, so that the topographic factors (i.e., loess slope morphology) that have changed significantly in coal mining subsidence have become the main controlling factors of soil erosion intensity in coal mining subsidence areas [8]. Therefore, the lack of research on the influence of coal mining subsidence on the surface morphology of the loess layer restricts the scientific understanding of the soil erosion effect derived from coal mining subsidence [20].

In view of this, this paper takes the typical loess coal mining area in northern Shaanxi as the research area and takes the overlying strata of the 2^{-2} main coal seam in the area as the geological prototype. Based on the numerical simulation test method, the influence of coal mining subsidence on the surface loess slope morphology (slope, slope length) under the same mining and geological conditions is studied and revealed. On this basis, the soil erosion effect of surface loess slope deformation is calculated and analyzed by using different scale soil erosion models, in order to enrich and deepen the study of the soil and water loss law in coal mining subsidence areas and to provide a scientific basis for soil and water conservation in coal mining subsidence areas in northern Shaanxi and ecological protection and high-quality development in the middle reaches of the Yellow River.

2. Overview of the Study Area

The study area is located in the loess hilly and gully area in Yulin City, northern Shaanxi Province. The ridges and hills are undulating, and the valleys are vertical and horizontal [29]. The slope gradient on both sides of the valley is generally within 5°~45° (Figure 1). The region belongs to the semi-arid continental climate, with perennial drought and little rain. The rainfall is concentrated from July to September, and the annual rainfall is between 194.7 mm and 531.6 mm. The ecological environment in the study area is fragile, the vegetation type is single, mainly drought-tolerant sand plants and shrubs, and the anti-interference ability is poor. The soil erosion in the area is serious, and the soil erosion modulus exceeds 4000 $t \cdot km^{-2} \cdot a^{-1}$ [30]. The strata in the study area from old to new are the Upper Triassic Yongping Formation (T_3y), the Lower Jurassic Fuxian Formation (J_1f), the Middle Jurassic Yan'an Formation (J_2y), the Zhiluo Formation (J_2z), the Anding Formation (J_2a), and the Tertiary and Quaternary periods. Among them, the 2^{-2} coal seam in the Yan'an Formation is the main mineable coal seam in the area, with a thickness of 3~10 m, generally buried at a depth of 0~300 m [31]. The overlying bedrock is characterized by sand–mudstone interbedding. Among them, there are many layers of sandstone, generally

12 layers; the thickness of single-layer sandstone is about 14 m, and the sand/mud ratio is about 80%. The mining method of the main coal seam is long-wall comprehensive mechanized mining, and the mining thickness is generally about 5 m.

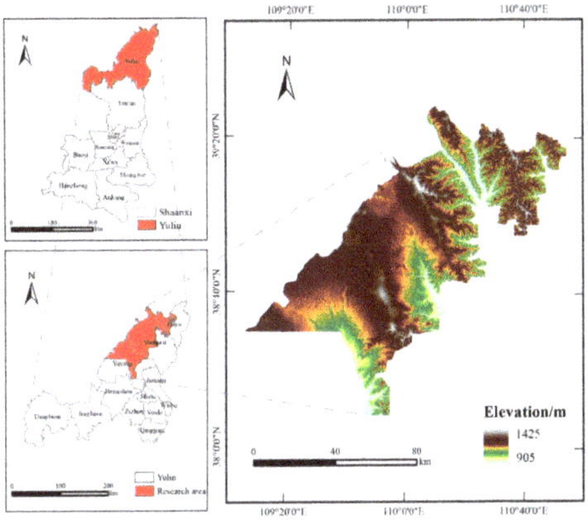

Figure 1. Location diagram of study area.

3. Research Method and Process

3.1. Numerical Model Construction and Experiment

In this study, FLAC^{3D} version 6.0 was selected as the numerical simulation software, mainly because of its powerful three-dimensional geological modeling and geotechnical mechanics simulation capabilities. FLAC^{3D} (Fast Lagrangian Analysis of Continua in 3 Dimensions) adopts the explicit finite difference method, which can effectively simulate the progressive failure and instability process of geological materials and is especially suitable for simulating the surface deformation caused by coal mining subsidence. It shows good adaptability in the dynamic simulation of surface displacement and related geological response caused by coal mining subsidence and can accurately reveal the change law of slope morphology and its relationship with coal mining subsidence. In addition, FLAC^{3D} supports dynamic adjustment of local material parameters and can flexibly simulate different engineering processes according to specific geological conditions and mining conditions. Its efficient three-dimensional modeling, flexible parameter modification and powerful post-processing function can extract displacement, stress, and strain data in real time and generate an intuitive and clear displacement field and stress distribution chart, which provides strong support for the study of the slope deformation law. As a widely recognized numerical simulation tool for geotechnical engineering at home and abroad, FLAC^{3D} has achieved remarkable results in a number of studies on coal mining subsidence and surface deformation [32,33]. However, the material model assumption of FLAC^{3D} is relatively simplified, which may not fully reflect the complex nonlinear deformation behavior of a loess layer. Further experimental verification and model adjustment are needed to optimize the simulation results.

3.1.1. Model Framework

According to the morphological characteristics of the loess slope in the study area and the burial structure of the 2^{-2} main coal seam from the bottom to the top 'floor-coal seam-bedrock layer-loess layer', a three-dimensional geological numerical model was constructed. The natural slope shape of the loess slope in the model is set to four types: linear, concave, convex, and compound. These types are based on the classification criteria of slope shape and cover all kinds of slope shapes in the study area. The natural slopes of the loess slope are set to 5°, 15°, 25°, 35°, 45°, and 5 other gradients [34], which can cover the slope range

erodibility K value is 0.306 $t \cdot hm^2 \cdot h \cdot (MJ \cdot hm^2 \cdot mm)^{-1}$; the slope length and slope factor (L, S) were calculated by the formula proposed by Desme et al. [44] and McCool D K et al. [45,46]. The vegetation coverage and biological measures factor B was obtained by the vegetation coverage equation proposed by Cai Chongfa et al. [47]. The factor E of soil and water conservation engineering measures was selected as 0.9, and the factor T of tillage measures was set as 1.

(2) Typical erosion rainfall scale

The empirical formula of Wu et al. [48] is based on the long-term rainfall erosion characteristics, slope morphology, and soil erosion dynamics in the Loess Plateau. After many field verifications, it is widely used in soil erosion analysis in the Loess Plateau and other erosion sensitive areas. The formula can effectively predict soil loss under different rainfall conditions by quantifying the interaction between rainfall, slope, slope length, and soil erosion. In this paper, three typical rainfall conditions (16.6 mm, 26.9 mm, 39.7 mm) in the empirical formula of Wu et al. were selected, and the soil erosion modulus was calculated by combining slope and slope length, as shown in Table 3. In the analysis of the model, It is understood that the main erosive rainfall in the Loess Plateau of northern Shaanxi is concentrated within the range of 15 to 45 mm. Rainfall within this range occurs frequently, accounting for 50-65% of the annual precipitation events, and the runoff generated from this rainfall contributes to 81.7% of the total runoff. It is proved that the application of the formula under this rainfall condition is reasonable, and it has good applicability to the typical erosive rainfall area of the Loess Plateau in northern Shaanxi. It should be noted that Wu's empirical formula is widely used in the Loess Plateau, but the model is mainly based on empirical data, which may not fully capture soil erosion under complex terrain or extreme weather conditions. Therefore, when applied in other regions, it may be necessary to make appropriate adjustments to the model.

Table 3. Comprehensive factors of rainfall and soil erosion in Loess Plateau relationship.

Rain Fall /(mm)	Rainfall Intensity /($mm \cdot h^{-1}$)	Raindrop Kinetic Energy /($J \cdot m^{-2}$)	Quantization of *S*, *L*, and *Ms*	Correlation Index R
16.6	12.73	255.64	$Ms = 39.727S^{0.741}L^{-0.577}$	0.970
26.9	122.28	766.11	$Ms = 213.955S^{0.992}L^{-0.310}$	0.997
39.7	14.44	634.80	$Ms = 40.333S^{0.847}L^{-0.432}$	0.990

(3) Calculation of the contribution of slope factor and slope length factor in soil erosion modulus

The contribution of slope factor and slope length factor to the soil erosion modulus was quantitatively analyzed by using the coefficients of each variable in the multivariate linear regression equation. The size and positive and negative values of the coefficient reflect the positive or negative effects of each factor on the soil erosion modulus, so as to clarify its importance in the process of soil erosion.

4. Results

Based on the numerical simulation test and data calculation, the slope, slope length, and soil erosion modulus of the surface loess slope of the 20 models before and after mining were obtained, respectively, as shown in Table 4.

Table 4. The change in slope gradient, slope length, and soil erosion modulus of the surface loess slope of each model after full mining.

Slope Shape	Natural Slope/°	Slope After Mining		Slope Length After Mining		Under the Scale of Annual Erosion Rainfall		Soil Erosion Modulus Under Typical Field Erosion Rainfall Scale					
								16.6 mm		26.9 mm		39.7 mm	
		Variable Quantity /°	Rate of Change /%	Variable Quantity /m	Rate of Change /%	Variable Quantity /t·hm^{-2}·a^{-1}	Rate of Change /%	Variable Quantity /t·hm^{-2}	Rate of Change /%	Variable Quantity /t·hm^{-2}	Rate of Change /%	Variable Quantity /t·hm^{-2}	Rate of Change /%
straight	5	0.59	11.80	25.67	3.73	33.23	10.99	1.93	9.64	22.93	12.12	2.89	10.61
	15	1.30	8.67	7.29	3.14	52.84	9.67	5.24	8.24	71.62	10.36	8.18	9.07
	25	1.59	6.36	4.35	3.06	48.39	6.42	6.46	5.80	92.37	7.06	10.17	6.27
	35	1.70	4.86	2.78	2.66	41.38	4.58	6.47	4.14	94.34	4.93	10.19	4.43
	45	1.78	3.96	1.24	1.46	40.30	3.95	5.58	2.85	85.96	3.48	8.99	3.09
concave	5	0.55	11.00	32.31	5.27	26.62	9.33	1.98	9.58	22.39	11.62	2.89	10.34
	15	1.43	9.53	9.87	4.90	50.49	9.91	6.44	9.73	84.28	11.93	9.82	10.56
	25	1.82	7.28	5.32	4.54	44.85	6.55	8.20	6.97	111.38	8.26	12.55	7.43
	35	1.91	5.46	3.03	3.15	41.30	4.76	7.41	4.63	105.93	5.46	11.54	4.93
	45	1.92	4.27	1.37	1.74	40.20	4.09	6.15	3.08	92.94	3.72	9.81	3.32
convex	5	0.63	12.60	21.28	2.94	37.50	12.11	1.95	9.87	23.74	12.64	2.96	10.98
	15	1.22	8.13	6.94	2.70	54.19	9.41	4.76	7.72	66.47	9.77	7.53	8.53
	25	1.48	5.92	4.13	2.54	48.20	5.98	6.54	6.10	97.41	7.60	10.52	6.69
	35	1.66	4.74	2.34	2.00	46.07	4.82	5.83	3.86	88.52	4.71	9.39	4.18
	45	1.61	3.58	1.01	1.03	43.01	3.92	5.38	2.87	87.07	3.61	8.91	3.16
composite	5	0.57	11.40	35.34	5.13	29.45	9.74	1.96	9.78	22.57	11.93	2.88	10.59
	15	1.30	8.67	10.20	4.37	51.45	9.39	5.68	8.94	75.95	11.00	8.76	9.73
	25	1.63	6.52	5.70	3.98	45.62	6.03	6.85	6.16	95.66	7.32	10.64	6.58
	35	1.70	4.86	3.21	3.02	40.02	4.38	6.60	4.25	95.20	4.99	10.34	4.51
	45	1.74	3.87	1.10	1.21	41.86	3.97	5.19	2.70	81.57	3.34	8.46	2.95

4.1. The Influence of Coal Mining Subsidence on the Slope of Loess Slope

4.1.1. Slope Variation Characteristics of Subsidence Slope Under the Same Natural Slope Shape

According to Table 4, the contrast diagram of the increment of and increase in the slope of the loess slope after coal mining subsidence under the same natural slope condition is drawn; this is shown in Figure 3.

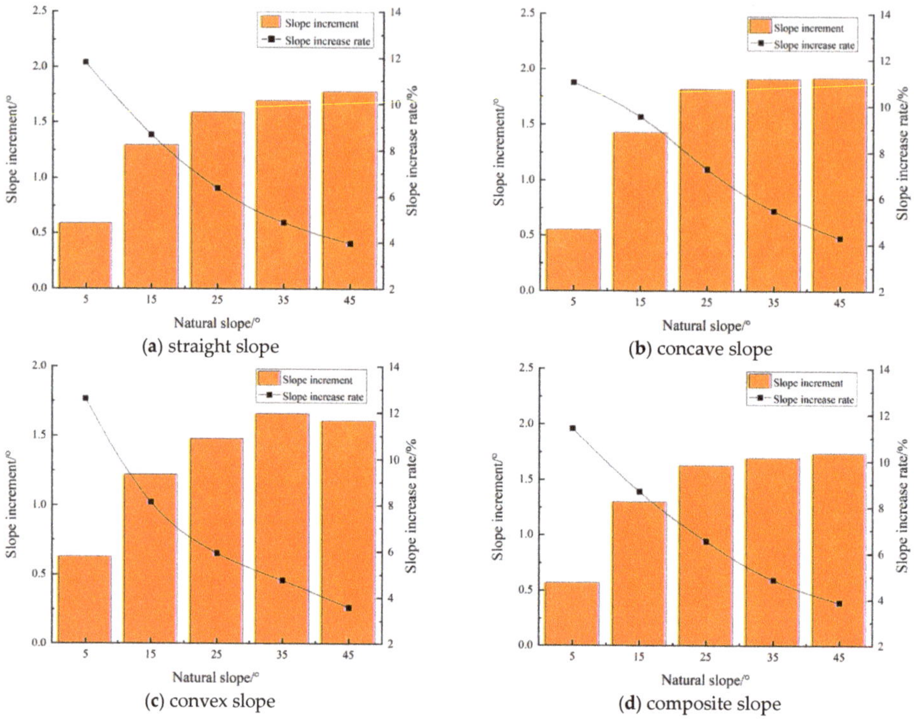

Figure 3. Under the same natural slope shape, the slope increment and increase after full mining.

First, under the condition of full mining, when the natural slope is [5°, 25°], the slope increments of the linear, concave, convex, and complex loess slopes increase with the increase in the slope. Among them, the concave loess slope has the largest slope increment when the natural slope is 25°, which is 1.27 times, 1.49 times, and 1.20 times that of the linear, convex, and complex slopes. When the natural slope is (25°, 45°], the change range of the slope increments of these four loess slopes is small, and the change value of the increments is within 0.1°. In general, coal mining subsidence will lead to the slope increments of linear, concave, convex, and complex loess slopes increasing first and then tending to be stable with the increase in natural slope. Secondly, under the condition of full mining, when the natural slope is [5°, 15°), the average increase in the slope gradient of the linear, concave, convex, and composite loess slopes is more than 10%, and the increase in the slope gradient of the convex loess slope is the largest when the natural slope is 5°, with an increase of more than 12%. When the natural slope is (15°, 25°], the average increase in the slope gradient of the linear, concave, convex, and composite loess slopes is 7.51%, 8.41%, 7.03%, and 7.59%, respectively. When the natural slope is (25°, 45°], the average increase in the slope of the linear, concave, convex, and composite loess slopes is 5.01%, 5.62%, 4.75%, and 5.03%, respectively. Among them, the slope of the three loess slopes of the linear, convex, and composite shapes is increased by less than 4% when the natural slope is 45°. It can be seen that coal mining subsidence will lead to the increase in the slope of the four slope shapes, which decreases with the increase in the natural slope. When the natural slope is 5°, the increase in the slope of the four slope shapes is the largest, and the change trend is highly consistent with the exponential function, as shown in Equations (1)–(4).

$$\text{Slope increase equation of straight slope}: \ Ci = 13.159e^{-0.028i} \quad R^2 = 0.9938 \quad (1)$$

$$\text{Slope increase equation of concave slope}: \ Ci = 13.064e^{-0.025i} \quad R^2 = 0.9894 \quad (2)$$

$$\text{Slope increase equation of convex slope}: \ Ci = 13.650e^{-0.031i} \quad R^2 = 0.9844 \quad (3)$$

$$\text{Slope increase equation of complex slope}: \ Ci = 13.016e^{-0.027i} \quad R^2 = 0.9985 \quad (4)$$

In the equation: *i*—The natural slope of four kinds of slope shape loess slope, °; *Ci*—Slope increase, %.

4.1.2. Slope Variation Characteristics of Subsidence Slope Under the Same Natural Slope

According to Table 4, the curve of the slope gradient increment of and increase in the loess slope with the linear, concave, convex, and compound shapes after coal mining subsidence under the same natural slope condition is drawn; this is shown in Figures 4 and 5.

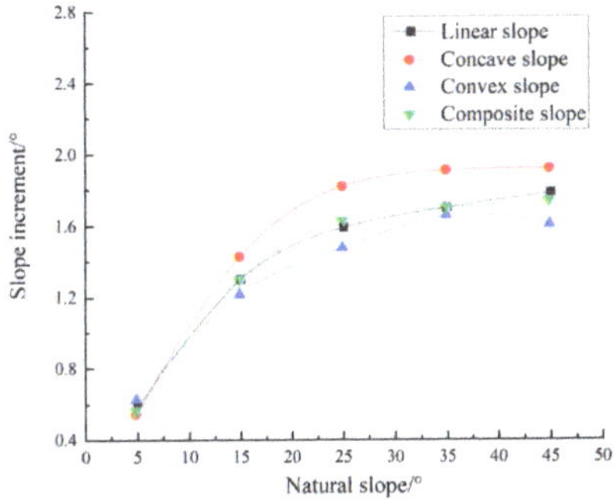

Figure 4. Slope increment after full mining on the same slope.

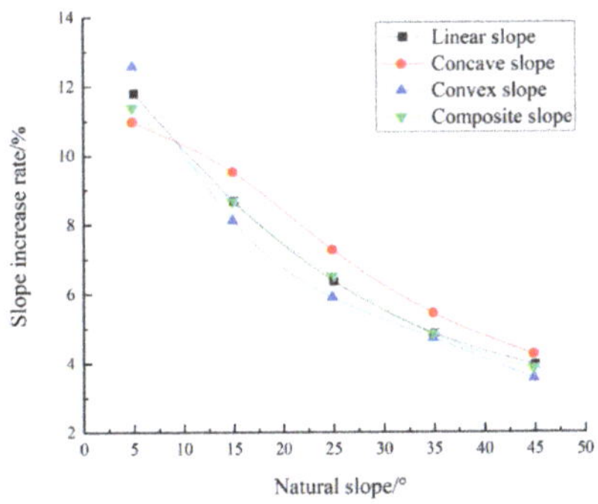

Figure 5. Slope increase after full mining on the same slope.

It can be seen from Figures 4 and 5 that under the same natural slope, the influence of coal mining subsidence on the slope of linear, concave, convex, and complex loess slopes is different. Specifically:

Firstly, under the condition of full mining, when the natural slope is [5°, 15°), the slope increments of the linear, concave, convex, and composite loess slopes are all below 0.7°, and there is no significant difference. When the natural slope is [15°, 45°], the slope increment of these four loess slopes is between 1.2° and 2°. Among them, the slope increment of the concave loess slope is always the largest, which is 11.11%, 11.86%, and 11.11% higher than that of the linear, convex, and composite slopes, respectively. It can be seen that the influence of coal mining subsidence on the slope of the concave loess slope is the most significant. Secondly, under the condition of full mining, when the natural slope is [5°, 15°), the slope increase of the linear, concave, convex, and composite loess slopes is the largest, which is more than 10%. The order of the slope increase of the loess slopes with different slope shapes caused by coal mining subsidence is convex slope > linear slope > composite slope > concave slope. When the natural slope is [15°, 45°], the order of the slope increase of the loess slope under different slope shapes caused by coal mining subsidence is concave slope > straight slope ≈ composite slope > convex slope. The slope increase of the concave loess slope is 11.32%, 18.61%, and 10.99% higher than that of the straight, convex, and composite slopes, respectively.

It can be seen, firstly, that the convex loess slope with a natural slope of less than 15° and the concave loess slope with a natural slope of greater than 15° are the most sensitive to coal mining subsidence. Second, the natural slope of 15° is the key demarcation point for the transformation of the sensitive slope shape of the loess slope in the coal mining subsidence area of northern Shaanxi.

4.2. The Influence of Coal Mining Subsidence on Slope Length of Loess Slope

4.2.1. The Variation Characteristics of Slope Length of Subsidence Slope Under the Same Natural Slope Shape

According to Table 4, the comparison chart of the reduction and the change in the slope length of the loess slope after the mining subsidence under the same natural slope condition is drawn; this is shown in Figure 6.

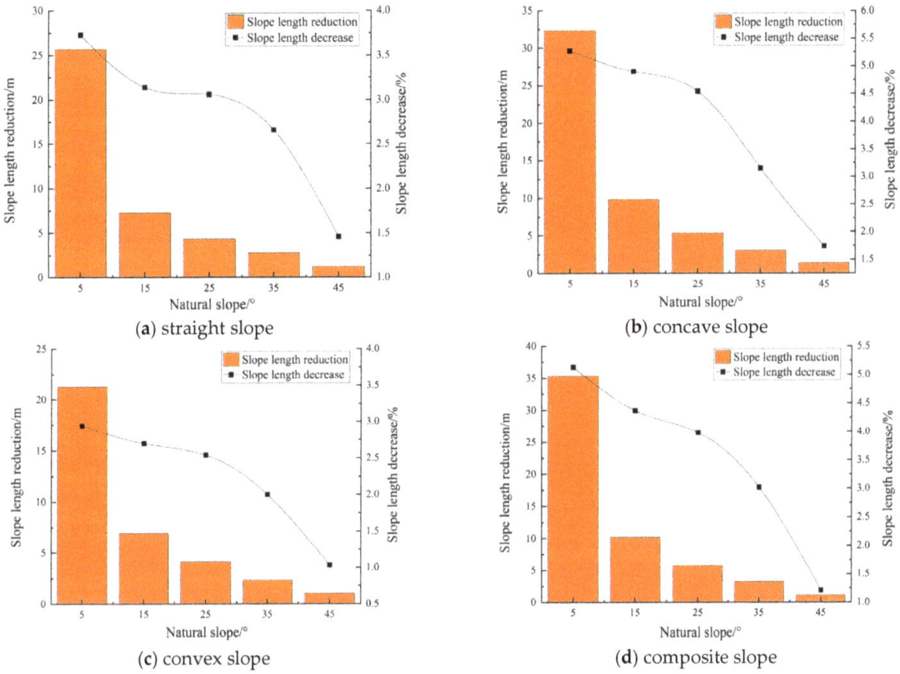

Figure 6. Under the same natural slope shape, the comparison chart of slope length reduction and decrease after full mining.

It can be seen from Figure 6 that no matter what kind of slope shape, coal mining subsidence will lead to the decrease in the slope length of the loess natural slope, and the smaller the natural slope, the greater the decrease in slope length. Specifically:

Firstly, under the condition of full mining, when the natural slope is [5°, 15°), the average decrease in the slope length of the linear, concave, convex, and composite loess slopes is 3.44%, 5.09%, 2.82%, and 4.75%, respectively. When the natural slope was [15°, 25°), the average decrease in the slope length of the four loess slopes was 3.10%, 4.72%, 2.62%, and 4.18%, respectively. When the natural slope was [25°, 35°), the average decrease in the slope length of the four loess slopes was 2.86%, 3.84%, 2.27%, and 3.50%, respectively. When the natural slope is [35°, 45°], the average decrease in the slope length of these four loess slopes is 2.06%, 2.44%, 1.51%, and 2.11%, respectively. It can be seen that coal mining subsidence will lead to the decrease in the slope length of the loess slope with the increase in the natural slope. Secondly, under the condition of full mining, the slope length of the linear, concave, convex, and composite loess slopes with a natural slope of 5° has the largest decrease, which is 1.44 times, 1.47 times, 1.43 times, and 1.63 times that of the average decrease in the slope length of the four slopes with a natural slope of 15°, 25°, 35°, and 45°. It can be seen that the influence of coal mining subsidence on the slope length of the loess slope is the most obvious when the natural slope is 5°, and its change trend conforms to the characteristics of a quadratic polynomial function, as shown in Equations (5)–(8).

$$\text{The slope length reduction amplitude equation of straight slope}: \ Li = -0.0011i^2 + 0.0048i + 3.597 \qquad R^2 = 0.9394 \quad (5)$$

$$\text{The slope length reduction amplitude equation of concave slope}: \ Li = -0.0022i^2 + 0.023\text{i} + 5.1784 \qquad R^2 = 0.9929 \quad (6)$$

$$\text{The slope length reduction amplitude equation of convex slope}: \ Li = -0.0013i^2 + 0.0196\text{i} + 2.8265 \qquad R^2 = 0.9873 \quad (7)$$

The slope length reduction equation of composite slope : $Li = -0.0019i^2 + 0.0025i + 5.0356 \quad R^2 = 0.9836$ (8)

In the equation: i—The natural slope of four kinds of slope shape of loess slope, °; Li—Reduction in slope length, %.

4.2.2. The Variation Characteristics of Slope Length of Subsidence Slope Under the Same Natural Slope

According to Table 4, under the same natural slope conditions, the curves of the decrease and decrease in the slope length of the loess slope after mining subsidence are drawn and are straight, concave, convex, and composite. See Figures 7 and 8.

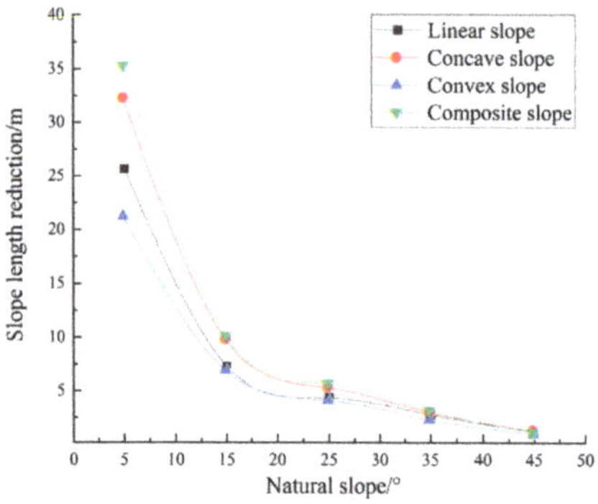

Figure 7. The change curve of slope length reduction after full mining under the same slope.

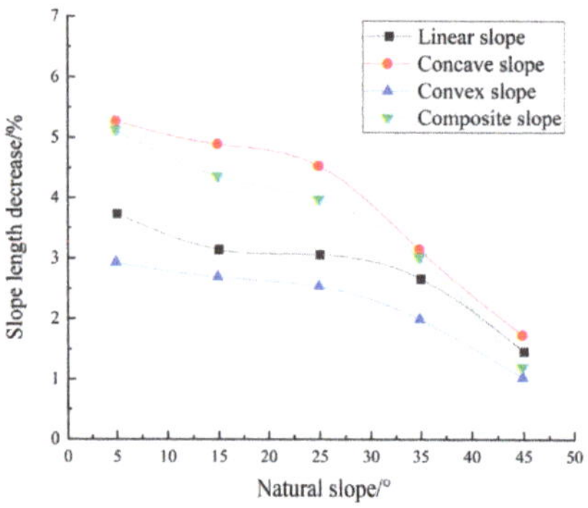

Figure 8. The change curve of slope length after full mining under the same slope.

It can be seen from Figures 7 and 8 that under the same natural slope, the influence of coal mining subsidence on the slope length of the linear, concave, convex, and composite loess slopes is different. Specifically:

Firstly, under the condition of full mining, when the natural slope is [5°, 25°], the average decrease in the slope length of the linear, concave, convex, and compound loess slopes caused by mining subsidence is 3.31%, 4.90%, 2.73%, and 4.49%, respectively. The order of the slope length decrease of the loess slope under different slope shapes is concave slope > compound slope > linear slope > convex slope. Among them, the decrease in the slope length of the concave loess slope is always the largest, which is 1.48 times, 1.80 times, and 1.09 times that of the average decrease in the slope length of the linear, convex, and compound loess slopes. When the natural slope is (25°, 45°], the order of the slope length reduction of the loess slope under the different slope shapes caused by coal mining subsidence is concave slope > composite slope ≈ straight slope > convex slope. It can be seen that the influence of coal mining subsidence on the slope length of the concave loess slope is the most significant.

Second, under the condition of full mining, when the natural slope is [5°, 25°], the slope length of the three loess slopes of the linear, concave, and convex slopes is relatively stable, and the variation range is below 0.8%. When the natural slope is (25°, 45°], the slope length of the linear, concave, convex, and composite loess slopes changes significantly, and the slope length of these four loess slopes with a natural slope of 25° is 2.10 times, 2.61 times, 2.47 times, and 3.28 times that of 45°. It can be seen, firstly, that the decrease in the slope length of the loess subsidence slope under different slope shapes shows a decreasing trend with the increase in natural slope. Second, the natural slope of 25° is the key point of the sudden change rate of the slope length of the loess slope in the coal mining subsidence area of northern Shaanxi.

4.3. Analysis of Soil Erosion Effect of Slope Deformation

4.3.1. Soil Erosion Effect Under Annual Erosion Rainfall Scale

(1) The variation characteristics of soil erosion modulus under the same natural slope shape

According to Table 4, the contrast diagram of the increment of and increase in the soil erosion modulus of the loess slope with the same natural slope shape after coal mining subsidence under the annual erosion rainfall scale is drawn; this is shown in Figure 9.

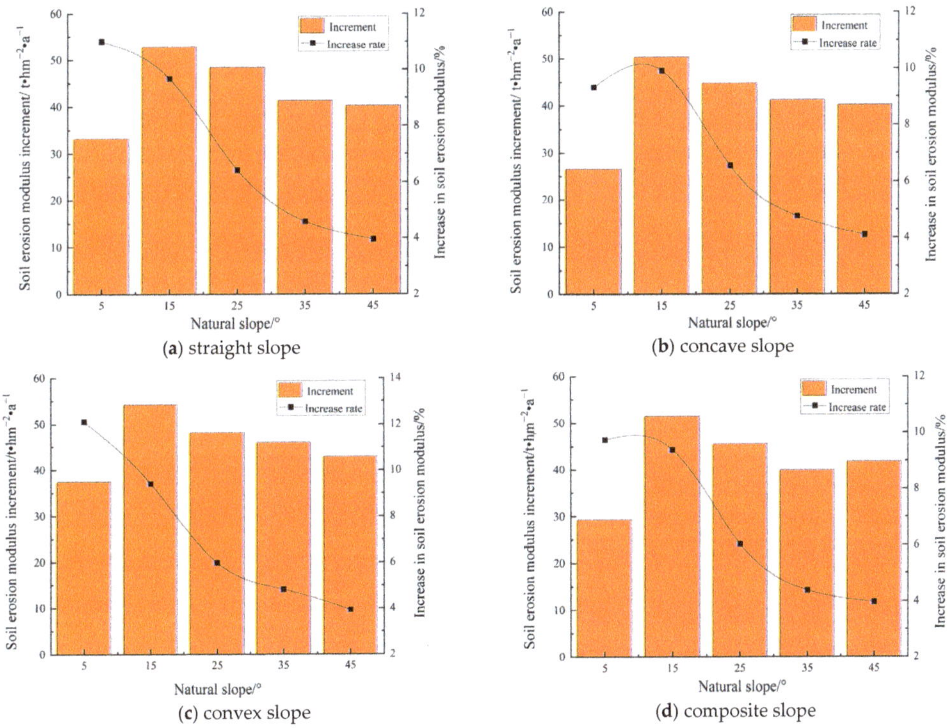

(**a**) straight slope (**b**) concave slope

(**c**) convex slope (**d**) composite slope

Figure 9. Comparison of soil erosion modulus increment and increase after full mining under CSLE model.

From Figure 9, it can be seen that no matter what the slope shape, coal mining subsidence will lead to an increase in the soil erosion modulus on the loess slope. Specifically:

First, under the condition of full mining, when the natural slope is [5°, 15°], the average increase in the soil erosion modulus of the linear, concave, convex, and composite loess slopes is 10.33%, 9.62%, 10.76%, and 9.56%, respectively. When the natural slope is (15°, 25°), the average increase in the soil erosion modulus of these four loess slopes is 6.42%, 6.55%, 5.98%, and 6.03%, respectively. When the natural slope was (25°, 45°], the average increase in the soil erosion modulus of the four loess slopes was 4.26%, 4.43%, 4.37%, and 4.18%, respectively. In addition, the increment of the soil erosion modulus of the linear, concave, convex, and complex loess slopes with a natural slope of 15° is the largest, which is 1.67 times, 1.12 times, 1.24 times, and 1.26 times that of the average increment of the soil erosion modulus of the four slopes with natural slopes of 5°, 25°, 35°, and 45°.

Second, according to the data of the slope increase, slope length decrease, and soil erosion modulus increase of the loess slope in different models, the quantitative relationship between the soil erosion modulus increase (R_{M_1}) and slope increase (R_s) and slope length decrease (R_l) under the annual erosion rainfall scale was constructed by using a multiple linear fitting method, as shown in Equation (9).

$$R_{M_1} = 1.007R_s - 0.279R_l + 0.644 \qquad R^2 = 0.9325 \tag{9}$$

In the formula: R_{M_1}—The increase in soil erosion modulus under the scale of one-year erosion rainfall, %; R_s—slope gradient increase, %; R_l—slope length decrease, %.

It can be seen, firstly, that under the scale of annual erosion rainfall, the increase in the soil erosion modulus of the loess subsidence slope with any slope shape shows a decreasing trend with the increase in natural slope. Secondly, the natural slope of 15° is the key point of the abrupt change in soil erosion intensity on the loess slope of the coal mining subsidence area in northern Shaanxi under the annual erosion rainfall scale. Thirdly, under the scale of annual erosion rainfall, the increase in the slope gradient has a positive effect on the increase in the soil erosion modulus, and the decrease in the slope length has a negative effect on the increase in the soil erosion modulus. The contribution of the slope increase was 92.9%, and the contribution of the slope length decrease was 7.1%. Therefore, the increase in the slope is the main factor of the increase in the soil erosion modulus.

(2) The variation characteristics of soil erosion modulus under the same natural slope angle According to Table 4, the curve of the soil erosion modulus increment of and the increase in the linear, concave, convex, and composite loess slopes with the same natural slope after coal mining subsidence under the annual erosion rainfall scale is drawn; this is shown in Figures 10 and 11.

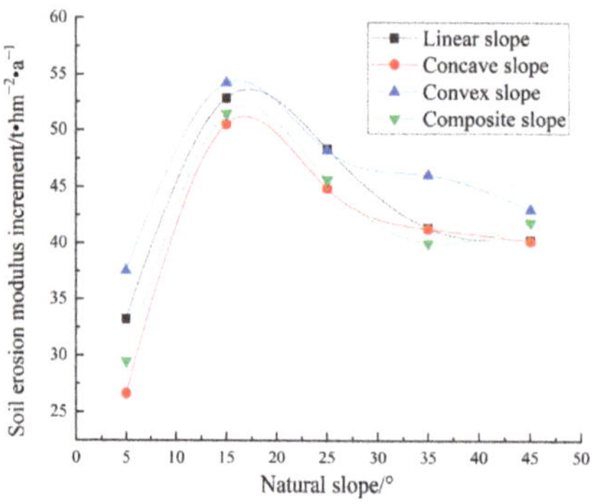

Figure 10. The curve of soil erosion modulus increment after full mining under CSLE model.

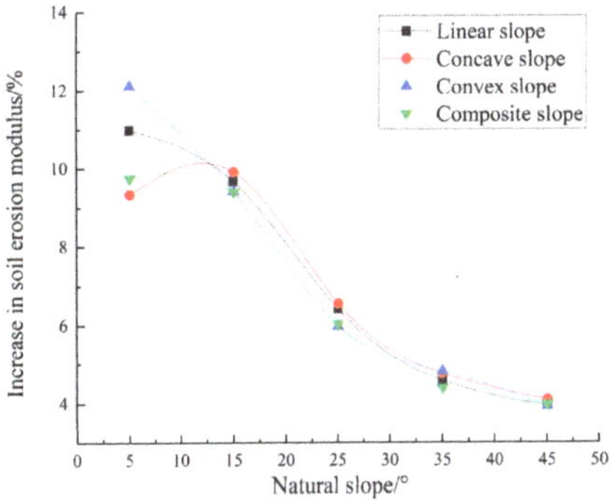

Figure 11. The curve of soil erosion modulus increase after full mining under CSLE model.

It can be seen from Figures 10 and 11 that under the same natural slope, the influence of coal mining subsidence on the soil erosion modulus of the linear, concave, convex, and complex loess slopes is different. Specifically:

First, under the condition of full mining, when the natural slope is [5°, 15°), the average increase in the soil erosion modulus of the linear, concave, convex, and composite loess slopes is 10.99%, 9.33%, 12.11%, and 9.74%, respectively. Among them, the increase in the soil erosion modulus of the convex loess slope is the largest, which is 1.10 times, 1.30 times, and 1.24 times of that of the linear, concave, and composite loess slopes. When the natural slope is [15°, 45°], the average increase in the soil erosion modulus of the linear, concave, convex, and composite loess slopes is 9.60%, 6.25%, 4.64%, and 3.98%, respectively. Under the same slope condition, there is no significant difference in the increase in the soil erosion modulus among the four slopes. It can be seen that under the annual erosion rainfall scale, coal mining subsidence has the most significant effect on the increase in the soil erosion modulus of the convex loess slope with slope < 15°.

4.3.2. Soil Erosion Effect Under Typical Field Erosion Rainfall Scale

(1) The variation characteristics of soil erosion modulus under the same natural slope shape

According to Table 4, the average increment of and average increase in the soil erosion modulus of the loess slope with the same natural slope shape after coal mining subsidence were compared under the typical rainfall scale of erosion, as shown in Figure 12.

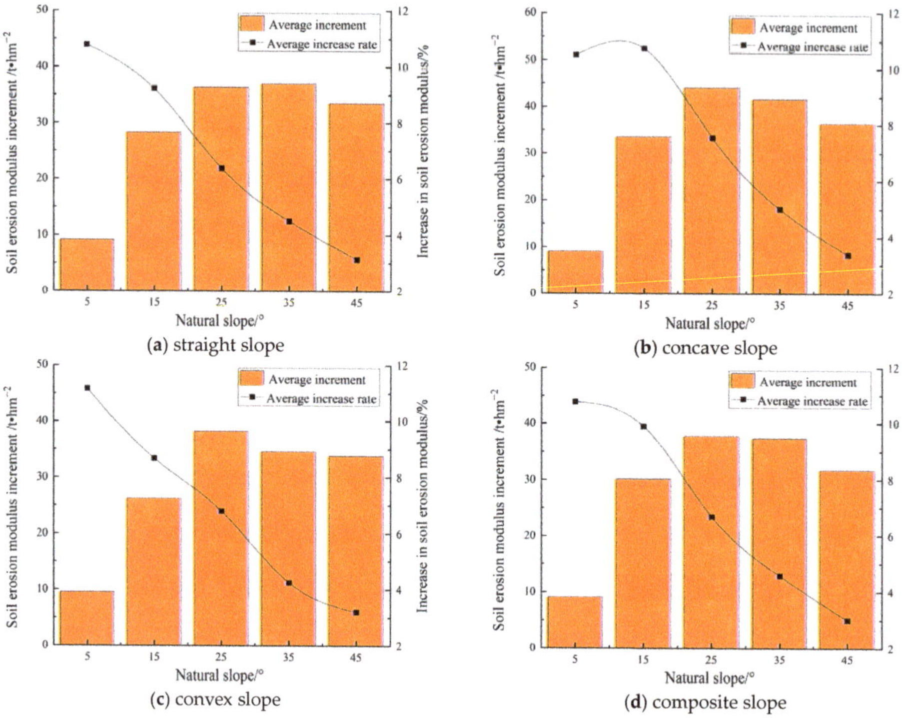

Figure 12. Under three typical rainfall conditions, the average increment of and average increase in soil erosion modulus after full mining are compared.

From Figure 12, it can be seen that coal mining subsidence will lead to an increase in soil erosion modulus on loess slopes under all slope shape and rainfall conditions, and the smaller the natural slope, the greater the increase in soil erosion modulus. Specifically:

Firstly, under the condition of full mining, when the natural slope is [5°, 25°], the average increase in the soil erosion modulus of the linear, concave, convex, and composite loess slopes under three typical erosive rainfall conditions is 8.80%, 9.60%, 8.88%, and 9.11%, respectively, and the increase in the soil erosion modulus is the largest when the natural slope is 25°, which is 3.93 times, 4.85 times, 4.00 times, and 4.13 times that of the increase in the soil erosion modulus when the natural slope is 5°. When the natural slope is (25°, 35°], the average increase in the soil erosion modulus of the linear, concave, convex, and composite loess slopes under three typical erosive rainfall conditions is 5.44%, 6.28%, 5.52%, and 5.64%, respectively. The average increment of the soil erosion modulus of the loess slopes with any slope shape in this natural slope range is not much different from that of the natural slope of 25°, and the average increment is not more than 1.5 t·hm^{-2}. When the natural slope is (35°, 45°], the average increase in the soil erosion modulus of the linear, concave, convex, and composite loess slopes under three typical erosive rainfall conditions is 3.82%, 4.19%, 3.73%, and 3.79%, respectively. In this natural slope range, the average increase in the soil erosion modulus of the loess slope with any slope shape is slightly lower than that of the natural slope of 35°, and the average decrease is not more than 3.9 t·hm^{-2}.

Secondly, according to the data of the slope increase, slope length decrease, and soil erosion modulus increase of the loess slope in the different models, the quantitative relationship between the soil erosion modulus increase (R_{M_2}) and slope increase (R_s) and slope length decrease (R_l) under a typical erosion rainfall scale was constructed by using the multiple linear fitting method, as shown in Equation (10).

$$R_{M_1} = 1.007R_s - 0.279R_l + 0.644 \; R^2 = 0.9325 \quad (10)$$

In the formula: R_{M_2}—The increase in soil erosion modulus under typical field erosion rainfall scale, %; R_s—Slope gradient increase, %; R_l—Slope length decrease, %.

It can be seen, firstly, that under the typical rainfall scale of erosion, the increment of the soil erosion modulus of the loess subsidence slope with an arbitrary slope shape increases first and then decreases with the increase in the natural slope. Secondly, the natural slope of 25° is the key point of the abrupt change in soil erosion intensity on the loess slope in the coal mining subsidence area of northern Shaanxi under the typical erosion rainfall scale. Thirdly, under the typical rainfall scale of erosion, the increase in slope and the decrease in slope length have a positive effect on the increase in the soil erosion modulus. The contribution of the slope increase is 79.1%, and the contribution of the slope length decrease is 20.9%. Therefore, the increase in the slope is the main factor of the increase in the soil erosion modulus.

(2) The variation characteristics of soil erosion modulus under the same natural slope angle

According to Table 4, the curves of the average increment of and average increase in the soil erosion modulus of the linear, concave, convex, and compound loess slopes with the same natural slope after coal mining subsidence under a typical rainfall scale are drawn, as shown in Figures 13 and 14.

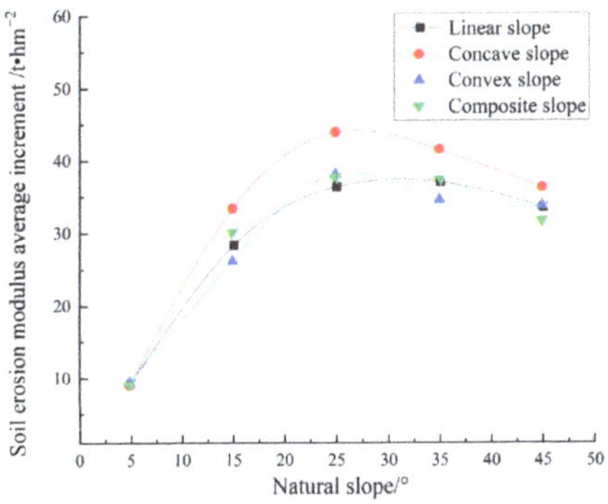

Figure 13. Under three typical rainfall conditions, the average increment curve of soil erosion modulus.

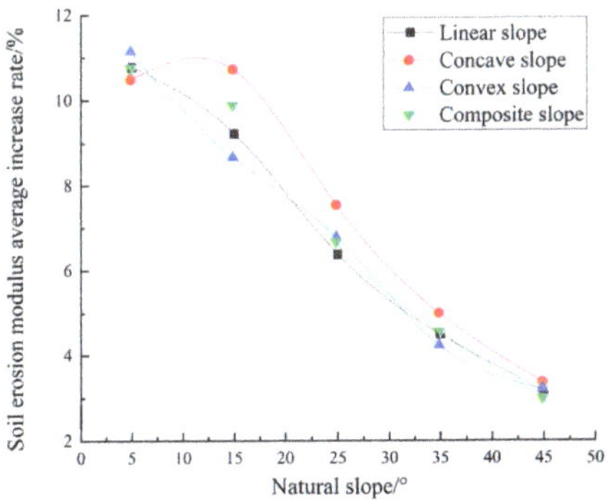

Figure 14. Under three typical rainfall conditions, the average increase curve of soil erosion modulus.

It can be seen from Figures 13 and 14 that under the same natural slope and three typical rainfall conditions, the influence of coal mining subsidence on the soil erosion modulus of the linear, concave, convex, and complex loess slopes is different. Specifically:

First, under the condition of full mining, when the natural slope is [5°, 15°), the average increase in the soil erosion modulus of the linear, concave, convex. and composite loess slopes under three typical rainfall conditions is 10.01%, 10.63%, 9.92%, and 10.33%, respectively. In this slope range, the average increase in the soil erosion modulus of the concave and composite loess slopes did not change significantly, and the increase value was below 0.8%. When the natural slope is [15°, 45°], the average increase in the soil erosion modulus of the linear, concave, convex, and composite loess slopes under three typical rainfall conditions is 9.63%, 6.85%, 4.59%, and 3.18%, respectively. In this slope range, the average increase in the soil erosion modulus of the loess slopes under different slope shapes is concave slope > composite slope ≈ linear slope ≈ convex slope. It can be seen that under the typical rainfall scale of erosion, the influence of coal mining subsidence on the average increase in the soil erosion modulus on the concave loess slope is the most obvious.

5. Discussion

5.1. Effects of Slope Gradient on Stability and Deformation of Loess Slope

Slope gradient is closely related to slope stability, and it is generally believed that the greater the slope, the worse the stability of the slope. Kokutse et al. [49] pointed out that the effect of the slope on slope stability is primarily achieved by altering the cohesion of the soil, and that different slopes play a crucial role in the variation in the safety factor. Zhang et al. [50] found that the slope gradient is closely related to stability. The unstable slope gradient is mostly concentrated above 35°, and the increase in slope gradient will change the stress distribution and reduce the slope safety factor. Zhu Boliang et al. [51] quantitatively calculated 76 loess slope models through the geotechnical slope stability analysis system. The results show that the increase in slope significantly reduces the stability coefficient of the loess slope, especially when the slope is less than 55°, the slope stability coefficient decreases sharply. Katz et al. [52] studied the negative effects of slope on slope stability and deformation from the perspective of numerical simulation and verified that the larger the slope, the larger the slope movement size, which in turn leads to the worse slope stability. Haijun Qiu et al. [53] used a combination of remote sensing images and field surveys. This study revealed the complex relationship between slope and slope movement size and further explained the influence of slope on loess slope deformation. Xi et al. [54] extracted the slope information of a coal mining subsidence area by using the digital image and multi-period digital elevation model (DEM) obtained by UAV and found that coal mining subsidence would lead to the overall increase in surface slope. Based on the law and model of mining subsidence, Huang et al. [55] studied the change in loess slope in a coal mining subsidence area through digital elevation model analysis and remote sensing image processing technology and found that coal mining subsidence led to the shortening of the surface slope length factor. This finding and the results of this study show that coal mining subsidence not only leads to the increase in surface slope, but also causes the shortening of slope length, which affects the change in slope morphology. In addition, this study also found that with the increase in the natural slope of the loess slope, the increase in the slope of the slope after the subsidence also increases, while the decrease in the slope length is relatively reduced, which provides a new research direction for the in-depth study of the law of slope movement and deformation in the subsidence area after coal mining.

5.2. Influence of Slope Shape on the Stability and Deformation Characteristics of Loess Slopes

The influence of slope shape on slope stability is still controversial in academia. Tang et al. [56] deeply explored the influence of different slope shapes on the stability of soil slopes through seismic simulation shaking table tests. The results show that the slope shape has a significant effect on the stability and deformation of the slope. The PGA (peak ground acceleration) amplification coefficient of the concave slope at the turning point of the slope is significantly higher than that of the convex slope and the straight slope, which makes the concave slope more prone to damage and deformation. Huang et al. [57] further confirmed the importance of slope shape to the stability and deformation of the loess slope through the combination of a small shaking table test and FLAC^{3D}numerical simulation. Their study found that under dynamic loading, convex slopes are more prone to instability, while concave slopes show stronger stability. In addition, Sharma et al. [58] revealed the phenomenon whereby the pore water pressure of the concave slope increased under the action of precipitation through three-dimensional numerical simulation, which led to a significant increase in slope deformation and an increase in landslide risk, once again highlighting the important influence of slope shape on slope stability. On the other hand, Gao J et al. [59] found that the concave slope with a medium height and steep slope is more likely to move and deform by using the method of remote sensing interpretation, which provides new evidence for the influence of slope shape on slope stability. This study reveals the close relationship between slope shape and slope under the special dynamic load of coal mining subsidence. Specifically, under the influence of the same coal mining subsidence, when the natural slope does not exceed 5°, the convex slope shows the largest movement deformation; when the natural slope exceeds 5°, the deformation of the concave slope becomes the most significant. This finding provides a new perspective for understanding the stability and deformation characteristics of the surface loess slope in the coal mining subsidence area.

5.3. Influence of Coal Mining Subsidence on Soil Erosion

The surface movement and deformation caused by coal mining subsidence will significantly change the original topography of the region, where the change in slope shape (slope gradient and slope length) is the most prominent. Slope and slope length characteristics are the key topographic factors affecting slope soil erosion, which has been the general consensus in the field of soil and water conservation [60]. Slope gradient has been repeatedly studied and confirmed as a key topographic factor affecting slope runoff and soil erosion. Farhan Y [61] studied the correlation between soil erosion and the topographic unit and slope in a watershed in northern Jordan. It was found that the soil erosion intensity on the slope was highly correlated with the slope, especially in the slope of 15–25°. The intensity of soil erosion on the slope was significantly enhanced; Xu Zhenjian et al. [62] carried out indoor experiments on loessial soil in the Loess Plateau and adjusted the experimental soil bin to three slopes of 5°, 10°, and 20°. The study found that the slope can affect the total runoff and soil erosion by affecting the runoff. Under the condition of rainfall intensity of 1.5 mm min^{-1}, the total erosion amount of the slope reached the maximum when the slope was 10°; Huang Jun et al. [63] conducted rainfall experiments on the typical red soil hilly area in the upper reaches of the Hanjiang River in the Pearl River Basin and used the PI10 index algorithm to calculate the rainfall erosion of each rainfall event. The study found that the soil erosion modulus variable and the slope under the condition of rainfall increased first and then decreased, and the critical slope value was about 10.75°. By selecting the typical small watershed of Qilong Bay in the middle and upper reaches of the Yellow River Basin as the research area, and based on the observation data of 10 runoff plots from 2008 to 2016, Wang Min et al. [64] analyzed that the amount of soil erosion increased with the increase in slope, and the amount of soil erosion was the largest when the slope was 30°. The above research results are highly consistent with the results of the study on the increase in slope gradient caused by coal mining subsidence under the scale of annual erosion rainfall and the scale of typical field erosion rainfall, which leads to the increase in the slope soil erosion modulus. It can be seen that no matter what the slope shape is, the greater the influence of the same coal mining subsidence on the loess slope with larger natural slope, the more obvious the change in slope after subsidence and the resulting change in soil erosion intensity on the slope.

As another important topographic factor, the effect of slope length on soil erosion is controversial. Liu Ran et al. [65] took the loess slope of Ansai in the loess hilly and gully region as the research object and studied the law of runoff and sediment yield on the loess slope under the two slope lengths of 5 m and 10 m through the indoor artificial simulated rainfall test. The results show that the increase in slope length will increase the rain-bearing area of the slope, resulting in an increase in the runoff rate per unit width and a corresponding increase in the total sediment yield. Therefore, the slope length and the slope erosion amount show a significant positive correlation. Based on the runoff and sediment yield data of eight runoff plots in Longfengling Soil and Water Conservation Science and Technology Demonstration Park in Mentougou District of Beijing from 2005 to 2009, Liu Dong et al. [66] concluded that slope length had a significant effect on runoff and sediment yield per unit area in a small runoff area and that the amount of soil erosion would increase with the increase in slope length. However, Kinnell [67] found that the slope length below 1 m has the effect of increasing soil erosion intensity through the rainfall erosion test of the plot below 10 m and the WEPP rill erosion model. However, with the increase in slope length, the energy of slope flow decreases, which leads to a decrease in soil erosion intensity. Based on the data of soil and water conservation monitoring points in Shiqiao small watershed, Gao et al. [68] studied the effect of slope length on soil erosion in karst slope farmland. The study showed that the soil erosion modulus increased with the increase in slope length when the rainfall was greater than 30 mm, and the soil erosion modulus increased sharply and then decreased with the increase in slope length when the rainfall was less than 30 mm. Gu Zhijia et al. [69] quantitatively analyzed the soil erosion status of Baiquan County by using the Chinese soil loss equation CSLE. The study found that with the increase in slope length, the soil erosion modulus generally showed a trend of increasing first and then decreasing, and the slope length of 100~200 m contributed the most to erosion. The above research results are basically consistent with the research results in this paper, which show that the decrease in the slope length of the surface loess slope caused by coal mining subsidence affects the increase in the slope soil erosion modulus. More interestingly, the effect of slope length reduction caused by coal mining subsidence on soil erosion under the annual erosive rainfall scale and the typical field erosive rainfall scale is manifested as a mitigation effect and an aggravation effect, respectively, and these two effects are mainly concentrated on the loess slope with a natural slope of 15–25°. This enriches and deepens the scientific understanding of the soil erosion effect caused by the change in slope length of the surface loess slope in the coal mining subsidence area.

Although this study provides valuable insights into the effects of coal mining subsidence on loess slope morphology and soil erosion, there are still some aspects worthy of further exploration in the future. First of all, this study focuses on the law of action in the process of mining subsidence and fails to obtain sufficient measured data for verification. In the future, the research will strengthen the collection of field data, especially in the field survey and soil erosion monitoring in the mining subsidence area, to further verify the prediction results of the model. Secondly, the numerical model used in the study is based on certain assumptions, such as soil uniformity and linear physical properties. However, the actual soil may have complex nonlinear and heterogeneous characteristics, which may affect the simulation results. Future studies can consider using more complex models and incorporating more changes in geological and soil properties to improve the accuracy and reliability of simulations. Finally, this study mainly focuses on the immediate or short-term impact of coal mining subsidence. This is based on the fact that high-intensity underground coal mining can cause a large-scale rock movement in the subsidence area in a short period of time and spread to the surface, significantly changing the original topography (the shape of the surface slope), which in turn has a profound impact on the characteristics and basic laws of soil erosion on the slope of the subsidence area. The coal mining subsidence in the Yellow River Basin generally takes 3–6 months from occurrence to basic stability, and it can reach complete stability in about 12 months. In such a short period of time, soil erosion factors such as precipitation and soil and water conservation measures will not change significantly. Although soil erosion factors such as soil and vegetation will deteriorate to a certain extent under the significant change in topographic factors, the impact on soil erosion will take a long time to appear. In view of this, the significant change in loess slope morphology caused by coal mining subsidence in the short term has become the main controlling factor of soil erosion intensity, which is also the focus of the study. However, long-term effects (such as vegetation regeneration, soil compaction, and climate change) are indeed important factors after coal mining subsidence. Future research can further explore these long-term effects and assess their role in the longer term after coal mining subsidence.

6. Conclusions

(1) Coal mining subsidence will lead to the increase in the slope of the loess slope, and the smaller the natural slope, the greater the increase in slope. Under the same natural slope, 'concave loess slope with natural slope of 15°' is the most sensitive to coal mining subsidence. It can be said that the natural slope of 15° is the key dividing point for the transformation of the sensitive slope shape of the loess slope in the coal mining subsidence area of northern Shaanxi.

(2) Coal mining subsidence will lead to the decrease in slope length of the loess natural slope, and the smaller the natural slope, the greater the decrease in slope length. Under the same natural slope, the slope length of the concave loess slope is always the largest, and the influence of coal mining subsidence on the slope length of the concave loess slope is the most significant. In addition, the natural slope of 25° is the key point of the sudden change rate of the slope length of the loess slope in the coal mining subsidence area of northern Shaanxi.

(3) Coal mining subsidence will lead to the increase in soil erosion modulus on the loess slope, and the smaller the natural slope, the greater the increase in the soil erosion modulus. In addition, when further analyzing the soil erosion intensity of the loess slope in the coal mining subsidence area of northern Shaanxi, we found two key natural slope thresholds. Under the scale of annual erosion rainfall, the 15° slope became the boundary point for the significant change in soil erosion intensity. Under the typical erosion rainfall scale, the slope of 25° is the key point of the abrupt change in soil erosion intensity.

(4) For the loess subsidence slope with any slope shape, the increase in slope gradient is the main factor of the increase in the soil erosion modulus. Under the annual erosion rainfall scale, the increase in slope had a significant positive effect on the increase in the soil erosion modulus, and the contribution rate was as high as 92.9%, while the decrease in slope length showed a weak negative effect. Under the typical erosion rainfall scale, the increase in slope and the decrease in slope length have a positive effect on the increase in the soil erosion modulus, but the increase in slope is still the dominant factor.

7. Recommendations and Future Actions

This study focused on the northern Shaanxi mining area in the middle reaches of the Yellow River. Considering the unique geological structure, climatic conditions, and coal mining practices in this area, the study revealed the impact of coal mining subsidence on loess slope morphology (such as slope and slope length) and the soil erosion modulus, which provided valuable basic data for further understanding the soil erosion process in coal mining subsidence areas. Although the conclusions of this study are mainly based on the specific conditions of the mining area in northern Shaanxi, the research results can be popularized by appropriately adjusting the model parameters, which can provide a reference for the study of coal mining subsidence in other areas. Future research should further verify the applicability of these conclusions in different regions and explore how to optimize the model according to the differences in geological, climatic, and mining conditions, so that it can be widely used in other mining areas. However, there are still some key factors that have not been explored, and future research can be further studied via the following aspects:

(1) The influence of coal mining subsidence on other geomorphological features.

Although this study focuses on the changes in slope gradient and slope length after coal mining subsidence, future research can consider the influence of coal mining subsidence on slope aspect, curvature, and surface roughness. These factors have a potential impact on the change in slope morphology and soil erosion, especially in coal mining subsidence areas, where the change in surface morphology may be more complex. Therefore, future research can combine remote sensing technology, GIS tools, and field surveys to further analyze the impact of these geomorphological factors on soil erosion, thereby providing a more comprehensive scientific basis for soil and water conservation measures in coal mining subsidence areas.

(2) Soil erosion characteristics of slope after coal mining subsidence under different conditions

Although this study analyzed multiple slopes (5°, 15°, 25°, 35°, 45°) through numerical simulation and obtained 15° and 25° slope angles as key critical points for soil erosion and slope stability, different slope angles may have different performances under different conditions. Especially in some specific precipitation patterns, geological conditions, or climatic conditions, the influence of slope on soil erosion may show different trends. The steeper slope may lead to higher soil erosion when the precipitation intensity is greater, while the gentler slope may have a significant effect on soil erosion under other conditions. Therefore, future research should further explore the influence characteristics of different slope angles under changing climate, precipitation, and soil conditions and identify other possible critical values or turning points, so as to provide a more comprehensive scientific basis for the optimization of soil and water conservation measures.

(3) Soil erosion characteristics of slope after coal mining subsidence under long-term effects

Long-term effects (such as vegetation regeneration, soil compaction, climate change, etc.) are undoubtedly important factors affecting coal mining subsidence. Over time, factors such as vegetation restoration, soil compaction, and climate change will change surface morphology, further exacerbating slope stability and soil erosion rates. Especially in the process of vegetation restoration, if the recovery speed is slow or the climate change is intensified, the slope stability may be further reduced, and the soil erosion problem may be more serious. Therefore, future research can further explore these long-term effects, assess their role in the longer term after coal mining subsidence, and consider how to take effective soil and water conservation measures on a long-term scale to mitigate these effects.

Author Contributions: S.S.: conceptualization, methodology, and writing—review and editing. R.N.: methodology, software drawing, data curation, writing—original draft, and experiment. S.Y.: investigation, methodology. X.C.: investigation, software drawing, and experiment. H.R.: investigation and writing—review and editing. B.C.: writing—review and editing. Y.L.: investigation and writing—review and editing; L.T.: investigation and writing—review and editing. All the authors reviewed the manuscript. All authors have read and agreed to the published version of the manuscript.

Funding: This paper has obtained the following: State Key Laboratory for Safe Mining of Deep Coal Resources and Environment Protection, Huainan Mining (Group) Co., Ltd., 232000 China (HNKY2024YB402); the Key Research and Development Program of Xianyang City (L2024-ZDYF-ZDYF-SF-0069); the Key Research and Development Program of Ningxia Hui Autonomous Region (2024BEG02005); Shaanxi Province Public Welfare Geological Survey Project (Project Number: 202412).

Institutional Review Board Statement: Not applicable.

Informed Consent Statement: Not applicable.

Data Availability Statement: The original contributions presented in the study are included in the article; further inquiries can be directed to the corresponding author.

Conflicts of Interest: The authors declare no conflicts of interest.

References

1. Huo, C.; Liu, T.; Fan, B.; Zhao, Y.; Wang, D.; Zhang, J.; Zheng, C. Study on national coal resources exploration and exploitation layout under carbon neutrality and emission peak settings. *Geol. Rev.* **2022**, *68*, 938–944.
2. National Bureau of Statistics of China. 2024 National Economic and Social Development Statistical Bulletin. Available online: https://www.stats.gov.cn/xxgk/sjfb/zxfb2020/202502/t20250228_1958817.html (accessed on 28 February 2025).
3. Xie, H.; Ren, S.; Xie, Y.; Jiao, X. Development opportunities of the coal industry towards the goal of carbon neutrality. *J. China Coal Soc.* **2021**, *46*, 2197–2211.
4. Wang, S.; Shen, Y.; Song, S.; Liu, L.; Gu, L.; Wei, J. Change of coal energy status and green and low-carbon development under the "dual carbon" goal. *J. China Coal Soc.* **2023**, *48*, 2599–2612.
5. Peng, S.; Bi, Y. Strategic consideration and core technology about environmental ecological restoration in coal mine areas in the Yellow River basin of China. *J. China Coal Soc.* **2020**, *45*, 1211–1221.
6. Wu, Q. One line and one game of chess boost the high-quality development of the Yellow River Basin. *Sci. Technol. Ind. China* **2023**, *4*, 1–2.
7. Wang, S.; Wei, J.; Song, S.; Wang, S.; Sun, T. Influence of thick sandstone on development of overburden mining fissures in northern Shaanxi coal mining area of Yellow River Basin and suggestions on water-preserved coal mining. *Coal Geol. Explor.* **2022**, *50*, 1–11.
8. Song, S.; Sun, T.; Zheng, B.; Niu, R.; Ruan, H.; Cheng, X. Effect of coal mining subsidence on loess slope morphology and soil erosion in loess gully region of Northern Shaanxi. *Coal Sci. Technol.* **2023**, *51*, 422–435.
9. Xu, X.; Sun, X. River Basin Green Quality Development and Its Environmental Law Guarantee—Reflections from the "Yellow River Basin Development Conference". *J. Henan Univ. Soc. Sci. Ed.* **2022**, *62*, 35–41+153.
10. Outline of the Yellow River Basin Ecological Protection and High-Quality Development Plan. Available online: http://www.gov.cn/gongbao/content/2021/content_5647346.htm (accessed on 17 February 2025).
11. Song, S.; Zhao, X.; Wang, S.; Zhang, Y. Analysis and numerical simulation on the influence of the overlying strata's rock-soil ratio on the mining subsidence. *China Coal* **2015**, *41*, 63–67.
12. Wang, X.; Zhu, Y.; Huang, X. Field tests on deformation property of self-weight collapsible loess with large thickness. *Int. J. Geomech.* **2014**, *14*, 04014001. [CrossRef]
13. Lian, X.; Zhang, Y.; Yuan, H.; Wang, C.; Guo, J.; Liu, J. Law of movement of discontinuous deformation of strata and ground with a thick loess layer and thin bedrock in long wall mining. *Appl. Sci.* **2020**, *10*, 2874. [CrossRef]
14. Drewnik, M.; Skiba, M.; Szymański, W.; Żyła, M. Mineral composition vs. soil forming processes in loess soils—A case study from Kraków (Southern Poland). *Catena* **2014**, *119*, 166–173. [CrossRef]
15. Tang, F.; Qiao, D.; Zhang, J. Influence of Collapsible Loess Layer on Mining Subsidence in Loess Covered Mining Area. *Coal Eng.* **2015**, *47*, 88–90+94.
16. Yao, Y.; Zhang, Y.; Gao, X.; Huang, H.; Liu, D.; Hui, X. Study on permeability and collapsibility characteristics of sandy loess in northern Loess Plateau, China. *J. Hydrol.* **2021**, *603*, 126883. [CrossRef]

17. Wang, L.; Li, X.A.; Li, L.C.; Hong, B.; Yao, W.; Lei, H.N.; Zhang, C. Characterization of the collapsible mechanisms of Malan loess on the Chinese Loess Plateau and their effects on eroded loess landforms. *Hum. Ecol. Risk Assess.* **2020**, *26*, 2541–2566. [CrossRef]
18. Li, P.; Vanapalli, S.; Li, T. Review of collapse triggering mechanism of collapsible soils due to wetting. *J. Rock Mech. Geotech. Eng.* **2016**, *8*, 256–274. [CrossRef]
19. Smalley, I.J.; Marković, S.B. Loessification and hydroconsolidation: There is a connection. *Catena* **2014**, *117*, 94–99. [CrossRef]
20. Song, S.; Feng, Z.; Sun, T.; Zheng, B.; Wei, J. Loess slope deformation and soil erosion effect in coal mining subsidence area of northern Shaanxi. *J. Xi'an Univ. Sci. Technol.* **2023**, *43*, 301–311.
21. Liu, N.; Zhao, X.; Zhou, W.; Yuan, Z. Effect of underground coal mining on slope morphology and erosion. *Coal Eng.* **2020**, *52*, 122–126.
22. Fan, Z.; Guo, Y.; Yang, K. Routinized Studying Mode on Land Surface Movement and Deformation Law of Mining Subsidence in Coal Mine. *Coal Sci. Technol.* **2014**, *42*, 252–255.
23. He, L.; Wu, D.; Ma, L. Numerical Simulation and Verification of Goaf Morphology Evolution and Surface Subsidence in a Mine. *Eng. Fail. Anal.* **2023**, *144*, 106918. [CrossRef]
24. An, C.; Long, J.; Hu, H.; Li, X.; Li, M.; Li, J. Study on Influence of Coal Mining under Ancient Building on Deformation Disturbance of Overlying Accumulation Layer on Slope. *Coal Sci. Technol.* **2024**, *52*, 169–180.
25. Wang, B.; Steiner, J.; Zheng, F.; Gowda, P. Impact of rainfall pattern on interrill erosion process. *Earth Surf. Process. Landf.* **2017**, *42*, 1833–1846. [CrossRef]
26. Babur, E.; Uslu, Ö.S.; Battaglia, M.L.; Diatta, A.; Fahad, S.; Datta, R.; Zafar-ul-Hye, M.; Hussain, G.S.; Danish, S. Studying soil erosion by evaluating changes in physico-chemical properties of soils under different land-use types. *J. Saudi Soc. Agric. Sci.* **2021**, *20*, 190–197. [CrossRef]
27. Zhou, P.; Luukkanen, O.; Tokola, T.; Nieminen, J. Effect of vegetation cover on soil erosion in a mountainous watershed. *Catena* **2008**, *75*, 319–325. [CrossRef]
28. Diyabalanage, S.; Samarakoon, K.K.; Adikari, S.B.; Hewawasam, T. Impact of soil and water conservation measures on soil erosion rate and sediment yields in a tropical watershed in the Central Highlands of Sri Lanka. *Appl. Geogr.* **2017**, *79*, 103–114. [CrossRef]
29. Wang, S.; Wei, J.; Song, S.; Sun, Q.; Yang, T. Experiment and numerical simulation of overburden and surface damage law in shallow coal seam mining under the gully. *Bull. Eng. Geol. Environ.* **2022**, *81*, 207.
30. Wang, S.; Du, L.; Song, S. Influence of mining ground fissures on soil erodibility in Northern Shaanxi coal mining area of Yellow River Basin. *J. China Coal Soc.* **2021**, *46*, 3027–3038.
31. Zhao, C.; Jin, D.; Li, Z.; Shen, X.; Wang, H.; Wang, S.; Xu, F.; Wang, Q. Analysis of overlying aquifer water inrush above mining seam in Yushen mining area. *J. China Coal Soc.* **2021**, *46*, 523–533.
32. Lokhande, R.D.; Murthy, V.M.S.R.; Singh, K.B.; Verma, C.P.; Verma, A.K. Numerical Modeling of Pot-Hole Subsidence Due to Shallow Underground Coal Mining in Structurally Disturbed Ground. *J. Inst. Eng. India Ser. D* **2018**, *99*, 93–101. [CrossRef]
33. Behera, B.; Yadav, A.; Singh, P.S.G.; Sharma, S.K. Numerical Modeling Study of the Geo-mechanical Response of Strata in Longwall Operations with Particular Reference to Indian Geo-mining Conditions. *Rock Mech. Rock Eng.* **2019**, *53*, 1827–1856. [CrossRef]
34. Zheng, B.; Song, S.; Cheng, X.; Niu, R.; Cheng, X.; Ruan, H.; Li, G. The influence of coal mining subsidence on the movement and deformation of loess slope in the loess gully area of Northern Shaanxi. *Front. Earth Sci.* **2023**, *11*, 1273389. [CrossRef]
35. Song, S.J.; Zhang, Y.J.; Zhang, Y.; Du, H.D.; Liu, N. *Influence of Coal Mining Subsidence on Surface Slope Morphology and Its Soil and Water Loss Effect in Typical Coal Mining Areas in the Middle Reaches of the Yellow River*; Geological Publishing House: Beijing, China, 2020.
36. Liu, B.; Guo, S.; Li, Z.; Xie, Y.; Zhang, K.; Liu, X. Sampling survey of water erosion in China. *Soil Water Conserv. China* **2013**, *10*, 26–34.
37. Liu, B.Y.; Zhang, K.L.; Xie, Y. An empirical soil loss equation. In *Proceedings of ISCO Pot*; Tsinghua Press: Beijing, China, 2022; pp. 139–143.
38. Zhang, H.; Zhang, R.; Qi, F.; Liu, X.; Niu, Y.; Fan, Z.; Zhang, Q.; Li, J.; Yuan, L.; Yang, S.; et al. The CSLE Model Based Soil Erosion Prediction: Comparisons of Sampling Density and Extrapolation Method at the County Level. *Catena* **2018**, *165*, 465–472. [CrossRef]
39. Budeba, M.D.; Joubert, J.W.; Webber-Youngman, R.; Shafiee, S. Predicting the Efficiency of a Surface Coal Mine for Competitiveness. *Int. J. Min. Reclam. Environ.* **2017**, *31*, 187–204. [CrossRef]
40. Wang, X.; Wang, L.; Ma, N. Temporal and Spatial Dynamic Analysis of Soil Erosion in Shendong Mining Area. *Yellow River* **2022**, *44*, 124–127.
41. Hu, L.; Su, J.; Sang, Y.Z.; Zhang, W.J.; Wang, Q. Spatial and temporal characteristics of rainfall erosivity in Shaanxi Province. *Arid Land Geogr.* **2014**, *37*, 1101–1107.

42. Wischmeier, W.H.; Smith, D.D. *Predicting Rainfall-Erosion Losses from Cropland East of the Rocky Mountains: Guide for Selection of Practices for Soil and Water Conservation*; Agricultural Research Service, US Department of Agriculture: Washington, DC, USA, 1965.
43. Song, S.; Wang, S.; Sun, T. Variation of soil erodibility on loess slope under various subsidence years in coal mining subsidence area located in Northern Shaanxi. *Coal Sci. Technol.* **2022**, *50*, 289–299.
44. Desmet, P.J.J.; Govers, G. A GIS procedure for automatically calculating the USLE LS factor on topographically complex landscape units. *J. Soil Water Conserv.* **1996**, *51*, 427–433.
45. McCool, D.K.; Foster, G.R.; Weesies, G.A. Slope length and steepness factors (LS). In *Predicting Soil Erosion by Water: A Guide to Conservation Planning with the Revised Universal Soil Loss Equation (RUSLE)*; United States Government Printing: Washington, DC, USA, 1997; pp. 101–141.
46. Liu, B.Y.; Nearing, M.A.; Risse, L.M. Slope gradient effects on soil loss for steep slopes. *Trans. ASAE* **1994**, *37*, 1835–1840. [CrossRef]
47. Cai, C.; Ding, S.; Shi, Z.; Huang, L.; Zhang, G. Study of applying USLE and geographical information system IDRISI to predict soil erosion in small watershed. *J. Soil Water Conserv.* **2000**, *2000*, 19–24.
48. Wu, F.; Zhao, X.; Liu, B. *Analysis of Erosion Environment and Dynamic Mechanism of Gentle Slope Cultivated Land*; Shaanxi Science Press: Xi'an, China, 2000.
49. Kokutse, N.K.; Temgoua, A.G.T.; Kavazović, Z. Slope stability and vegetation: Conceptual and numerical investigation of mechanical effects. *Ecol. Eng.* **2016**, *86*, 146–153. [CrossRef]
50. Zhang, M.; Sun, C.; Xiao, P. A demonstration project for detailed geo-hazard survey in the Baota district, Yan'an. *Northwest Geol* **2007**, *40*, 29–55.
51. Zhu, B.; Wang, Y.; Wang, J. The variation law of stability factor of loess slopes with different slopes. *Soil Water Conserv. China* **2020**, *08*, 42–43.
52. Katz, O.; Morgan, J.K.; Aharonov, E.; Dugan, B. Controls on the size and geometry of landslides: Insights from discrete element numerical simulations. *Geomorphology* **2014**, *220*, 104–113. [CrossRef]
53. Qiu, H.; Cui, P.; Regmi, A.D.; Hu, S.; Wang, X.; Zhang, Y. The effects of slope length and slope gradient on the size distributions of loess slides: Field observations and simulations. *Geomorphology* **2018**, *300*, 69–76. [CrossRef]
54. Xi, B.; Gao, Y.; Yang, B.; Li, L.; Kong, W. Analysis of terrain change characteristics and disaster causing points under mining disturbance in western mountainous area. *Metal Mine* **2021**, *7*, 172–178.
55. Huang, Y.; Wang, Y.; Wang, M.; Tian, F.; Ao, J. Effect of mining subsidence on soil erosion in mountainous area of the Loess Plateau. *Trans. Chin. Soc. Agric. Eng.* **2014**, *30*, 228–235.
56. Tang, W.; Ma, S.; Liu, X.; Zhao, X. Influence of topographic and geomorphic conditions on the dynamic response of slope acceleration. *J. Yangtze River Sci. Res. Inst.* **2019**, *36*, 98.
57. Huang, Z.W. Stability Analysis of Concave and Convex Slopes Under Earthquake Action. Master's Thesis, Chongqing University, Chongqing, China, 2017.
58. Sharma, R.H. Evaluating the effect of slope curvature on slope stability by a numerical analysis. *Aust. J. Earth Sci.* **2013**, *60*, 283–290. [CrossRef]
59. Gao, J. Identification of topographic settings conducive to landsliding from DEM in Nelson County, Virginia, USA. *Earth Surf. Process. Landf.* **1993**, *18*, 579–591. [CrossRef]
60. Sun, C.-J.; Lin, R.-J.; Zheng, Z.-J.; Wang, J.-R.; Sun, J.-L. Characteristic analysis of soil and water loss in typical small watersheds of the Middle Yellow River based on RUSLE model. *Southwest China J. Agric. Sci.* **2022**, *35*, 200–208.
61. Farhan, Y.; Zreqat, D.; Nawaysa, S. Assessing the influence of physical factors on spatial soil erosion risk in northern Jordan. *J. Am. Sci.* **2014**, *10*, 9–29.
62. Xu, Z.; Quan, X.; Shi, H.; He, J.; Cai, Q.; Sun, L. Erosion process and spatial distribution characteristics of erosion-deposition on the loess slope. *J. Shaanxi Norm. Univ. Nat. Sci. Ed.* **2021**, *49*, 98–105.
63. Huang, J.; Jin, P.; Jiang, X.; Lin, L.; Kou, X.; Xu, Z.; Lin, B.; Fang, Z. Model construction and verification of vegetation cover and management factor in southern red soil region of China. *Trans. Chin. Soc. Agric. Eng.* **2020**, *36*, 106–114.
64. Wang, M.; Jin, H.; Gao, Z.; He, Y.; Zhang, W.; Song, X.; Zhang, R. Analysis of the influencing factors of soil erosion in Qilong Bay. *Water Saving Irrig.* **2023**, *8*, 17–25.
65. Liu, R.; Yu, X.; Cai, Q.; Sun, L.; Fang, H.; Jia, G.; He, J. Erosion process of loess slope and influencing factors in the loess hilly-gully region, China. *Chin. J. Appl. Ecol.* **2021**, *32*, 2886–2894.
66. Liu, D. Benefits of runoff and sediment reductions of different soil and water conservation measures in Mentougou area of Beijing City. *Bull. Soil. Water Conserv.* **2015**, *35*, 107–110.
67. Kinnell, P.I.A. A review of the design and operation of runoff and soil loss plots. *Catena* **2016**, *145*, 257–265. [CrossRef]
68. Gao, R.; Gao, H.; Song, J.; Li, H. Influence of slope length on soil erosion in yellow sloping farmland of karst. *Res. Soil Water Conserv.* **2018**, *25*, 53–57.

69. Gu, Z.; Xie, Y.; Li, A.; Liu, G.; Shi, Y. Assessment of soil erosion in rolling hilly region of Northeast China using Chinese Soil Loss Equation (CSLE) model. *Trans. Chin. Soc. Agric. Eng.* **2020**, *36*, 49–56.

minerals

MDPI

Article

Enhanced Flame Retardancy of Silica Fume-Based Geopolymer Composite Coatings Through In Situ-Formed Boron Phosphate from Doped Zinc Phytate and Boric Acid

Yachao Wang [1,2,3,*], Yufei Qu [3], Chuanzhen Wang [2] and Juan Dou [1]

1 State Key Laboratory of Green and Low-Carbon Development of Tar-Rich Coal in Western China, Xi'an University of Science and Technology, Xi'an 710054, China
2 Anhui Engineering Research Center for Coal Clean Processing and Carbon Reduction, Huainan 232001, China
3 School of Resources Engineering, Xi'an University of Architecture & Technology, Xi'an 710055, China
* Correspondence: wangyachao@xauat.edu.cn

Abstract

Silica fume-based geopolymer composite coatings, an approach to using metallurgical solid waste, exert flame retardancy with ecological, halogen-free, and environmentally friendly advantages, but their fire resistance needs to be improved further. Herein, a silica fume-based geopolymer composite flame-retardant coating was designed by doping boric acid (BA), zinc phytate (ZnPA), and melamine (MEL). The results of a cone calorimeter demonstrated that appropriate ZnPA and BA significantly enhanced its flame retardancy, evidenced by the peak heat release rate (p-HRR) decreasing from 268.78 to 118.72 $kW \cdot m^{-2}$, the fire performance index (FPI) increasing from 0.59 to 2.83 $s \cdot m^2 \cdot kW^{-1}$, and the flame retardancy index increasing from 1.00 to 8.48, respectively. Meanwhile, the in situ-formed boron phosphate (BPO_4) facilitated the residual resilience of the fire-barrier layer. Furthermore, the pyrolysis kinetics indicated that the three-level chemical reactions governed the pyrolysis of the coatings. BPO_4 made the pyrolysis E_α climb from 94.28 (P5) to 127.08 (B3) $kJ \cdot mol^{-1}$ with temperatures of 731–940 °C, corresponding to improved thermal stability. Consequently, this study explored the synergistic flame-retardant mechanism of silica fume-based geopolymer coatings doped with ZnPA, BA, and MEL, providing an efficient strategy for the high-value-added recycling utilization of silica fume.

Keywords: flame retardancy; zinc phytate; boric acid; silica fume; boron phosphate

1. Introduction

Wood is a biological material that has been widely used in the fields of decoration, architecture, and furniture because of its renewable, environmental, and abundant resources [1,2]. However, wood is highly flammable [3], and its extensive use in daily life inevitably holds fire risk. The most direct and effective method to mitigate this risk is by brushing flame-retardant coatings on its surface [4], which minimizes the fire risk of flammable materials, benefits personnel evacuation and rescue, and reduces fire losses. Building coatings with environmental protection, ecology, low toxicity, and efficient fire resistance have become an important direction for the research and development of flame-retardant coatings.

Recently, inorganic geopolymer has emerged as a promising candidate for flame-retardant coatings due to its durability, halogen-freeness, environmental friendliness, and low-cost advantages [5,6]. Silica fume is a metallurgical by-product [7] with high emissions but low utilization efficiency that has been used as a flame-retardant coating. Shahidi et al. [8] investigated the flame retardancy of an intumescent flame retardant enhanced by silica fume and graphene oxide/talc. However, silica fume-based geopolymer coatings exhibit certain limitations in practical applications, such as insufficient flame retardancy [9] and low residual structural strength [10], failing to protect plywood effectively. Therefore, it is necessary to optimize its flame retardancy through a synergistic flame-retardant mechanism.

Currently, one of the most commonly used flame retardants is melamine (MEL) due to its lower cost and non-toxicity during combustion, without secondary pollution. During combustion, the MEL takes away heat and releases N_2, NH_3, and other non-flammable gases, diluting the concentration of O_2 and toxic gases. However, the flame retardancy of composite coating is not effectively improved by adding MEL alone. In recent years, numerous flame-retardant systems consisting of a variety of materials have been developed to seek synergistic flame-retardant effects [11,12]. For instance, combinations such as phosphorus/boron [13] and phosphorus/boron/nitrogen [14] have been demonstrated to achieve synergistic flame retardancy. Zhang et al. [15] improved the thermal stability and flame retardancy of furfurylated wood by introducing a multifunctional catalyst system comprising BA and ammonium dihydrogen phosphate. Huo et al. [16] incorporated hyperbranched oligomer-containing phosphorus/nitrogen/boron (BDHDP) in epoxy resin (EP), achieving a UL-94 V-0 level with 1.5 wt.% doped BDHDP. Consequently, it is a good strategy to seek a multi-element synergistic effect for improving the flame retardancy of geopolymer coatings.

Zinc phytate (ZnPA) is formed by chelating phytic acid and zinc ions, serving as an eco-friendly and efficient flame retardant due to its high phosphorus content [17]. Phosphorus-containing flame retardants facilitate catalytic charring and combustion prevention and generate phosphorus radicals capable of capturing active radicals, thereby quenching the combustion chain reaction [18]. Zhang et al. [19] used Zn^{2+} ions in ZnPA to chelate with components such as dopamine, DOPO, and fly ash, preparing an organic–inorganic hybrid flame-retardant coating. In this sense, ZnPA could form a zinc ion chelate cross-linked structure with the silica fume-based geopolymer to enhance the stability and continuity of the residues during fire.

Additionally, boric acid (BA) is another flame retardant widely used across various industries; it acts as an effective flame-retardant by forming a protective B_2O_3 glassy layer during combustion, which shields material from oxygen and heat [20]. Savas et al. [21] investigated the impact of zinc borate (ZnB) on the thermal and flame retardancy of polyamide 6 (PA6) composites containing aluminum hypophosphite (AlHP), where the as-formed BPO_4 enhanced the flame retardancy. Therefore, the synergism between BA and ZnPA could further form a zinc-chelated silicon phosphorus boron composite structure within the silica fume-based geopolymers during combustion in theory, thereby enhancing the flame retardancy.

Consequently, this research designed flame-retardant coatings integrating silica fume-based geopolymer with ZnPA/BA/MEL, characterized by X-ray diffraction (XRD), a cone calorimeter (CC), and scanning electron microscopy (SEM), respectively. It demonstrates that in situ-formed BPO_4 significantly enhances the residue strength and improves the flame retardancy of the condensed phase. Generally, the novelty of this thesis is in proposing a recycling strategy for high-value-added utilization of silica fume, probing the synergistic flame-retardant mechanism of ZnPA/BA/MEL-containing silica fume-based geopolymer coatings, and exploring the design of a halogen-free Si-P-N-C multi-element composite flame retardant system.

2. Experiment and Methods

2.1. Raw Materials

The silica fume was a gray powder with a SiO_2 content > 86 wt.% (mass percentage concentration), a density of 1.62 g/cm^3, and a Blaine-specific surface area of 25 m^2/g, purchased from Xi'an LinYuan company (Xi'an, China). ZnPA was procured from Guangzhou Jiale Chemical company (Guangzhou, China). Melamine (MEL) was obtained from Wuxi Yatai United Chemical company (Yantai, China). BA was supplied by Zhengzhou Del Boron Chemical company (Zhengzhou, China). Analytically pure polyacrylamide (PAM) was purchased from Tianjin Fuchen chemical reagent factory (Tianjin, China). Analytical pure sodium silicate ($Na_2SiO_3{\cdot}9H_2O$) was purchased from Sinopharm chemical reagent company (Jinan, China). Potassium hydroxide (KOH) was purchased from Hongyan chemical reagent company (Tianjin, China). The silane coupling agent (KH-550, CAS no.: 919-30-2) was provided by Shandong Yousuo chemical technology company (Jinan, China). Polydimethylsiloxane (PDMS) was purchased from Dongguan Tianyu chemical company (Dongguan, China). The plywood was supplied by the timber processing plant of Xi'an in China with secondary flame retardancy.

2.2. Preparation of Geopolymer Composite Coating

Figure 1 illustrates the preparation steps of silica fume-based geopolymer composite coatings via the multi-step sol–gel method. Firstly, 0.5 g KH-550, 1 g MEL, and varying dosages of ZnPA were dissolved in 15 g of water, according to our previous research [22], as shown in Table 1. Magnetic stirring at 700 $r{\cdot}min^{-1}$ at 60 °C was applied for pretreating ZnPA for approximately 30 min.

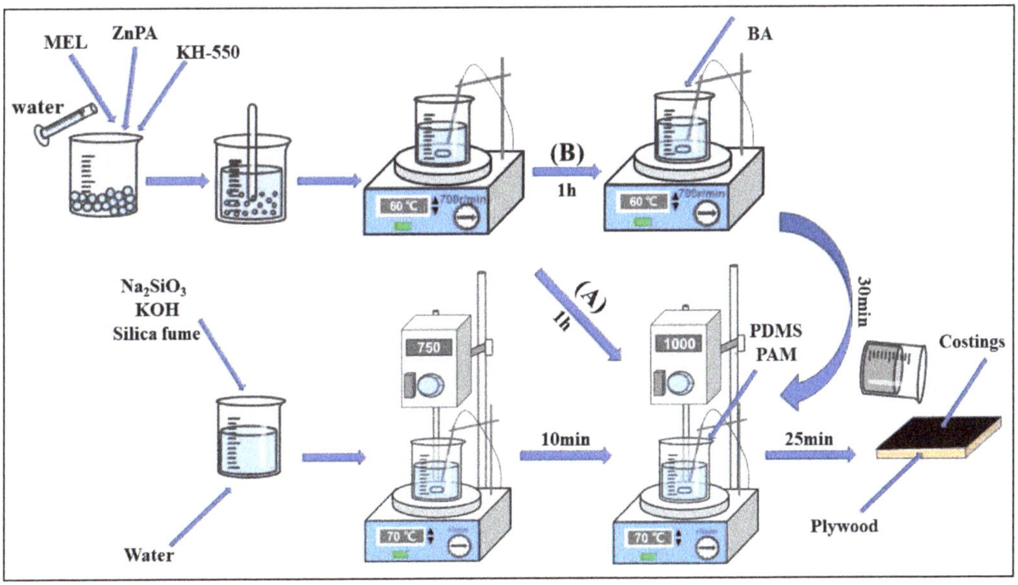

Figure 1. The preparation diagrams of the geopolymer composite coatings.

Secondly, 14.21 g $Na_2SiO_3{\cdot}9H_2O$, 5.61 g KOH, 30 g silica fume, and 30 g water were poured into another glass beaker [22]. The silica sol mixture was obtained by 10 min of stirring at 70 °C with 750 $r{\cdot}min^{-1}$. Then, a pretreated ZnPA solution was injected into the silica sol and mixing was continued for another 20 min, with the speed increased to 1000 $r{\cdot}min^{-1}$.

Table 1. The composition of coating samples.

Samples	$Na_2SiO_3·9H_2O$/g	KOH/g	Silica Fume/g	KH-550/g	MEL/g	PAM/g	H_2O/g	PDMS/g	ZnPA/g	BA/g
P0	14.21	5.61	30	0.5	1	0.2	45	0.25	0.48	0.0
P1	14.21	5.61	30	0.5	1	0.2	45	0.25	0.95	0.0
P2	14.21	5.61	30	0.5	1	0.2	45	0.25	1.43	0.0
P3	14.21	5.61	30	0.5	1	0.2	45	0.25	1.9	0.0
P4	14.21	5.61	30	0.5	1	0.2	45	0.25	2.38	0.0
P5	14.21	5.61	30	0.5	1	0.2	45	0.25	2.85	0.0
P6	14.21	5.61	30	0.5	1	0.2	45	0.25	2.38	0.0
B1	14.21	5.61	30	0.5	1	0.2	45	0.25	2.38	0.5
B2	14.21	5.61	30	0.5	1	0.2	45	0.25	2.38	1.0
B3	14.21	5.61	30	0.5	1	0.2	45	0.25	2.38	1.5
B4	14.21	5.61	30	0.5	1	0.2	45	0.25	2.38	2.0
B5	14.21	5.61	30	0.5	1	0.2	45	0.25	2.38	2.5
B6	14.21	5.61	30	0.5	1	0.2	45	0.25	2.38	3.0

Finally, PAM as a thickener and PDMS as an antifoaming agent were doped into ZnPA/MEL-containing silica fume-based geopolymer coatings by stirring for 5 min. The dosages of ZnPA were 0 g (0.0 wt.%), 0.48 g (0.5 wt.%), 0.95 g (1.0 wt.%), 1.43 g (1.5 wt.%), 1.9 g (2.0 wt.%), 2.38 g (2.5 wt.%), and 2.85 g (3.0 wt.%). Then, the as-formed composite coating was covered evenly on the surface of the plywood (100 × 100 × 5 mm^3) 3 times with an interval of about 20 min.

Based on the results of the ZnPA-containing geopolymer coatings, further addition of BA was carried out to prepare BA/ZnPA/MEL-containing silica fume-based geopolymer composite coatings. The preparation process was identical to that of the ZnPA-containing coating, as shown in Figure 1B. The dosages of BA were 0.5 g (0.5 wt.%), 1.0 g (1.0 wt.%), 1.5 g (1.50 wt.%), 2 g (2.0 wt.%), 2.5 g (2.5 wt.%), and 3.0 g (3.00 wt.%). The samples with a ZnPA dosage of 0.5 wt.%–3 wt.% were numbered as P1–P6. BA dosages of 0.5 wt.%–3 wt.% were referred to as B1-B6, respectively, and the coating without ZnPA or BA was denoted as the control P0.

2.3. *Characterizations*

2.3.1. Flame Retardancy Testing

The combustion performance of the sample was assessed by a cone calorimeter (CC, ZY6243, Zhongnuo Instruments Company, Dongguan, China) according to ISO 5660-1:2015 [23]. The irradiative heat flux of CC was 50 $kW·m^{-2}$ (approximately 715 °C), with a distance between the coating and the ignition needle of 25 mm. The following parameters were recorded in real time by CC, such as time to ignite (TTI, s), heat release rate (HRR, $kW·m^{-2}$), peak heat release rate (p-HRR, $kW·m^{-2}$), time to p-HRR (T_P, s), total heat release (THR, $MJ·m^{-2}$), and weight loss (WL, g). Meanwhile, the following four parameters were used to assess the flame retardancy of the sample. The fire growth index (FGI) reflected the potential growth and intensity of a fire, which was calculated by Formula (1). The fire performance index (FPI) evaluated the fire performance of the materials, which was calculated by Formula (2). The average effective heat of combustion (AEHC) was the ratio of THR to WL. The flame retardancy index (FRI) was used to quantify the flame resistance of the materials, which was calculated by Formula (3). It was divided into three grades: "FRI < 1", "1 < FRI < 10", and "10 < FRI < 100", which represented "poor", "good", and "excellent" flame retardancy, respectively [22].

$$FGI = \frac{p - HRR}{T_p} \tag{1}$$

$$FPI = \frac{TTI}{p - HRR} \tag{2}$$

$$FRI = \frac{THR * (\frac{p-HRR}{TTI})_{Control}}{THR * (\frac{p-HRR}{TTI})_{Composite}} \quad (3)$$

2.3.2. Microstructure Testing

Infrared spectra of the samples before combustion were captured in the spectral range of 4000–500 cm^{-1} by an infrared spectrometer Nicolet iS50 (Thermo, Singapore). The composition of the residual layer was analyzed by an X-ray diffractometer (D/MAX-2400) with Cu K_α (λ = 1.54056 Å) radiation at 40 kV and 40 mA within the range of 2θ = 10~50°, with a scanning step of 2°, which was preferentially ground into powder. The morphological structures of the barrier layer were observed using a scanning electron microscope (Gemini 500). Each sample needed to be sprayed with gold before scanning electron microscopy to improve the conductivity of the sample. Thermogravimetric analysis (TGA) of samples was conducted by the Mettler (Germany) thermal–gravimetric analyzer, from 30 to 1000 °C at a rate of 20 °C·min^{-1} in a N_2 atmosphere.

3. Results

3.1. Flame Retardancy of Samples

Figure 2a shows the HRR curve for the ZnPA-containing samples, which gradually shifted to the right with the increasing content of ZnPA. When the ZnPA dosage was 2.38 g, p-HRR exerted the lowest value, decreasing from 268.78 (P0) to 156.35 kW·m^{-2} (P5), and T_P was delayed to 404 s, indicating that an appropriate ZnPA content imparted greater flame retardancy. However, doping in excess of 2.85 g ZnPA led to a rise in p-HRR and the early emerged peak at 144 s because the excess ZnPA accelerated the coating catalytic decomposition and uneven dispersion in the coatings.

Figure 2a′ displays the HRR curve for the BA-containing samples. The HRR curve decreased at first and then increased. Compared with P5, B3 exhibited a lower p-HRR value of 118.73 kW·m^{-2} and T_P was delayed from 404 s to 440 s. When the BA content exceeded 1.50 wt.%, p-HRR gradually rose, indicating that the excess BA led to a decrease in the flame retardancy due to the uneven dispersion of the doped BA in the coatings.

Figure 2b,b′ show the smoke temperature of the samples, which gradually decreased with the increasing ZnPA content, as shown in Figure 2b. Among these, P5 presented the lowest smoke temperature, reaching a peak value of 88.66 °C at 438 s. This indicated that the doped ZnPA gave enhanced flame retardancy to the composite coating. Regarding the 3.00 wt.% ZnPA-containing sample, the smoke temperature rose to 92.39 °C, corresponding to poor flame retardancy. As shown in Figure 2b′, B3 displayed the lowest smoke temperature at 494 s, decreasing from 88.66 (P5) to 84.68 °C. This was attributed to BA dehydrating during combustion, which had an endothermic and cooling effect. However, samples with a BA content exceeding 1.50 wt.% exhibited a higher smoke temperature in comparison to that of B3. The smoke temperature increased with the increasing BA content, which was completely consistent with the results of HRR.

Figure 2c,c′ present the residual weight value of the samples under an externally constant heat flux of 50 kW·m^{-2}. The lowest final residual weight was observed for P0 (35.29%). As the content of ZnPA increased, the residual weight also increased. P5 displayed the maximum final residual weight of 46.09% at 600 s, indicating that the incorporated ZnPA caused the composite coating to have a reduced weight loss rate. As shown in Figure 2c′, the residual weight increased and then dropped with the increasing BA dosage. Compared to P5, B3 exhibited a lower weight loss of 50.97% at 650 s. The residual weight decreased sequentially for B4, B5, and B6, indicating that a higher BA content was detrimental to improving the flame retardancy of the composite coating.

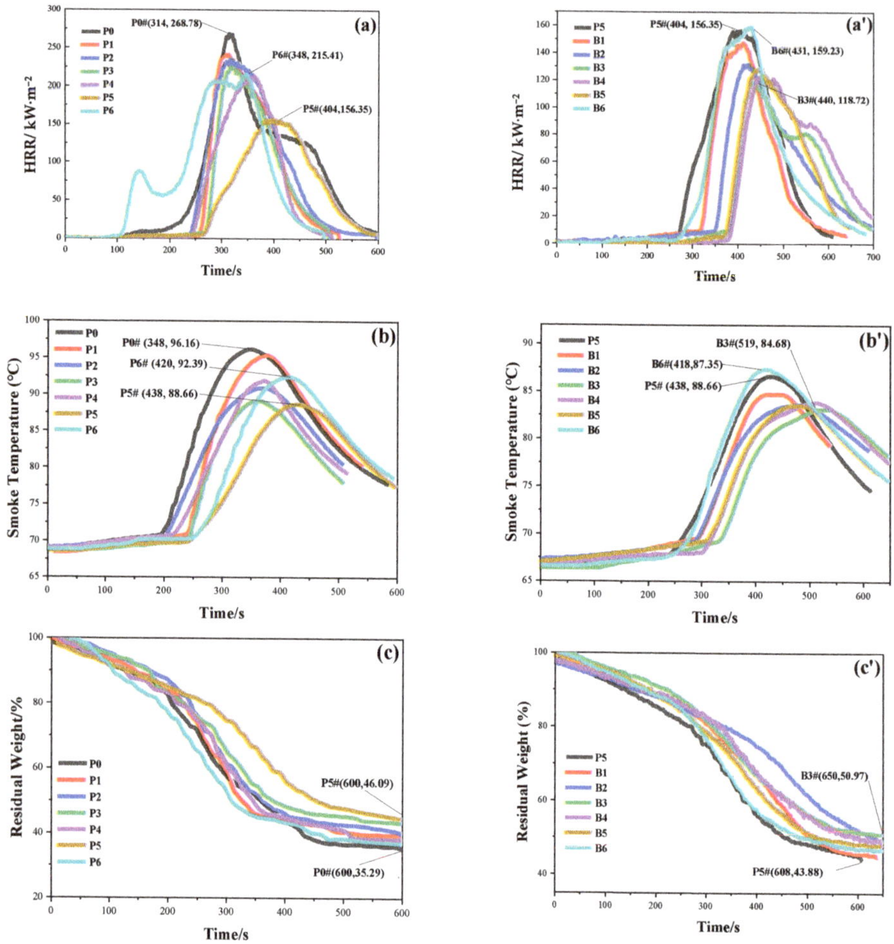

Figure 2. Flame retardancy of samples including HRR in (**a**) and (**a′**), smoke temperature in (**b**) and (**b′**), and residual weight in (**c**) and (**c′**), respectively.

Table 2 summarizes the combustion parameters of the samples. The doped ZnPA and BA extended the TTI. Among all samples, B3 exhibited the longest TTI of 336 s and the highest flame retardancy, evidenced by the highest FPI of 2.83 s·m^2·kW^{-1} and the lowest FGI of 0.27 kW·m^2·s^{-1}. FRI was 8.48, indicating a 'good' flame retardant grade compared to P0. Conversely, when the content of BA exceeded 1.50 wt.%, the parameters exhibited a contrary trend. For instance, the TTI of B5 shrank to 318 s compared to B3, the FPI declined to 2.48 s·m^2·kW^{-1}, and FGI climbed to 0.29 kW·m^{-2}·s^{-1} with the lower FRI of 7.48, corresponding to diminished flame retardancy. Therefore, the appropriate amount of doped ZnPA and BA generated synergism, leading to the formation of a dense and resilient siliceous layer, corresponding to enhanced flame retardancy.

Table 2. Combustion parameters of samples in CC.

Samples	TTI/s	T_P/s	p-HRR/kW·m^{-2}	FPI/s·m^2·kW^{-1}	FGI/kW·m^{-2}·s^{-1}	WL/g	THR/MJ·m^{-2}	AEHC/kW·kg^{-1}	FRI
P0	158	314	268.78	0.59	0.86	30.93	36.67	1.19	1.00
P1	219	308	242.31	0.90	0.79	29.16	31.27	1.07	1.80
P2	201	319	234.90	0.86	0.74	29.93	35.93	1.20	1.49

Table 2. *Cont.*

Samples	TTI/s	T_P/s	p-HRR/kW·m^{-2}	FPI/s·m^2·kW^{-1}	FGI/kW·m^{-2}·s^{-1}	WL/g	THR/MJ·m^{-2}	AEHC/kW·kg^{-1}	FRI
P3	232	321	223.93	1.04	0.70	26.04	27.92	1.07	2.31
P4	205	364	213.97	0.96	0.59	24.56	27.79	1.13	2.15
P5	228	404	156.35	1.46	0.39	27.12	27.79	1.02	3.27
P6	78	348	215.41	0.36	0.62	31.37	40.44	1.29	0.56
B1	266	414	147.49	1.80	0.36	26.84	22.42	0.84	5.02
B2	294	423	131.89	2.23	0.31	29.46	23.44	0.80	5.93
B3	336	440	118.72	2.83	0.27	24.30	20.83	0.86	8.48
B4	327	453	123.31	2.65	0.27	23.90	22.62	0.95	7.31
B5	318	443	128.29	2.48	0.29	25.01	20.68	0.83	7.48
B6	260	431	159.23	1.63	0.37	26.51	27.63	1.04	3.69

3.2. FTIR Spectra of Samples

Figure 3 presents the obtained FTIR spectra of the composite coatings before burning. The wide peak at 3398 cm^{-1} was assigned to stretching vibrations of -OH and N-H [24]. Among them, -OH was mainly derived from geopolymer and BA and N-H were mainly derived from MEL. The broad peak at 1600 cm^{-1} was assigned to the C=N absorption peak of the triazine ring [25]. The asymmetric vibrations of the Si-O-Si bond were detected at 1120 cm^{-1}, whereas the symmetric vibrations were located at 765 cm^{-1} [26], attributed to the hydrolysis of silane coupling agents and alkali-activated geopolymer sols. P-O-C symmetric vibrations occurred at 980 cm^{-1} [27]. After the addition of BA, the absorption peaks of the B3 and B6 samples at 617 cm^{-1} showed a strengthening, attributed to the deformation vibration of the atoms in the B-O bond in BA [28,29]. Generally, the spectra of samples exhibited similar curves without obvious differences, except for the peaks at 980 cm^{-1} and 617 cm^{-1}. Although the dosage of ZnPA and BA was very low, presenting tiny peaks, they could improve the flame retardancy, as evidenced by the results of the CC. The specific reasons will be explained later in Section 4.

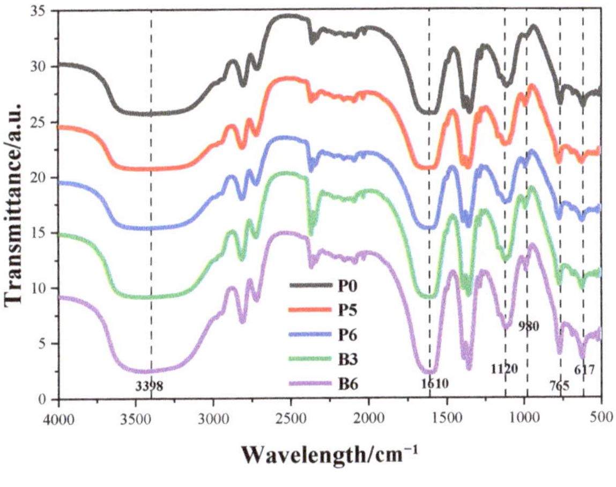

Figure 3. FTIR spectra before burning.

3.3. XRD Analysis

Figure 4 depicts the X-ray diffraction (XRD) patterns of the raw materials. Figure 5 presents the XRD of the residual silica fume-based geopolymer coatings after combustion. A broad hump observed between 15 and 35° at 2θ suggested a high content of amorphous silicates, which contributed to enhanced flame retardancy. Overlaid upon this broad hump, discrete peaks corresponded to quartz (SiO_2, PDF no. 46-1441, no. 04-0379), graphite (G, PDF no. 05-0625), zinc metaphosphate ($Zn(PO_3)_2$, PDF no. 01-0587), boric oxide (B_2O_3, PDF no. 46-1045), and boron phosphate (BPO_4, PDF no. 12-0380). Notably, the diffraction peaks at 29.12° and 40.56° were attributed to the hexagonal structure of BPO_4. However, peaks corresponding to melamine and its derivatives were not detected, possibly due to their complete decomposition at high temperatures.

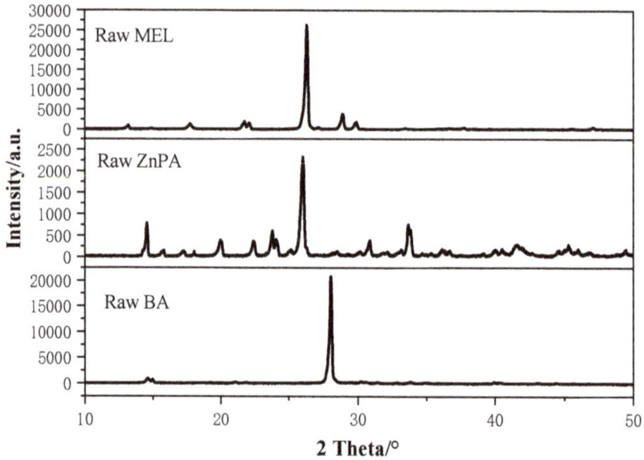

Figure 4. XRD of raw materials.

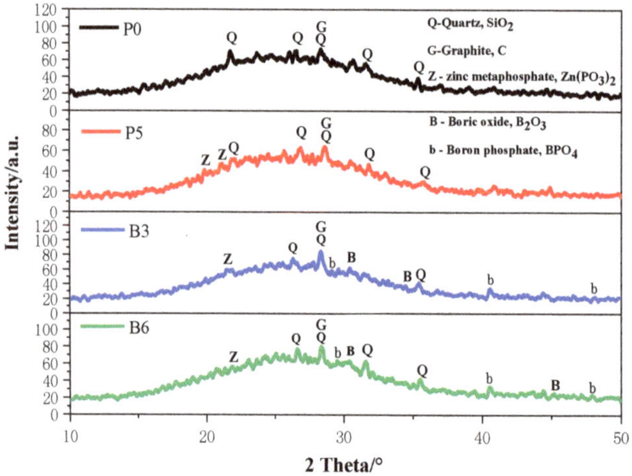

Figure 5. XRD of coating residues after burning.

Essentially, ZnPA decomposed into phosphoric acid and $Zn(PO_3)_2$ while releasing CO_2 through a dehydration process [17], as shown in reaction (4). Subsequently, the partially generated phosphoric acid dehydrated into polyphosphoric acid, which then further dehydrated into P_2O_5 [30], as shown in reactions (5) and (6). BA underwent dehydration at high temperatures, converting into B_2O_3 [31], as shown in reaction (7). The generated P_2O_5 further reacted with B_2O_3 and transformed into BPO_4 [32], as shown in reaction (8), holding a high melting point of 1200 °C and excellent stability [33]. The XRD analysis confirmed that the doped ZnPA and BA generated in situ reactions.

$$C_6H_6O_{24}P_6Zn_6 \rightarrow H_3PO_4 + Zn(PO_3)_2 + CO_2 + H_2O \tag{4}$$

$$nH_3PO_4 \rightarrow H(PO_3)_n + nH_2O \tag{5}$$

$$H(PO_3)_n \rightarrow nP_2O_5 + H_2O \quad (6)$$

$$2B(OH)_3 \rightarrow B_2O_3 + 3H_2O \quad (7)$$

$$P_2O_5 + B_2O_3 \rightarrow 2BPO_4 \quad (8)$$

3.4. Appearance

Figures 6 and 7 illustrate the appearance of the composite coatings before burning. The coating without ZnPA and BA appears slightly whitened in Figure 6a. Because MEL is an alkaline compound [34], it resulted in excessive alkali content on the coating. With the increasing dosage of ZnPA, the whitening phenomenon diminished, as shown in Figure 6b–g, and Figure 6f presents a flat and smooth surface. The excess ZnPA caused a slightly rough surface, as shown in Figure 6g, due to its uneven dispersion. After the addition of BA, the white powders were dispersed on the surface, as shown in Figure 7a–e. This phenomenon became obvious with the increasing BA dosage. When the content of BA reached 3.0 wt.%, a rough surface with white small particle aggregation appeared, as shown in Figure 7e.

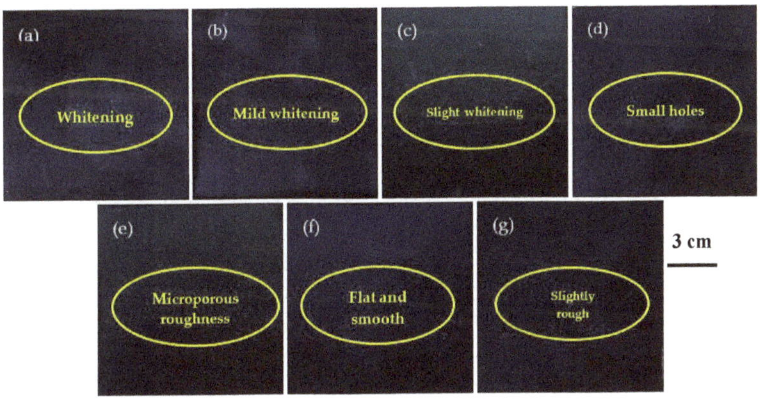

Figure 6. Appearance of samples including (**a**) P0, (**b**) P1, (**c**) P2, (**d**) P3, (**e**) P4, (**f**) P5, and (**g**) P6.

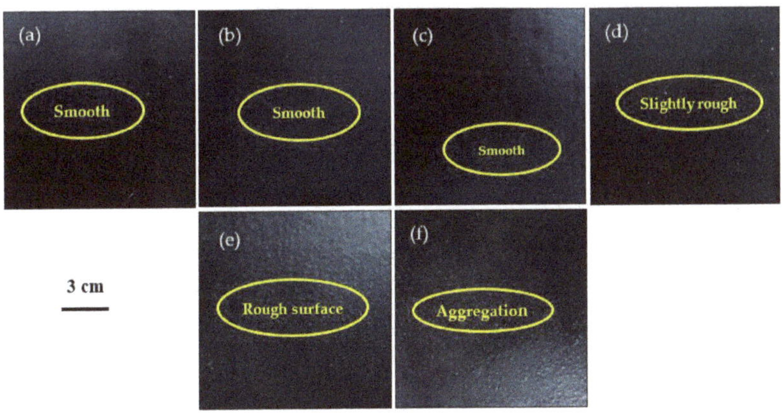

Figure 7. Appearance of samples including (**a**) B1, (**b**) B2, (**c**) B3, (**d**) B4, (**e**) B5, and (**f**) B6.

3.5. Residual Appearance of Samples

All samples formed a siliceous layer [35], as shown in Figures 8 and 9. This layer effectively inhibited flame propagation and trapped the transfer of heat and mass, thereby protecting the underlying plywood against complete combustion. As shown in Figure 8a, the siliceous layer was inherently brittle and fragile. As shown in Figure 8b–e, the coating hardly swelled with a low content of ZnPA.

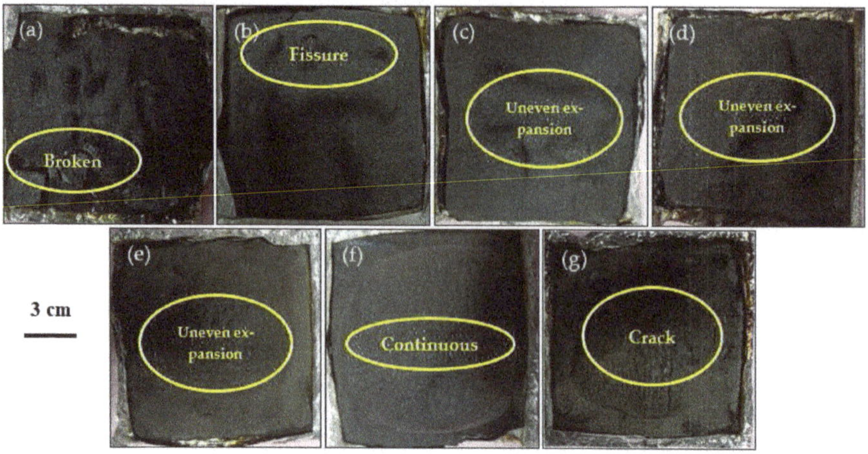

Figure 8. Residual appearance of the sample in CC including (**a**) P0, (**b**) P1, (**c**) P2, (**d**) P3, (**e**) P4, (**f**) P5, and (**g**) P6.

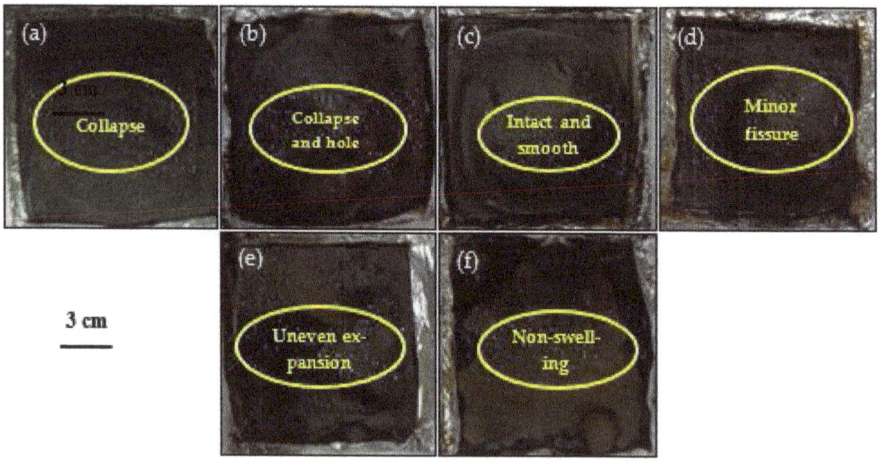

Figure 9. Residual appearance of the sample in CC including (**a**) B1, (**b**) B2, (**c**) B3, (**d**) B4, (**e**) B5, and (**f**) B6.

The surface of the layer was unevenly convex, which affected the flame retardancy of the samples and their heat release rate. A continuous swelling layer was observed with the increasing content of ZnPA. Especially for the P5 in Figure 8f, the siliceous layer was uniformly expanded, continuous, and intact, but small cracks could be observed on the surface of the layer. When the content of ZnPA reached 3.0 wt.% in Figure 8g, cracks were observed, which were averse to the formation of the intact layer.

However, the integrity of the siliceous layer gradually improved when the BA was doped into the coatings, as shown in Figure 9a,b. The unbroken, smooth, and robust homogeneous barrier layer was generated for B3 with the addition of 1.5 wt.% BA, as shown in Figure 9c, and a smooth surface was observed, indicating that a resilient swelling siliceous layer was formed. When the BA content exceeded 1.5 wt.%, an inhomogeneous barrier layer emerged, as shown in Figure 9d–f, leading to diminished flame retardancy.

3.6. SEM of Residues

Figure 10 illustrates the microscopic morphology (500× and 2000×) of the residues after the combustion of the composite coatings. Figure 10a,a′ reveal sparse fibrous barriers covering the fractured and discontinuous siliceous layer. This layer was easily penetrated by gases produced from pyrolysis and could not effectively prevent the transfer of heat and flammable gases during firing.

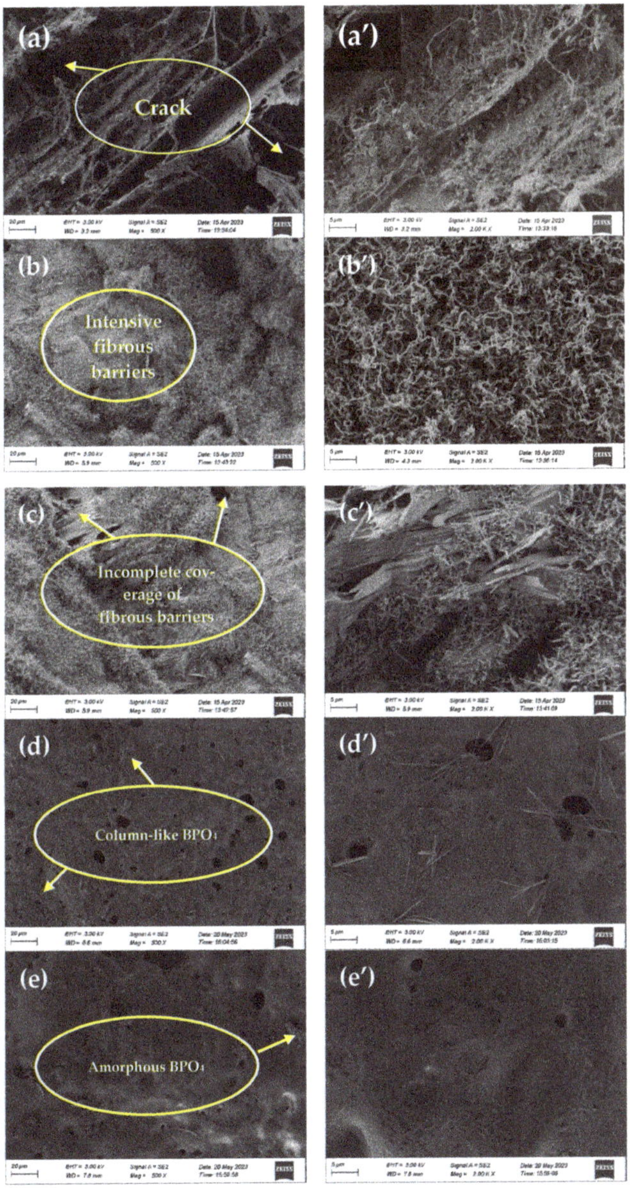

Figure 10. SEM of the surface of residues including (**a**,**a′**) P0, (**b**,**b′**) P5, (**c**,**c′**) P6, (**d**,**d′**) B3, and (**e**,**e′**) B6.

When ZnPA was doped into the coating, as shown in Figure 10b,b′, dense fibrous barriers appeared on the surface of the barrier layer, delaying flame propagation and preventing volatile gases and small liquid molecules from entering the combustion zone. When adding an excessive dosage of ZnPA, the coverage of the fibrous decreased, as shown in Figure 10c,c′, which adversely affected the flame retardancy.

Figure 10d,d′ show the formation of column-like BPO_4 on the surface of the B3 siliceous layer. During combustion, BA and ZnPA decomposed into phosphate and borate, respectively, which further reacted to form column-like BPO_4. The integration of BPO_4 with the siliceous layer enhanced the quality of the barrier layer. Notably, BPO_4 typically adopts a cristobalite-like structure [36]. According to our results, the formed BPO_4 exhibited a column-like structure, a discrepancy that may stem from varying conditions during BPO_4 crystal growth, leading to a diversity of crystal structures [37]. The column-like BPO_4 increased the surface area of the siliceous layer, hindering heat transfer and protecting the underlying plywood from combustion. Meanwhile, the porous surface of the barrier layer facilitated the volatiles' release into the surrounding air. However, an excess BA resulted in reduced BPO_4 on the surface of the siliceous layer, exhibiting an irregular and amorphous structure, leading to poor flame retardancy. Therefore, the appropriate amount of ZnPA (2.5 wt.%) and BA (1.5 wt.%) led to the in situ formation of BPO_4, enhancing the toughness of the barrier layer, presenting higher flame retardancy.

3.7. Thermal Performance Analysis

According to the thermal performance analysis, 5% weight loss temperature ($T_{d5\%}$), 10% weight loss temperature ($T_{d10\%}$), 20% weight loss temperature ($T_{d20\%}$), the initial degradation temperature (T_i), the final degradation temperature (T_f), and the corresponding maximum weight loss temperature (T_{max}) and carbon residue rate of the samples are expressed in Table 3. The comparison reveals that B3 had better thermal stability because the reaction of ZnPA and BA formed BPO_4 with high thermal stability [38], thereby blocking the heat transfer further.

Table 3. Pyrolysis parameters of samples.

Samples	[a] $T_{d5\%}$/°C	[b] $T_{d10\%}$/°C	[c] $T_{d20\%}$/°C	[d] T_i	[e] T_f	[f] T_{max}/°C	Char Residue/wt.%
P0	133.3	210.0	718.0	30	998.3	250.7	78.61
P5	151.3	243.7	774.3	30	999.3	245.0	78.40
P6	113.3	194.0	515.0	30	998.3	244.3	76.81
B3	80.7	160.7	393.0	30	997.0	252.3	74.14
B6	125.0	243.0	721.3	30	999.0	258.7	77.64

[a] Temperature of 5% weight loss ($T_{d5\%}$). [b] Temperature of 10% weight loss ($T_{d10\%}$). [c] Temperature of 20% weight loss ($T_{d20\%}$). [d] Initial decomposition temperature (T_i). [e] Final decomposition temperature (T_f). [f] Temperature of the maximum rate of weight loss (T_{max}).

Figure 11a shows the TGA curves for the geopolymer coatings, depicting temperature variations at a constant heating rate of 20 °C/min. With the addition of ZnPA and BA, continuous weight loss was observed in all samples during the heating process. This may be attributed to the thermal decomposition of ZnPA, which produced phosphorus compounds, accelerating the dehydration of the composite coatings [39]. The formation of BPO_4 was catalyzed [40], prompting the decomposition of the siliceous layer, consequently reducing the weight loss of the residues.

Figure 11b presents the DTG of the sample coatings. The whole pyrolysis process could be categorized into four stages: (I) 30–198 °C, (II) 198–439 °C, (III) 439–731 °C, and (IV) 731–1000 °C. In the first stage of 30–198 °C, weight loss was primarily attributed to the loss of free water and bound moisture from the coating. In the temperature range of

198–439 °C, the primary reactions involved the decomposition of ZnPA and MEL. MEL released inert gases such as NH_3 [41]. ZnPA decomposed within this temperature range, forming pyrophosphate, metaphosphate, and metal phosphate [42]. After adding BA, BA dehydrated and formed vitreous boron oxide. The third stage occurred at 439–731 °C, primarily due to the thermal decomposition of polyphosphoric acid and the formation of the blocking layer. The final stage occurred between 731 and 1000 °C, mainly attributed to the decomposition of the blocking layer and the formation of BPO_4. Moreover, some unreacted boride and pyrolysis products of ZnPA volatilized at high temperatures. In summary, ZnPA and BA formed BPO_4 at high temperatures within the siliceous coatings, improving the flame retardancy of the composite coating.

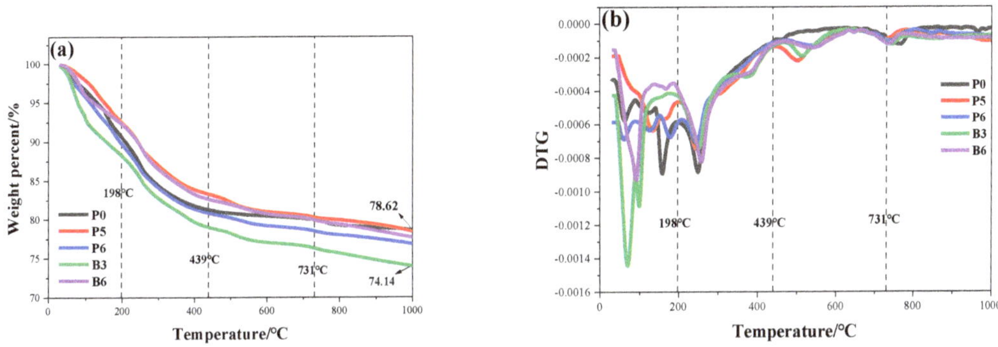

Figure 11. TG/DTG curves of samples including (**a**) TG and (**b**) DTG.

3.8. Pyrolysis Kinetics

Pyrolysis kinetics evaluates a material that undergoes decomposition or chemical reactions during heating to obtain the reaction mechanism, activation energy, and pyrolysis kinetics parameters. Using the modified Coats–Redfern integral method, the thermal decomposition kinetics of the coating were calculated using Formulas (9)–(13). A plot of the $\ln[G(\alpha)/T^2] - 1/T$ relationship was generated. The activation energy E_α and correlation coefficient R^2 were obtained by the linear regression fitting of the curve.

$$f(\alpha) = (1-\alpha)^n \tag{9}$$

$$G(\alpha) = (1-\alpha)^{-2} \tag{10}$$

$$\frac{d\alpha}{dT} = \frac{A}{\beta}\exp(-\frac{E_\alpha}{RT})f(\alpha) \tag{11}$$

$$\ln[\frac{G(\alpha)}{T^2}] = \ln(\frac{AR}{\beta E_\alpha}) - \frac{E_\alpha}{RT} \tag{12}$$

$$\alpha = \frac{m_0 - m_t}{m_0 - m_f} \tag{13}$$

n is the reaction order of combustion. A is the preexponent factor in units of min^{-1}. $E\alpha$ is the catalytic activation energy, measured in kilojoules per mole ($kJ \cdot mol^{-1}$). R is the ideal gas constant, 8.3145 $J \cdot mol^{-1} \cdot K^{-1}$. β is the heating rate of 20 $°C \cdot min^{-1}$. T is the thermodynamic temperature in units of K. α is the weight conversion rate of the sample during pyrolysis, %. m_0 represents the initial mass of the sample. m_t is the real-time mass of the coating at time. m_f corresponds to the final remaining mass.

By performing calculations and screening among the 29 commonly used mechanism functions [43], a three-level chemical reaction model (F3) was chosen to plot the curves, as illustrated in Figure 12. This meant that the reaction process could be controlled by regulating reactant design, condition control, equivalence ratio, etc., and the doped ZnPA and BA could affect the pyrolysis. The pyrolysis was mainly divided into the following four stages, as shown in Table 4: (IV) 60–198 °C, (III) 198–439 °C, (II) 439–731 °C, and (I) 731–940 °C. Stage IV was mainly the volatilization of small molecular substances. In comparison to P5, the E_α of B3 decreased to 15.94 kJ·mol^{-1} in stage III and 16.04 kJ·mol^{-1} in stage II. This was because the thermal decomposition of ZnPA and BA catalyzed the thermal degradation and carbonization of the coating. The E_α of B3 was 127.08 kJ·mol^{-1} in the range of 731–940 °C, indicating that ZnPA and BA reacted with the geopolymer to form a dense siliceous barrier layer with BPO_4, preventing plywood from coming into contact with air and enhancing the flame retardancy.

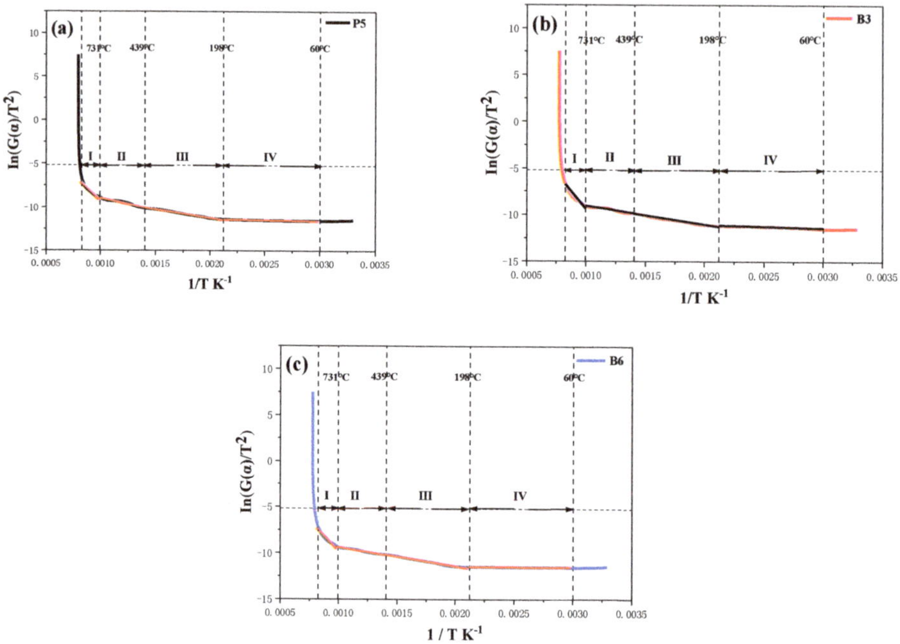

Figure 12. Fitting pyrolysis kinetics of sample coatings including (**a**) P5, (**b**) B3, and (**c**) B6. (The red lines are the fitted lines, the same to the black line in (**b**). The fitted lines cover the original curves).

Table 4. Kinetic parameters of samples calculated using F3.

Sample	Temperature	Intercept	Slope	Adj.R^2	E$_\alpha$/kJ·mol^{-1}
P5	60–198 °C	−11.02	−190.19	0.95	1.58
	198–439 °C	−7.35	−1948.92	0.99	16.20
	439–731 °C	−6.13	−2800.53	0.95	23.28
	731–940 °C	2.20	−11,339.87	0.94	94.28
B3	60–198 °C	−10.40	−333.17	0.95	2.77
	198–439 °C	−7.19	−1916.77	0.99	15.94
	439–731 °C	−7.13	−1929.49	0.93	16.04
	731–940 °C	5.91	−15,284.84	0.93	127.08
B6	60–198 °C	11.32	−61.67	0.94	0.51
	198–439 °C	−7.24	−2062.03	0.99	17.14
	439–731 °C	−7.15	−2172.26	0.98	18.06
	731–940 °C	3.42	−12,986.68	0.91	107.98

4. Discussion

In this research, an alkali-activated silica fume-based geopolymer coating was optimized by incorporating MEL, ZnPA, and BA to enhance its flame retardancy. Various characterization techniques, including CC, XRD, SEM, and TGA, were employed to analyze the coating. The results indicate that the optimal combination of BA and ZnPA significantly improved the flame retardancy. The flame-retardant mechanism of ZnPA/BA/MEL-containing silica fume-based geopolymer coatings is summarized as follows.

Firstly, the condensation reactions and the formation of coordination bonds were crucial in enhancing flame retardancy. During the combustion, a condensation reaction occurred, which transformed Si[OH]$_4$ into the Si-O-Si network structure [44], as shown in reaction (14). This silicon network structure contributed to the heat dissipation of the coating.

Secondly, zinc ions from ZnPA reacted with the Si-O-Si structure to form crosslinking networks [45], as shown in reaction (15). This crosslinking enhanced the strength of the residues [19]. Additionally, ZnPA enhanced flame retardancy by generating phosphorus-containing free radicals that quenched the pyrolysis process [46]. The production of phosphoric acid, polyphosphoric acid, and metal phosphates catalyzed the formation of residual layers.

Furthermore, BA helped in reducing the surface temperature of the coating by producing B_2O_3 and water. Furthermore, zinc ions and boron could enhance charring [47,48]. Moreover, MEL accelerated the production of polyphosphoric acid [49], releasing inert gases, such as NH_3. These gases diluted the combustible gases and O_2 in the vicinity, reducing the combustion intensity.

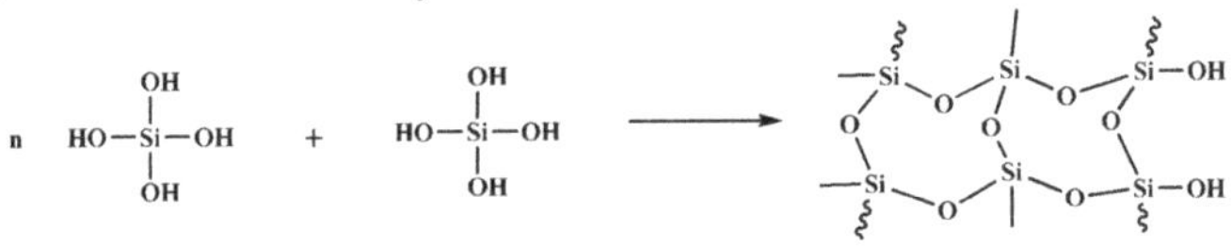

(14)

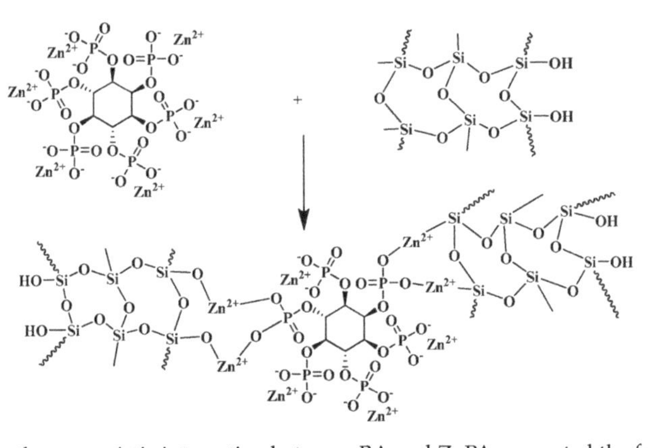

(15)

Thirdly, the synergistic interaction between BA and ZnPA promoted the formation of highly thermally stable BPO_4. BPO_4 formed a ceramic shielding layer atop the siliceous layer, as shown in Figure 10. This additional layer significantly enhanced the physical barrier properties of the siliceous layer. Moreover, BPO_4 had a filling effect, strengthening the structural integrity of the residual layer and forming a dense residual layer [50]. This layer acted as a barrier to heat transfer, thereby protecting the underlying plywood. In addition, BPO_4 increased the surface area of the siliceous layer. This was similar to the findings of Decsov et al. [51], who changed the specific surface area of 2-hydroxypropyl-β-cyclodextrin (HP-β-CD) microfibers, leading to an increase in the flame-retardant index of the composites. Thus, BPO_4 introduced more active sites on the surface of the barrier layer. It adsorbed flammable organic volatiles, preventing their release and diffusion during combustion and therefore reducing the rate of flame spread.

Consequently, the synergistic flame-retardant mechanism of silica fume-based composite coatings, as illustrated in Figure 13, primarily involved the quenching effect, dilution action, heat absorption, and catalytic charring of coatings. The interpenetrating network structure and BPO_4 promoted the generation of a continuous and robust siliceous layer that acted as a barrier. However, the complex reaction mechanism among ZnPA, BA, MEL, and geopolymer, as well as the smoke suppression performance of the composite coating, will be investigated in our subsequent research.

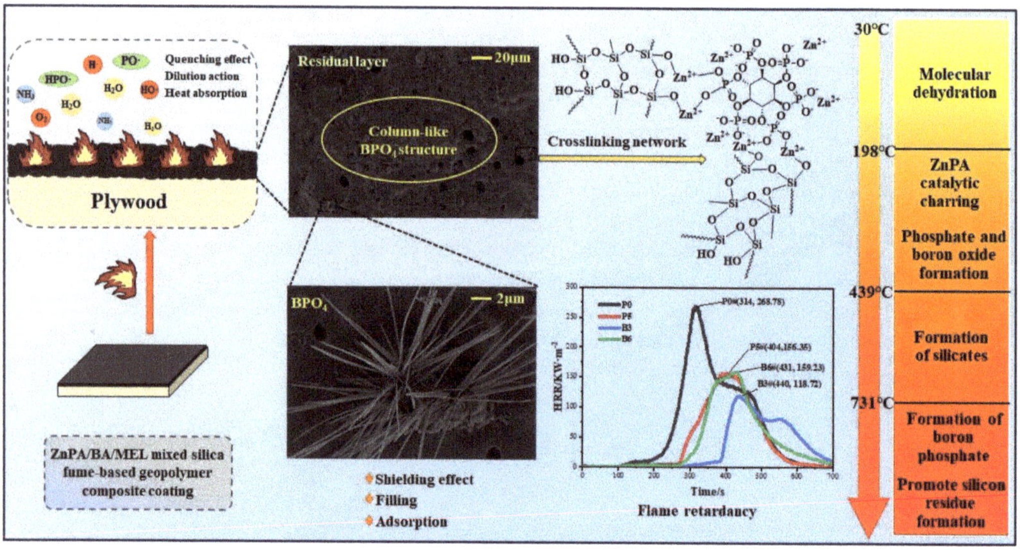

Figure 13. Flame-retarding mechanism of composite coatings.

5. Conclusions

In this study, the flame retardancy of ZnPA/BA/MEL-containing silica fume-based geopolymer coatings was preliminarily investigated. The following conclusions were drawn:

(1) An appropriate amount of ZnPA (2.5 wt.%) and BA (1.5 wt.%) enhanced the flame retardancy of silica fume-based geopolymer coating, evidenced by the FPI increasing from 0.59 (P0) to 2.83 (B3) $s\cdot m^{-2}\cdot kW^{-1}$, the FGI decreasing from 0.86 (P0) to 0.27 (B3) $kW\ m^{-2}\cdot s^{-1}$, and the FRI of 8.48.

(2) The in situ-formed boron phosphate (BPO_4) derived from the doped ZnPA and BA facilitated interpenetrating networks within siliceous gels, improving the residual resilience of the fire-barrier layer, which could shield, fill, and adsorb heat during firing.

(3) According to pyrolysis kinetic calculation, in situ-formed BPO_4 made the E_α climb from 94.28 (P5) to 127.08 $kJ\cdot mol^{-1}$ (B3) with temperatures of 731–940 °C, corresponding to improved flame retardancy.

Author Contributions: Methodology, Y.W.; Software, J.D.; Formal analysis, C.W.; Investigation, Y.W. and Y.Q.; Resources, Y.W., C.W. and J.D.; Data curation, Y.Q.; Writing—original draft, Y.W.; Writing—review & editing, Y.W. All authors have read and agreed to the published version of the manuscript.

Funding: This work was supported by the Opening Projects of State Key Laboratory of Green and Low-carbon Development of Tar-rich Coal in Western China (SKLCRKF23-08), the Natural Science Foundation of Shaanxi Province (2025JC-YBMS-560), and the open fund from Anhui Engineering Research Center for Coal Clean Processing and Carbon Reduction y (CCCE-2024007).

Data Availability Statement: The original contributions presented in this study are included in the article. Further inquiries can be directed to the corresponding author.

Conflicts of Interest: The authors declare no conflict of interest.

References

1. Dong, X.; Gan, W.; Shang, Y.; Tang, J.; Wang, Y.; Cao, Z.; Xie, Y.; Liu, J.; Bai, L.; Li, J.; et al. Low-value wood for sustainable high-performance structural materials. *Nat. Sustain.* **2022**, *5*, 628–635. [CrossRef]
2. Huang, Y.; Ma, T.; Wang, Q.; Guo, C. Synthesis of biobased flame-retardant carboxylic acid curing agent and application in wood surface coating. *ACS Sustain. Chem. Eng.* **2019**, *7*, 14727–14738. [CrossRef]
3. Al-Kaseasbeh, Q.; Al-Qaralleh, M. Valorization of hydrophobic wood waste in concrete mixtures: Investigating the micro and macro relations. *Results Eng.* **2023**, *17*, 100877. [CrossRef]
4. Sykam, K.; Hussain, S.S.; Sivanandan, S.; Narayan, R.; Basak, P. Non-halogenated UV-curable flame retardants for wood coating applications: Review. *Prog. Org. Coat.* **2023**, *179*, 107549. [CrossRef]
5. He, Z.; Shen, A.; Guo, Y.; Lyu, Z.; Li, D.; Qin, X.; Zhao, M.; Wang, Z. Cement-based materials modified with superabsorbent polymers: A review. *Constr. Build. Mater.* **2019**, *225*, 569–590. [CrossRef]
6. Ding, Z.; Xu, M.-R.; Dai, J.-G.; Dong, B.-Q.; Zhang, M.-J.; Hong, S.-X.; Xing, F. Strengthening concrete using phosphate cement-based fiber-reinforced inorganic composites for improved fire resistance. *Constr. Build. Mater.* **2019**, *212*, 755–764. [CrossRef]
7. Bajpai, R.; Choudhary, K.; Srivastava, A.; Sangwan, K.S.; Singh, M. Environmental impact assessment of fly ash and silica fume based geopolymer concrete. *J. Clean. Prod.* **2020**, *254*, 120147. [CrossRef]
8. Shahidi, S.S.; Mohammadi, S. Synergistic effect of nano hybrid multi-layered graphene oxide/talc and silica fume on the fire and water-resistance of intumescent coatings. *Prog. Org. Coat.* **2023**, *183*, 107736. [CrossRef]
9. Zhang, D.R.; Zhu, H.J.; Wu, Q.S.; Yang, T.; Yin, Z.F.; Tian, L. Investigation of the hydrophobicity and microstructure of fly ash-slag geopolymer modified by polydimethylsiloxane. *Constr. Build. Mater.* **2023**, *369*, 130540. [CrossRef]
10. Zhang, D.; Wang, Y.; Zhang, T.; Yang, Q. Engineering and microstructural properties of carbon-fiber-reinforced fly-ash-based geopolymer composites. *J. Build. Eng.* **2023**, *79*, 107883. [CrossRef]
11. Xu, Y.; Hu, H.; Tao, B.; Yin, R.; Liu, L.; Li, B. Safe and economical preparation of amino acid-derived bio-based triazine char-forming agent for efficient intumescent flame retardant polypropylene. *Constr. Build. Mater.* **2025**, *484*, 141876. [CrossRef]
12. Shi, Y.; Xu, Y.; Xu, K.; Yan, C.; Qin, A.; Du, C.; Xu, M.; Wang, C.; Li, B.; Liu, L. Fire-resistant and thermal-insulating alginate aerogel with intelligent bionic armor for exceptional mechanical and fire early-warning performance. *Chem. Eng. J.* **2024**, *498*, 155181. [CrossRef]
13. Huo, S.; Wang, C.; Shi, Q.; Yu, L.; Liu, Z.; Fang, Z.; Wang, H. A novel hyperbranched phosphorus-boron polymer for transparent, flame-retardant, smoke-suppressive, robust yet tough epoxy resins. *Compos. Part B Eng.* **2021**, *227*, 109395. [CrossRef]
14. Peng, H.; Mao, Y.; Wang, D.; Fu, S. B-N-P-linked covalent organic frameworks for efficient flame retarding and toxic smoke suppression of polyacrylonitrile composite fiber. *Chem. Eng. J.* **2022**, *430*, 133120. [CrossRef]
15. Zhang, L.; Zhang, W.; Peng, Y.; Wang, W.; Cao, J. Thermal behavior and flame retardancy of poplar wood impregnated with furfuryl alcohol catalyzed by boron/phosphorus compound system. *Ind. Crops Prod.* **2022**, *176*, 114361. [CrossRef]
16. Huo, S.; Sai, T.; Ran, S.; Guo, Z.; Fang, Z.; Song, P.; Wang, H. A hyperbranched P/N/B-containing oligomer as multifunctional flame retardant for epoxy resins. *Compos. Part B Eng.* **2022**, *234*, 109701. [CrossRef]
17. Zhang, S.; Wang, X.; Ding, M.; Huang, Y.; Li, L.; Wang, M. In-situ incorporation of metal phytates for green and highly efficient flame-retardant wood with excellent smoke-suppression property. *Ind. Crops Prod.* **2022**, *187*, 115287. [CrossRef]
18. Jiang, G.; Xiao, Y.; Qian, Z.; Yang, Y.; Jia, P.; Song, L.; Hu, Y.; Ma, C.; Gui, Z. A novel phosphorus-, nitrogen- and sulfur-containing macromolecule flame retardant for constructing high-performance epoxy resin composites. *Chem. Eng. J.* **2023**, *451*, 137823. [CrossRef]
19. Zhang, H.; Wang, Y.; Li, F.; Zhao, J. In-situ polymerized zinc phytate chelated Si-C-P geopolymer hybrid coating constructed by incorporating chitosan oligosaccharide and DOPO for flame-retardant plywood. *Constr. Build. Mater.* **2023**, *397*, 132416. [CrossRef]
20. Wang, K.; Wang, S.; Meng, D.; Chen, D.; Mu, C.; Li, H.; Sun, J.; Gu, X.; Zhang, S. A facile preparation of environmentally-benign and flame-retardant coating on wood by comprising polysilicate and boric acid. *Cellulose* **2021**, *28*, 11551–11566. [CrossRef]
21. Savas, L.A.; Dogan, M. Flame retardant effect of zinc borate in polyamide 6 containing aluminum hypophosphite. *Polym. Degrad. Stab.* **2019**, *165*, 101–109. [CrossRef]

22. Wang, Y.; Kou, X.; Deng, J.; Zhao, J.P.; Shi, H. Ammonium polyphosphate/expandable graphite/TiO_2 blended silica fume-based geopolymer coating for synergistically flame-retarding plywood. *Constr. Build. Mater.* **2022**, *317*, 125941. [CrossRef]
23. *ISO 5660-1:2015*; Reaction-to-Fire Tests—Heat Release, Smoke Production and Mass Loss Rate. Part 1: Heat Release Rate (Cone Calorimeter Method) and Smoke Production Rate (Dynamic Measurement). ISO: Geneva, Switzerland, 2015.
24. Yuan, X.; Luo, K.; Zhang, K.; He, J.; Zhao, Y.; Yu, D. Combinatorial Vibration-Mode Assignment for the FTIR Spectrum of Crystalline Melamine: A Strategic Approach toward Theoretical IR Vibrational Calculations of Triazine-Based Compounds. *J. Phys. Chem. A* **2016**, *120*, 7427–7433. [CrossRef]
25. Ong, H.R.; Prasad, R.; Khan, M.M.R.; Chowdhury, M.N.K. Effect of Palm Kernel Meal as Melamine Urea Formaldehyde Adhesive Extender for Plywood Application: Using a Fourier Transform Infrared Spectroscopy (FTIR) Study. *Appl. Mech. Mater.* **2012**, *121–126*, 493–498. [CrossRef]
26. Lei, J.; Song, H.; Wei, Y.; Zhao, S.; Qi, H. A novel strategy to enhance hydrothermal stability of Pd-doped organosilica membrane for hydrogen separation. *Microporous Mesoporous Mater.* **2017**, *253*, 55–63. [CrossRef]
27. Carli, L.; Schnitzler, E.; Ionashiro, M.; Szpoganicz, B.; Rosso, N. Equilibrium, thermoanalytical and spectroscopic studies to characterize phytic acid complexes with Mn (II) and Co (II). *J. Braz. Chem. Soc.* **2009**, *20*, 1515–1522. [CrossRef]
28. Elbeyli, İ.Y. Production of crystalline boric acid and sodium citrate from borax decahydrate. *Hydrometallurgy* **2015**, *158*, 19–26. [CrossRef]
29. Yu, K.; Wang, Y.; Li, F.; Zhao, J. In-situ grown SiC whiskers enhance flame retardancy of alkali-activated gold tailings geopolymer composite coatings by incorporating expanded graphite. *Constr. Build. Mater.* **2023**, *392*, 131936. [CrossRef]
30. Yang, F.; Yao, Y.; Xu, Y.; Wang, C.; Wang, M.; Ren, J.; Zhang, C.; Wu, F.; Lu, J. Evolution of the porous structure for phosphoric acid etching carbon as cathodes in Li-O_2 batteries: Pyrolysis temperature-induced characteristics changes. *Carbon Energy* **2023**, *6*, e372. [CrossRef]
31. Larsson, E.; Donzel-Gargand, O.; Heinrichs, J.; Jacobson, S. Tribofilm formation of a boric acid fuel additive-Material characterization; challenges and insights. *Tribol. Int.* **2022**, *171*, 107541. [CrossRef]
32. Hu, J.; Xia, H.; Hou, X.; Yang, T.; Si, K.; Wang, Y.; Wang, L.; Shi, Z. Enhanced thermal management performance of nanofibrillated cellulose composite with highly thermally conductive boron phosphide. *J. Mater. Chem. A* **2021**, *9*, 27049–27060. [CrossRef]
33. Ullah, S.; Ahmad, F.; Shariff, A.M.; Bustam, M.A.; Gonfa, G.; Gillani, Q.F. Effects of ammonium polyphosphate and boric acid on the thermal degradation of an intumescent fire retardant coating. *Prog. Org. Coat.* **2017**, *109*, 70–82. [CrossRef]
34. Tong, X.-M.; Zhang, T.; Yang, M.-Z.; Zhang, Q. Preparation and characterization of novel melamine modified poly (urea–formaldehyde) self-repairing microcapsules. *Colloids Surf. A Physicochem. Eng. Asp.* **2010**, *371*, 91–97. [CrossRef]
35. Yan, M.; Pan, Y.; Cheng, X.; Zhang, Z.; Deng, Y.; Lun, Z.; Gong, L.; Gao, M.; Zhang, H. "Robust–soft" anisotropic nanofibrillated cellulose aerogels with superior mechanical, flame-retardant, and thermal insulating properties. *ACS Appl. Mater. Interfaces* **2021**, *13*, 27458–27470. [CrossRef]
36. Wang, X.; Li, L.; Hong, W.B.; Yan, H.; Wu, S.Y.; Chen, X.M. Preparation and microwave dielectric properties of BPO_4 ceramics with ultra-low dielectric constant. *J. Mater. Sci. Mater. Electron.* **2021**, *32*, 6660–6667. [CrossRef]
37. Schmidt, M.; Ewald, B.; Prots, Y.; Cardoso-Gil, R.; Armbrüster, M.; Loa, I.; Zhang, L.; Huang, Y.-X.; Schwarz, U.; Kniep, R. Growth and characterization of BPO_4 single crystals. *Z. Anorg. Und Allg. Chem.* **2004**, *630*, 655–662. [CrossRef]
38. Chen, Y.; Duan, H.; Ji, S.; Ma, H. Novel phosphorus/nitrogen/boron-containing carboxylic acid as co-curing agent for fire safety of epoxy resin with enhanced mechanical properties. *J. Hazard. Mater.* **2021**, *402*, 123769. [CrossRef]
39. Kong, D.; Liu, J.; Zhang, Z.; Wang, S.; Lu, Z. Preparation of synergistic silicon, phosphorus and nitrogen flame retardant based on cyclosiloxane and its application to cotton fabric. *Cellulose* **2021**, *28*, 8115–8128. [CrossRef]
40. Liu, X.; Wang, J.-Y.; Yang, X.-M.; Wang, Y.-L.; Hao, J.-W. Application of TG/FTIR TG/MS and cone calorimetry to understand flame retardancy and catalytic charring mechanism of boron phosphate in flame-retardant PUR–PIR foams. *J. Therm. Anal. Calorim.* **2017**, *130*, 1817–1827. [CrossRef]
41. Bifulco, A.; Parida, D.; Salmeia, K.A.; Nazir, R.; Lehner, S.; Stämpfli, R.; Markus, H.; Malucelli, G.; Branda, F.; Gaan, S. Fire and mechanical properties of DGEBA-based epoxy resin cured with a cycloaliphatic hardener: Combined action of silica, melamine and DOPO-derivative. *Mater. Des.* **2020**, *193*, 108862. [CrossRef]
42. Guo, H.Z.; Lukovic, M.; Mendoza, M.; Schleputz, C.M.; Griffa, M.; Xu, B.W.; Gaan, S.; Herrmann, H.; Burgert, I. Bioinspired Struvite Mineralization for Fire-Resistant Wood. *Acs Appl. Mater. Interfaces* **2019**, *11*, 5427–5434. [CrossRef] [PubMed]
43. Ma, Z.; Wang, J.; Yang, Y.; Zhang, Y.; Zhao, C.; Yu, Y.; Wang, S. Comparison of the thermal degradation behaviors and kinetics of palm oil waste under nitrogen and air atmosphere in TGA-FTIR with a complementary use of model-free and model-fitting approaches. *J. Anal. Appl. Pyrolysis* **2018**, *134*, 12–24. [CrossRef]
44. Xiao, Y.; Liu, S.; Hu, Y.; Zhang, S.; Li, Z.; Li, L.; Feng, J. Excellent antioxidizing, thermally insulating and flame resistance silica-polybenzoxazine aerogels for aircraft ablative materials. *J. Appl. Polym. Sci.* **2022**, *139*, e52499. [CrossRef]
45. Rath, S.S.; Singh, S.; Rao, D.S.; Nayak, B.B.; Mishra, B.K. Adsorption of heavy metals on a complex Al-Si-O bearing mineral system: Insights from theory and experiments. *Sep. Purif. Technol.* **2017**, *186*, 28–38. [CrossRef]

46. Zhang, W.; Wu, W.; Meng, W.; Xie, W.; Cui, Y.; Xu, J.; Qu, H. Core-shell graphitic carbon nitride/zinc phytate as a novel efficient flame retardant for fire safety and smoke suppression in epoxy resin. *Polymers* **2020**, *12*, 212. [CrossRef]
47. Piao, J.; Lai, Y.; Ren, J.; Wang, Y.; Feng, T.; Wang, Y.; Liu, W.; Dong, H.; Chen, W.; Jiao, C.; et al. Zn-doped carbon microspheres as synergist in intumescent flame-retardant thermoplastic polyurethane composites: Mechanism of char residues layer regulation. *Compos. Commun.* **2022**, *32*, 101173. [CrossRef]
48. Han, G.; Zhao, X.; Feng, Y.; Ma, J.; Zhou, K.; Shi, Y.; Liu, C.; Xie, X. Highly flame-retardant epoxy-based thermal conductive composites with functionalized boron nitride nanosheets exfoliated by one-step ball milling. *Chem. Eng. J.* **2021**, *407*, 127099. [CrossRef]
49. Song, F.; Zhao, Q.; Zhu, T.; Bo, C.; Zhang, M.; Hu, L.; Zhu, X.; Jia, P.; Zhou, Y. Biobased coating derived from fish scale protein and phytic acid for flame-retardant cotton fabrics. *Mater. Des.* **2022**, *221*, 110925. [CrossRef]
50. Yu, R.; Wen, X.; Zhu, Y.; Lou, S.; Li, Y.; Wang, S.; Liu, J.; Tang, T. Boron-doped copper phenylphosphate as temperature-response nanosheets to fabricate high fire-safety polycarbonate nanocomposites. *Compos. Part A* **2023**, *175*, 107812. [CrossRef]
51. Decsov, K.; Takács, V.; Marosi, G.; Bocz, K. Microfibrous cyclodextrin boosts flame retardancy of poly (lactic acid). *Polym. Degrad. Stab.* **2012**, *191*, 109655. [CrossRef]

mining

MDPI

Article

Assessing the Systemic Impact of Heat Stress on Human Reliability in Mining Through FRAM and Hybrid Decision Models

Ana Carolina Russo

Departamento de Engenharia de Minas e de Petróleo da Escola Politécnica, Universidade de São Paulo, São Paulo CEP 05508-030, SP, Brazil; anacarolinarusso@usp.br

Abstract

Occupational heat stress represents an increasing challenge to safety and operational performance in underground mining, where elevated temperatures, humidity, and limited ventilation are common. This study proposes an integrated framework to analyze the systemic impact of heat stress on human reliability in mining operations. We conducted a systematic literature review to identify empirical studies addressing thermal exposure, extracting key operational functions for modeling. These functions were structured using the Functional Resonance Analysis Method (FRAM) to reveal interdependencies and performance variability. Human reliability was evaluated using Fuzzy CREAM, which quantified the degree of contextual control associated with each function. Finally, we applied the Gaussian Analytic Hierarchy Process (AHP) to prioritize functions based on thermal impact, contextual reliability, and systemic connectivity. The results showed that functions involving subjective or complex judgment, such as assessing thermal stress or identifying psychophysiological indicators, exhibited lower reliability and higher vulnerability. In contrast, monitoring and control functions based on standardized procedures were more stable and resilient. This combined approach identified critical points of systemic fragility and offers a robust decision-support tool for prioritizing thermal risk mitigation. The findings contribute to advancing the scientific understanding of heat stress impacts in mining and support the development of targeted interventions to enhance human performance and safety in extreme environments.

Keywords: heat stress; human reliability; underground mining; FRAM; Fuzzy CREAM

1. Introduction

In mining environments, workers are frequently exposed to extreme environmental conditions—marked by high temperatures, elevated humidity, and limited ventilation—that impose significant physiological and cognitive demands. Such conditions lead to fatigue, human error, and workplace accidents, compromising both occupational health and the resilience of sociotechnical systems [1].

Heat stress refers to the total internal and external thermal loads that challenge the body's thermoregulatory capacity, leading to homeostatic imbalance and, in severe cases, heatstroke or death [2]. Research has shown that human performance under thermal stress declines non-linearly, affecting both physical workload capacity and decision-making quality [3,4]. Prolonged exposure to excessive heat is also associated with neurological and cardiovascular disorders, reinforcing the need for effective prevention and monitoring strategies [5,6].

While indices such as the WBGT (Wet Bulb Globe Temperature) and PHS (Predicted Heat Strain) are widely used for assessing thermal risks, their predictive capability in real-world operational contexts remains limited because they often neglect the behavioral, organizational, and cognitive aspects of human work [7,8]. Moreover, these traditional tools fall short in capturing the dynamic variability of sociotechnical systems, particularly under high thermal stress, where minor fluctuations in performance and context can escalate into critical failures. This highlights the gap for systemic approaches that integrate physiological responses to heat with human and functional variability—elements critical to the reliability of complex operations.

Recently, hybrid models have emerged as promising tools for addressing the multidimensional complexity of operations under heat stress. The Functional Resonance Analysis Method (FRAM) enables the modeling of organizational and operational variability, revealing how minor fluctuations can amplify into safety-critical failures [9,10]. Meanwhile, fuzzy adaptations of the Cognitive Reliability and Error Analysis Method (CREAM) allow quantifying human reliability under uncertainty by translating qualitative context conditions into fuzzy numerical scores [11–15].

Moreover, multi-criteria decision-making techniques such as the Analytic Hierarchy Process (AHP)—especially when extended with probabilistic modeling (Gaussian AHP)—enable the prioritization of critical functions based on multiple dimensions, including thermal impact, contextual reliability, and system connectivity [16,17]. However, no prior study has combined these three methods in an integrated framework to model and quantify systemic human reliability degradation under heat stress.

To our knowledge, no prior study has integrated the FRAM, fuzzy CREAM, and AHP in a unified model to evaluate thermal risk in mining systems. While hybrid models have been explored in other domains (e.g., urban transport) [18], this paper offers a novel contribution by adapting and extending such methods to the high-risk, heat-exposed context of underground mining.

This study stands out from previous works by uniquely integrating the FRAM, fuzzy CREAM, and Gaussian AHP into a unified framework that captures the multifactorial and systemic dimensions of heat stress in mining operations. Unlike traditional assessment tools such as the WBGT, PHS, or isolated applications of the CREAM, our approach models emergent variability, quantifies human reliability under contextual uncertainty, and systematically prioritizes critical operational functions—offering a more comprehensive and decision-oriented analysis of thermal risk.

We propose an integrated methodological framework that combines the Functional Resonance Analysis Method (FRAM), fuzzy Cognitive Reliability and Error Analysis Method (CREAM), and Gaussian Analytic Hierarchy Process (AHP) to evaluate the systemic impact of heat stress on human reliability in mining environments. By jointly incorporating operational dynamics, cognitive demands, and organizational structures, this model provides a novel lens to understand and manage human performance under thermal stress, supporting evidence-based strategies for improving safety and resilience in extreme occupational conditions.

2. Materials and Methods

This study integrates three complementary methodological approaches: functional system modeling using the Functional Resonance Analysis Method (FRAM), human reliability assessment through the fuzzy extension of the Cognitive Reliability and Error Analysis Method (Fuzzy CREAM), and prioritization of critical functions using the Gaussian Analytic Hierarchy Process (Gaussian AHP).

2.1. Functional Modeling Using FRAM

The first stage involved the identification of critical functions associated with heat stress in mining environments. A systematic review was conducted in the Scopus database using Boolean operators combining the terms "heat stress" OR "thermal stress" OR "occupational heat exposure" OR "high temperature" OR "WBGT" OR "physiological strain index" AND "mining industry" OR "mining operations" OR "underground mining" OR "surface mining" OR "mine workers". The search was restricted to scientific articles (document type: "ar") written in English.

To process the selected literature, we implemented an automated text-extraction procedure using Python 3.10.12. Full-text articles in PDF format were parsed with the PyPDF2 library, version 3.0.1, which enables structured analysis of document content. The extracted text was automatically translated into English using the deep_translator package (version 1.11.4, released on 28 June 2023) and the GoogleTranslator service, ensuring terminological consistency for further analysis.

Once compiled into a unified corpus, the text was filtered using regular expressions and part-of-speech tagging to detect sentences representing operational functions. Specifically, the pattern targeted action–object verb sequences associated with thermal exposure and performance degradation (e.g., "monitor core temperature", "evaluate thermal comfort"), drawing on previous methodological approaches by [19,20]. The regex algorithm was calibrated to identify both operational verbs (e.g., monitor, assess, manage) and domain-specific terms (e.g., heat stress, the WBGT, hydration, fatigue).

The extracted sentences were stored in a structured pandas DataFrame (version 1.5.3) and then subjected to manual review to ensure semantic consistency and contextual relevance. Only functions explicitly or implicitly linked to thermally induced variability or human–system interaction were retained for modeling.

The full source code used in this step is presented below.

```
# Install required libraries (run once in Colab or Jupyter)
!pip install PyPDF2 matplotlib pandas
# Import libraries
from PyPDF2 import PdfReader
import re
import pandas as pd
from collections import Counter
import matplotlib.pyplot as plt
# Define PDF file paths (uploaded to/content/in
Google Colab)
pdf_paths = [
"/content/0002889748507014.pdf",
"/content/mem035.pdf",
"/content/s40033-022-00389-z.pdf",
"/content/Vol.67,+No.11,+NOVEMBER+2019-15-
19.pdf"
]
# Function to extract full text from all PDFs
def extract_text_from_pdfs(paths):
full_text = ""
for path in paths:
reader = PdfReader(path)
for page in reader.pages:
text = page.extract_text()
if text:
full_text += text + "\n"
return full_text
# Define technical contexts associated with
thermal stress
contexts = [
"heat stress", "thermal comfort", "WBGT",
"hydration", "acclimatization",
"workload", "temperature", "body heat",
"fatigue", "performance",
"core temperature", "dehydration", "cooling",
"rest breaks", "TWL", "exhaustion"
]
# Regular expression to identify function-like
sentences (verb + context)
pattern =
rf"((?:{'|'.join(verbs)}).*?(?:{'|'.join(contexts)}).*?\.)"
# Extract sentences that match the pattern
raw_phrases = re.findall(pattern, full_text,
flags=re.IGNORECASE)
# Create DataFrame with extracted functional
sentences
df_functions = pd.DataFrame(raw_phrases,
columns=["Function (textual description)"])
# Count frequency of key context terms
used_terms = [ctx.lower() for phrase in
raw_phrases for ctx in contexts if ctx in
phrase.lower()]
```

```
# Extract text from selected papers
full_text = extract_text_from_pdfs(pdf_paths)

# Define action-related verbs (used as indicators of functional behavior)
verbs = [
"monitor", "evaluate", "assess", "estimate",
"measure", "record", "analyze",
"control", "regulate", "adjust", "implement",
"detect", "observe", "apply",
"interrupt", "determine", "manage", "reduce",
"prevent"
]
term_freq = Counter(used_terms)
# Generate frequency graph of terms
plt.figure(figsize=(10, 5))
plt.bar(term_freq.keys(), term_freq.values())
plt.xticks(rotation=45, ha='right')
plt.title("Frequency of heat-stress-related terms in extracted functions")
plt.xlabel("Operational topics")
plt.ylabel("Frequency")
plt.tight_layout()
plt.savefig("function_term_frequency.png")
plt.show()
# Display top 10 extracted functions
df_functions.head(10)
```

Each retained function was then structured according to the six canonical FRAM aspects (Table 1): Input (I), Output (O), Precondition (P), Resource (R), Control (C), and Time (T), as defined by [9]. The visual representation was created using the FRAM Model Visualizer (FMV), version 2.1.4 (October 2020), which supported the identification of functional couplings and potential variability propagation. This allowed for a non-linear, emergent interpretation of socio-technical dynamics.

Table 1. Six core aspects.

Core Aspects	Description
Input (I)	The trigger or starting point for the function
Output (O)	The result or outcome produced by the function
Preconditions (P)	The conditions that must be satisfied before the function can be executed
Resources (R)	The materials, tools, or information required to perform the function
Control (C)	The mechanisms or guidelines regulating the function's execution
Time (T)	The temporal aspect that defines when the function must be completed

Finally, a frequency analysis of domain-specific terms was performed to visualize the most recurrent technical elements within the extracted corpus, with prominent emphasis on concepts such as temperature, hydration, fatigue, and performance degradation.

To enhance methodological transparency in the modeling process, we provide additional detail on the semantic validation of functional elements. Manual review was conducted by the lead researcher, whose background in safety systems and thermal ergonomics guided a structured and consistent evaluation. Each function-like sentence identified through automated extraction was assessed against three inclusion criteria: (i) clear reference to thermally induced variability or human–system performance degradation, (ii) presence of an explicit action–object structure (e.g., "monitor temperature"), and (iii) alignment with at least one of the six canonical aspects defined in the FRAM framework (version 2.1.4).

This manual validation process followed a predefined protocol, and ambiguous or borderline cases were refined iteratively to ensure conceptual coherence within the system model.

To support the visual FRAM representation, a detailed summary table of all modeled functions, including their descriptions, mapped FRAM aspects, and key couplings, is provided as Supplementary Material (Table S1). Importantly, in the FRAM Model Visualizer (FMV), red interface dots represent undefined or unlinked couplings. These do not reflect modeling errors but instead signal potential variability propagation routes that warrant further specification or empirical verification.

In compliance with FRAM methodological standards [9,10], particular attention was given to the correct establishment of couplings between functions. Each input (I), control (C), resource (R), precondition (P), and time (T) assigned to a function was explicitly traced back to the output (O) of another function, ensuring that no input element exists without a functional source. This coupling process respects the principle that "nothing exists in isolation" within the model, a key tenet of functional resonance.

Furthermore, we revised the model to distinguish Foreground functions (those representing core operational activities) from Background functions, which act as systemic enablers with output-only behavior [21]. The Background functions are now shown in the Modern view as rectangular elements, as recommended for clarity (FRAM Model Visualizer, version 2.1.6).

In the final version, three terminal functions ("To Establish Control Measures Based on Heat Stress Data," "To Assess Thermal Comfort During Shifts," and "To Implement Procedures to Reduce Occupational Heat Stress") were reviewed. Their outputs were removed, as these functions conclude the modeling sequence. This adjustment eliminates methodological inconsistencies and adheres to the FRAM rule that terminal functions may legitimately lack outputs, as no further function depends on them [9,21].

The final FRAM model is presented in (Modern View), where all functional couplings were manually validated using the FRAM Model Visualizer (FMV). A complete list of functions, their descriptions, and associated aspects is provided in Table S1.

To enhance visual clarity and facilitate interpretation, Modern View adopts a color-coded border scheme to distinguish function types:

- Yellow-bordered functions denote human actions, such as perception, decision-making, or manual task execution.
- Blue-bordered functions represent technological processes, including sensor-based measurements and automated operations.
- Green-bordered functions correspond to organizational activities, typically involving planning, coordination, or institutional control.

This visual coding supports functional differentiation and helps identify the interaction between human, technological, and organizational components within the system.

2.2. *Human Reliability Assessment Using Fuzzy CREAM*

To evaluate human reliability under thermally stressful mining conditions, we implemented a fuzzy-logic extension of the Cognitive Reliability and Error Analysis Method (CREAM) using Python 3.10.12 and the scikit-fuzzy library version 0.5.0, released on 22 August 2024. This hybridization enables quantitative modeling of contextual variability while maintaining the qualitative reasoning structure of the classical CREAM.

The assessment followed five structured steps, as detailed below:

Step 1—Selection of Relevant CPCs

From the nine original Common Performance Conditions (CPCs) in the CREAM, we selected three contextually relevant dimensions for heat-exposed mining environments:

- CPC_1: Adequacy of organization
- CPC_2: Working conditions
- CPC_3: Time available

These CPCs were chosen based on domain-specific literature and expert consultation, focusing on their critical role in shaping operator performance under heat stress.

The selection of these three CPCs was guided not only by prior literature [12] but also by their direct contextual relevance to heat-stressed mining environments. While standard CREAM applications often consider additional factors such as training and experience, man–

machine interface (MMI), or crew collaboration, these dimensions were not prioritized in this study for specific reasons. In underground mining operations affected by heat, immediate performance degradation is most strongly influenced by organizational support structures (e.g., scheduling, staffing), environmental working conditions (e.g., temperature, humidity, ventilation), and time constraints on decision-making. Other CPCs such as "adequacy of training" or "crew collaboration" are undoubtedly relevant but are either indirectly represented through organizational structures or difficult to quantify reliably under thermal stress without direct field data. Therefore, the selection focused on the CPCs with the most observable and quantifiable impact on heat-related human reliability, aligning the modeling effort with the available evidence and the scope of literature-based functional analysis.

Step 2—Fuzzification of CPC Inputs

Each CPC was modeled using triangular fuzzy membership functions defined by

$$\mu_{poor}(x) = \begin{cases} 1, & x \leq a \\ \frac{b-x}{b-a}, & a < x < b \\ 0, & x \geq b \end{cases}$$

where a, b, and c define the lower limit, peak, and upper limit of each fuzzy set.

Each CPC is represented as a triplet

$$y = f(x_1, x_2, x_3) \in [0, 10]$$

The linguistic values {Adequate, Average, Poor} were associated with normalized fuzzy inputs and interpreted via linguistic rules.

Step 3—Fuzzy Rule Base and Inference

A set of fuzzy "IF–THEN" rules was constructed to relate CPC inputs to control modes. For example:

- IF CPC_1 is "Adequate" AND CPC_2 is "Adequate" AND CPC_3 is "Adequate" $\rightarrow$ THEN Control Mode = "Efficient"
- IF CPC_1 is "Poor" OR CPC_2 is "Poor" $\rightarrow$ THEN Control Mode = "Scrambled"

The fuzzy inference mechanism used Mamdani-type composition with the minimum operator for conjunction (AND) and maximum for disjunction (OR). The rules were aggregated using

$$\mu_R(y) = max_i(min(\mu_{Ai}(x), (\mu_{Bi}(y)))$$

where $\mu_{Ai}(x)$ is the degree of membership for the input condition and $\mu_{Bi}(x)$ the consequent fuzzy set for output.

Step 4—Defuzzification

To convert the fuzzy output into a crisp reliability score, we applied the centroid method

$$y^* = \frac{\int y \cdot \mu(y)\, dy}{\int \mu(y)\, dy}$$

This results in a continuous control level score $y^* \in [0,10]$, interpreted as the contextual reliability level for each FRAM function.

Step 5—Control Mode Classification

The defuzzified output y^* was mapped to four discrete control modes, as defined by Hollnagel (Table 2).

Priority vectors wk for each criterion were derived using the Gaussian extension of the geometric mean method

$$w_i^{(k)} = \left(\prod_{j=1}^{n} m_{ij}^{(k)}\right)^{1/n}, \text{ normalized such that } \sum_{i=1}^{n} w_i^{(k)} = 1$$

The global priority vector w was obtained by aggregating the individual vectors wk, weighted by coefficients αk, with ∑αk = 1

$$w = \sum_{k=1}^{m} a_k.w_k$$

The weights used were

α1 = 0.4 for Thermal Impact
α2 = 0.35 for Human Reliability
α3 = 0.25 for FRAM Connectivity

These coefficients reflect the relative emphasis on physiological risk, cognitive vulnerability, and systemic criticality, respectively.

Step 4—Consistency Verification

A Probabilistic Consistency Index (PCI) was computed for each matrix, following [16], to ensure that the preference structures satisfied the transitive logic. Matrices with PCI < 0.20 were considered acceptably consistent under uncertainty.

Step 5—Final Ranking

The resulting global scores enabled the prioritization of the nine functions modeled via the FRAM and evaluated through the Fuzzy CREAM. This probabilistic formulation mitigates the limitations of crisp AHP, reduces bias from rigid hierarchies, and enhances the interpretability of prioritization in real-world mining applications. The Gaussian AHP used here adapts the approach validated by [24].

3. Results

The initial phase of the study involved a systematic literature search conducted in the Scopus database, using a comprehensive Boolean query that combined descriptors related to heat stress and mining operations. This search returned five scientific articles. After full-text analysis, one article was excluded for not meeting the predefined inclusion criteria—specifically, it lacked sufficient functional detail relevant to heat-related performance in mining environments.

As a result, four articles formed the analytical corpus used for the extraction of operational functions associated with heat stress.

Table 5 presents a structured summary of the four scientific articles selected for this study. The summarized information includes authorship, research objectives, methodologies, study sites, main findings, and applied relevance. These studies served as the empirical and conceptual foundation for the extraction of functions modeled using the Functional Resonance Analysis Method (FRAM), supporting the functional characterization and systemic mapping proposed in this research.

Table 5. Summary of selected articles on occupational conditions and green mining.

Author	Research Location and Context	Objective	Methodology	Key Results	Relevance
[25]	This study was conducted in the South African gold mining sector, specifically within the deep-level mines of the Witwatersrand region.	The primary aim was to evaluate how interdisciplinary research in human sciences—particularly physiology, psychology, and sociology—could be applied to improve health, safety, and productivity in underground mining, with an emphasis on heat acclimatization and worker resilience in hot environments.	The study utilized a combination of field experiments, physiological monitoring (including rectal temperature, oxygen consumption, and heart rate), and controlled environmental exposure in climatic rooms. From 1965 onwards, new recruits underwent an 8-day acclimatization program in chambers maintained at 32 °C wet-bulb temperature. Exercises were calibrated to simulate moderate mining workloads, gradually increasing metabolic demand over the acclimatization period.	The controlled acclimatization protocol proved highly effective. By 1970, over 280,000 workers had participated across 28 acclimatization facilities. The program led to a significant reduction in heat stroke cases and fatalities, particularly due to improvements in early detection (via oral temperature checks) and immediate on-site cooling of affected individuals. Mortality rates dropped drastically compared to earlier years, when unrecognized or untreated heat stroke had resulted in up to 50% fatality among affected miners.	The study demonstrated that large-scale, structured acclimatization significantly enhances thermal safety in mining. It also allowed for partial work shift reallocation to include safety and productivity training, while reducing the supervisory burden. Wyndham's findings contributed to the foundational practices in occupational heat stress management and laid the groundwork for modern protocols in thermal risk mitigation in mining environments.
[26]	The study included both controlled laboratory trials and field observations at mining and construction sites in Northwest Australia, where summer conditions regularly exceed conventional heat stress thresholds.	The main objective was to evaluate the Thermal Work Limit (TWL) index as a practical, accurate, and robust tool for managing occupational heat stress in both controlled environments and real-world outdoor work settings. The TWL was compared to the widely used but often overly conservative Wet Bulb Globe Temperature (WBGT) index.	Twelve physically active male participants underwent trials in a climate chamber simulating hot environments (38–40 °C dry bulb, ~28 °C wet bulb). They performed repeated 30 min work intervals at increasing workloads (40–60 W) interspersed with rest, while heart rate, core temperature (via ingestible sensors), and hydration status were monitored. Field studies involved continuous environmental and physiological monitoring of mine and construction workers across three outdoor sites, using heart rate monitors, tympanic thermometers, and urine specific gravity measurements. Environmental conditions were assessed using both TWL and WBGT indices.	In the controlled trials, the TWL accurately predicted the metabolic threshold at which physiological strain (elevated core temperature >38.2 °C or heart rate >115 bpm) would occur. Ten of the twelve subjects stayed within the predicted limits; the two outliers were likely impacted by poor acclimatization or excess body fat. In the field, the TWL provided realistic thresholds that permitted safe continuous labor even when WBGT values suggested mandatory work-rest cycles. Despite consistently high WBGTs (>30 °C), workers maintained stable heart rates, hydration, and core temperatures when the TWL exceeded 140 W/m^2. Management protocols based on the TWL were proposed, allowing unrestricted work when the TWL >220 W/m^2 and recommending restrictions and monitoring when the TWL <140 W/m^2.	This study validated the TWL as a more reliable and applicable index for heat stress management in hot workplaces compared to the WBGT. It demonstrated that the TWL better reflects the cooling potential of environmental factors like air movement, thus avoiding unnecessary productivity losses. The authors recommend TWL-based protocols for industry-wide adoption, highlighting its benefits in protecting health while sustaining operational efficiency.

Table 5. *Cont.*

Author	Research Location and Context	Objective	Methodology	Key Results	Relevance
[27]	The study was carried out in a fully mechanized underground coal mine operated by Eastern Coalfields Limited (ECL), a subsidiary of Coal India Ltd. This mine, located in eastern India, produces 3.4 million tons annually and represents a key example of mechanized underground mining in the country.	The study aimed to evaluate the ergonomic conditions of underground machine operators by analyzing their working postures and physical workload. The goal was to identify risk factors for musculoskeletal disorders (MSDs) and fatigue, which are common in deep mining due to constrained workspaces and prolonged exposure to poor environmental conditions.	Work posture was assessed using the Ovako Working Posture Analysis System (OWAS), which classifies postural stress based on joint angles and time spent in each posture. Physical workload was evaluated using pulse rate measurements captured by a Polar M430 activity monitor, and energy expenditure was estimated via the Varghese formula. Four types of machinery operators were evaluated: continuous miner (CM), universal drilling machine (UDM), road header, and load haul dumper (LHD).	Postural analysis revealed that most operators regularly assumed slightly to distinctly harmful postures. For example, CM operators spent 70% of their time with the neck bent backward (distinctly harmful), while UDM, road header, and LHD operators exhibited prolonged bent or twisted trunk positions. In terms of workload, CM operators had a light workload (average pulse: 96 bpm; energy expenditure: 6.5 kJ/min), whereas the UDM, road header, and LHD operators showed moderately heavy workloads (pulse rates ~112–114 bpm; energy expenditure ~9.1–9.4 kJ/min). These findings suggest a substantial risk of fatigue and MSDs, particularly for UDM and road header operators.	This study emphasizes the urgent need for ergonomic interventions in Indian underground mines. The authors advocate for better postural training, workload management, and the adoption of ergonomic design principles (e.g., anthropometry and participatory ergonomics) to reduce occupational health risks. The methodology and findings may be extended to other mining contexts, including opencast and non-coal operations.
[28]	The study was conducted in two underground coal mines in India (referred to as Mine A and Mine B), both operated under subsidiaries of Coal India Limited. While the mines shared similar ventilation and geological conditions, they differed in depth—150 m for Mine A and 320 m for Mine B.	The objective of the study was twofold: (1) to assess the thermal comfort status of underground coal miners using standardized thermal stress indices, and (2) to examine the correlations between thermal comfort and physiological and environmental stress factors.	Twenty male participants (10 per mine) were selected based on strict inclusion criteria: healthy BMI and body surface area, no cardiovascular or physiological disorders, acclimatized to underground work, and with normal resting ECG. Thermal comfort was evaluated using the Predicted Mean Vote (PMV) and Predicted Percentage of Dissatisfied (PPD) indices derived from environmental data collected hourly during the work shift. The Pearson correlation and Student's t-tests were applied to determine the strength of association between the PMV/PPD and various environmental and physiological variables such as heart rate, metabolic heat generation, air velocity, convection and evaporation heat exchange.	Both mines showed high thermal stress levels with PMV and PPD values exceeding comfort thresholds—Mine A had a PMV of 3.74 and PPD of 99.28%, while Mine B showed a PMV of 4.09 and PPD of 99.97%. WBGT (Wet Bulb Globe Temperature) values were also elevated compared to OSHA's recommended limits. A strong positive correlation was found between PMV and metabolic heat production ($r = 0.99$), while strong negative correlations were found with evaporative heat loss ($r = -0.99$) and convective heat exchange in breathing ($r = -0.88$). Medium-strength correlations existed between the PMV and PPD ($r = 0.60$ in Mine A and $r = 0.53$ in Mine B). These findings indicate that thermal comfort in Indian underground mines is governed by multiple interacting factors, and even relatively shallow mines can present substantial heat stress risks.	The study underscores the importance of systematically identifying and monitoring the various determinants of thermal comfort in underground environments. As Indian mining shifts toward deeper operations, understanding the interplay between physiological strain, environmental heat stress, and mine design becomes increasingly critical. The authors advocate for better air velocity management, targeted work–rest schedules, and enhanced environmental monitoring to reduce discomfort and maintain productivity in heat-stressed mines.

These articles were subjected to an automated processing procedure designed to identify function-like sentences related to thermal exposure, workload, fatigue, hydration, and human–system interaction. The process yielded a curated set of functional elements which served as the foundation for the FRAM modeling stage.

3.1. Functional Modeling Using FRAM

The automated extraction process applied to the selected corpus of scientific articles yielded a set of operational functions directly or indirectly associated with heat stress in mining contexts. Using regular expressions calibrated to detect action–object verb patterns, the analysis captured function-like sentences containing technical terms related to thermal exposure, performance degradation, and physiological stress.

The extracted content was filtered to retain only those functions that demonstrated semantic alignment with human reliability under heat conditions, resulting in a refined dataset suitable for FRAM modeling. A frequency analysis of the key technical terms identified across the extracted functions is presented in Figure 1.

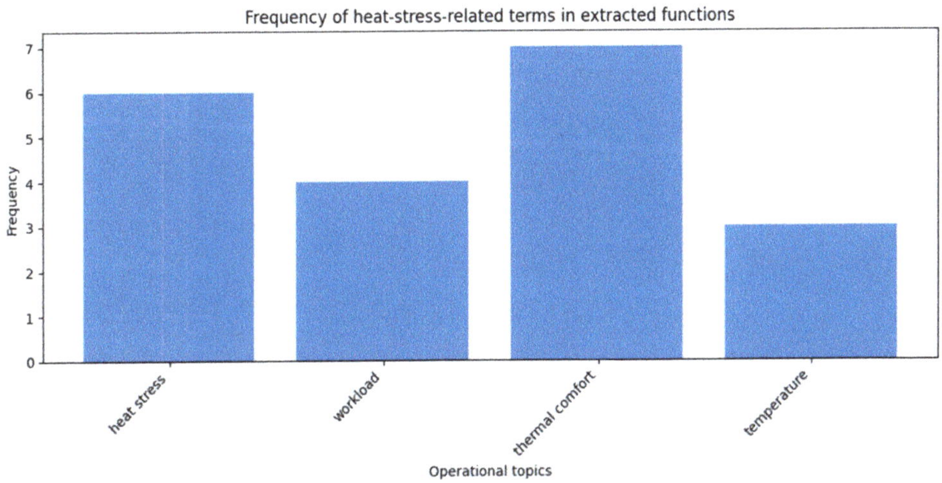

Figure 1. Related terms in extracted functions.

As shown, the most recurrent topic was "thermal comfort", appearing in seven functional descriptions, followed by "heat stress" (six occurrences), "workload" (four occurrences), and "temperature" (three occurrences). This distribution reflects a predominant concern in the literature with the perceptual and physiological dimensions of heat exposure, emphasizing the impact of environmental thermal conditions on human performance and task execution.

The prevalence of the term "thermal comfort" suggests a strong emphasis on subjective perception and environmental adaptation mechanisms, often linked to productivity and fatigue. Meanwhile, "heat stress" and "workload" represent objective dimensions of physiological strain and operational demand, commonly associated with increased risk of error and diminished reliability. "Temperature", while present with lower frequency, remains a fundamental parameter underlying all thermal risk assessments.

Figure 2 presents the functional model resulting from the application of the FRAM methodology, comprising nine critical functions extracted from the specialized literature on heat stress in mining. These functions were structured according to the six canonical FRAM aspects proposed by [9]: Input, Output, Preconditions, Resources, Control, and

Time. The model highlights key functional couplings and identifies potential points of systemic variability.

The function "Evaluate thermal comfort during shifts" directly links individual perception with operational decision-making, highlighting the subjective dimension of heat stress management. This aligns with [29], who argue for the critical role of contextualized ergonomics in underground mining operations.

In addition, the function "Quantify physical workload under thermal load conditions" emerged as a critical node due to its high degree of functional connectivity. It directly influences control mechanisms (e.g., rest breaks, task rotation), and variability in this function may trigger cascading effects on performance and safety. These findings are consistent with [30], who underscore the importance of physiological metrics in socio-technical systems under heat stress.

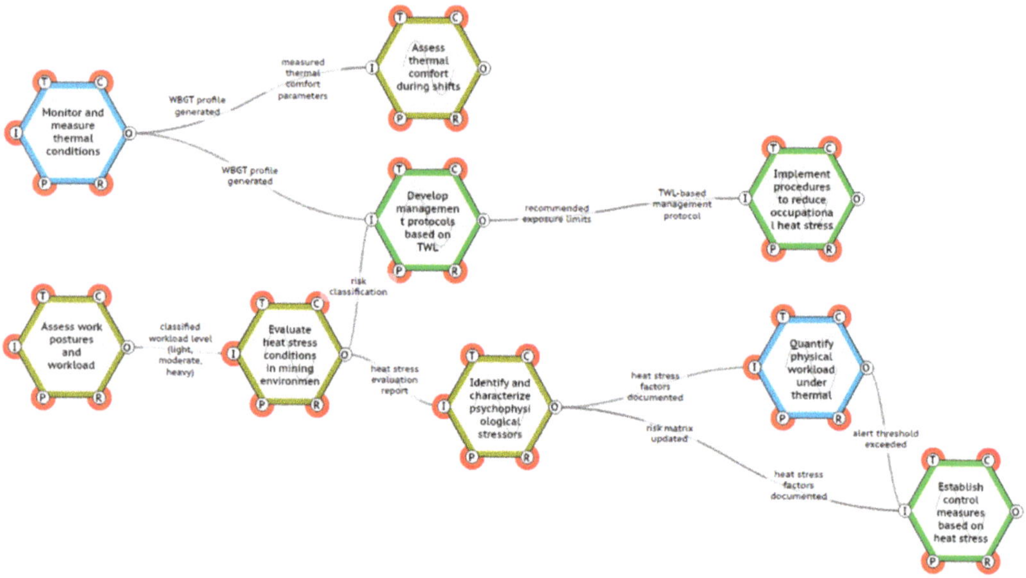

Figure 2. The functional model resulting from the application of the FRAM. The analysis revealed that the function "Monitor and measure thermal conditions" acts as a primary entry point and trigger for organizational response mechanisms. It plays a pivotal role in activating adaptive strategies and serves as a foundation for thermal risk mitigation. As emphasized by [10], identifying central functions in FRAM models helps to map elements that are most likely to amplify systemic variability.

The inclusion of functions such as "Implement control measures" and "Develop management protocols based on TWL" indicates an attempt to institutionalize operational learning. As observed by [31], resilient systems require mechanisms that support both anticipatory and responsive adjustments.

This model structure allows not only the visualization of information and resource flows under heat stress conditions but also the identification of bottlenecks and vulnerable functions likely to collapse if poorly managed. Such analysis is fundamental for guiding preventive and ergonomic interventions in extreme mining environments, as suggested by [12] when integrating human reliability with operational variability.

Beyond mapping functional structure, the FRAM framework emphasizes the propagation of performance variability as the core explanatory mechanism for emergent outcomes. In this model, everyday fluctuations in how a function is performed—due to time pressure, degraded environmental conditions, or organizational gaps—can interact non-linearly with variabilities in other functions, potentially reinforcing each other. This phenomenon, known as functional resonance, can amplify deviations in system behavior and lead to unexpected events or degraded performance [9]. For example, minor inconsistencies in evaluating heat stress may combine with reduced time availability or inadequate resources, making downstream functions (e.g., decision-making or emergency response) increasingly fragile. These dynamic interactions highlight the system's sensitivity to context, especially in thermally stressful environments like underground mining. Modeling this through the FRAM allows us not only to describe function couplings, but to identify where variability accumulates and propagates—a critical insight for anticipating systemic fragility under operational stress [10].

3.2. *Human Reliability Assessment Using Fuzzy CREAM*

Building upon the functional structure modeled via FRAM, a fuzzy-based adaptation of the Cognitive Reliability and Error Analysis Method (CREAM) was applied to quantify the human reliability associated with each critical function. The FRAM outputs provided the functional scope and interdependencies necessary to parameterize the context of control modes in the fuzzy logic system.

The fuzzy CREAM enables the translation of qualitative control context factors into reliability scores by modeling uncertainty and partial truth through fuzzy membership functions. In this study, each of the nine key functions extracted from the literature and structured via the FRAM model were assessed in terms of human reliability, resulting in a fuzzy score ranging from 0 to 10. These scores were categorized into four control modes—scrambled (0–2.5), inefficient (2.5–5.0), tolerable (5.0–7.5), and efficient (7.5–10)—based on the original CREAM framework proposed by [32] and further adapted through fuzzy extensions for industrial environments [12,13]. The thresholds were contextualized for mining operations, reflecting typical decision-making structures and cognitive demands encountered under thermal stress.

The resulting scores varied widely across functions, reflecting how contextual conditions and task characteristics influence operator performance under heat stress. The integration of the FRAM and fuzzy CREAM thus provides a layered analysis: while the FRAM reveals systemic variability and functional interconnections, the Fuzzy CREAM quantifies the cognitive demands and reliability potential of each function under real-world thermal stress conditions.

As shown in Figure 3, the function achieved a reliability score of 8.64, classifying it within the efficient control mode.

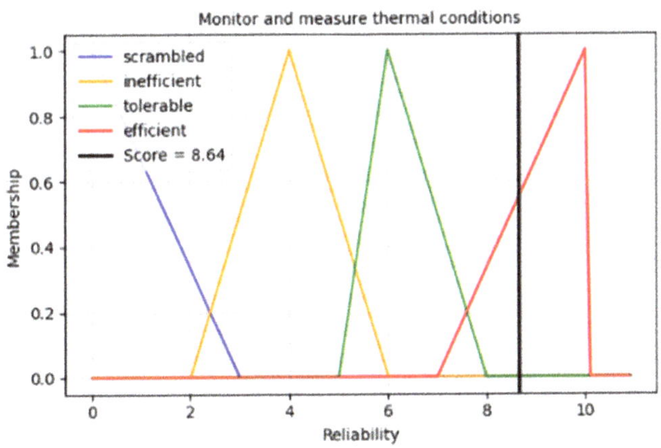

Figure 3. Monitor and measure thermal conditions.

This high level of reliability reflects the procedural and instrument-based nature of the task, which is typically supported by standardized protocols and real-time environmental monitoring tools. The stability of this function reinforces its role as a critical enabler of system feedback and adaptation. As highlighted by [9], functions with clear input–output relationships and strong control structures tend to exhibit high consistency, especially when automation assists in execution.

The function in Figure 4 achieved a score of 4.76, at the upper limit of the inefficient control range.

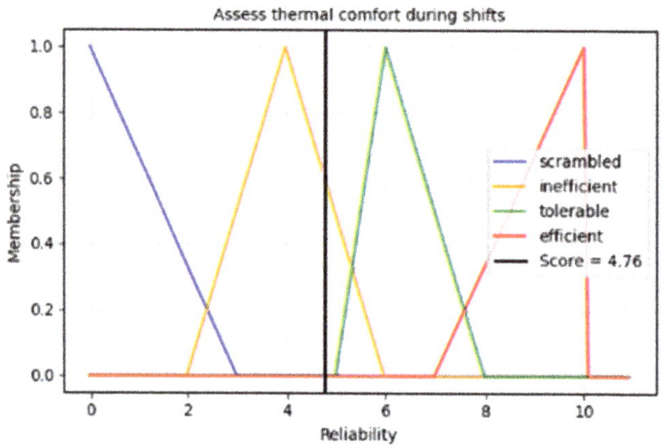

Figure 4. Assess thermal comfort during shifts.

Thermal comfort is inherently subjective and influenced by numerous personal and environmental variables, including clothing, humidity, acclimatization, and metabolic rate. This result indicates that while some structured procedures exist (e.g., surveys, observation protocols), the reliance on individual perception introduces inconsistency, corroborating the findings of [29], who emphasize the challenge of operationalizing subjective experience in safety-critical environments.

In contrast, the function shown in Figure 5 received a much lower score of 1.22, placing it in the scrambled mode.

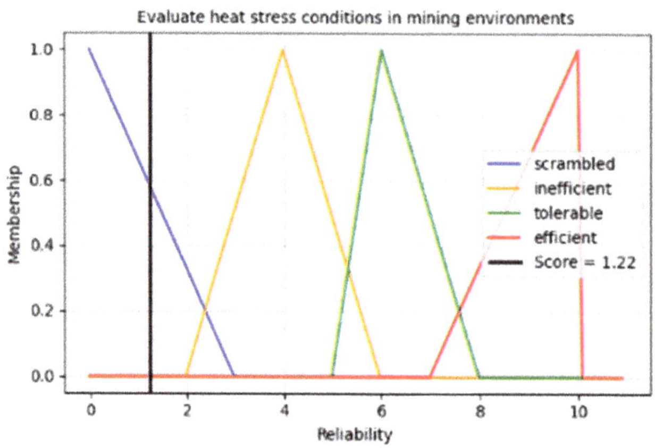

Figure 5. Evaluate heat stress conditions in mining environments.

This suggests a high degree of uncertainty and cognitive overload, likely stemming from the need to synthesize multiple data sources (e.g., thermal readings, physiological signals, workload classification) in dynamic conditions. The function's centrality in the FRAM model further amplifies its risk potential. According to [3], diagnostic functions that depend on complex judgment under stress conditions often present low reliability when context factors (e.g., fatigue, time pressure) are unfavorable.

Similarly, the function in Figure 6 scored 1.11, indicating scrambled control. This result reinforces the systemic vulnerability of this function, which often depends on indirect signals and non-observable indicators (e.g., cognitive fatigue, emotional stress, reduced alertness).

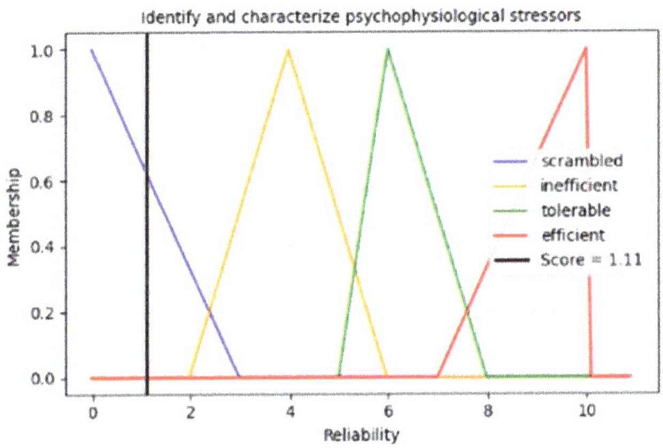

Figure 6. Identify and characterize psychophysiological stressors.

As observed by [30], psychophysiological factors are difficult to quantify in real time and are rarely integrated systematically into operational decision-making, increasing the likelihood of undetected performance degradation.

The function in Figure 7 scored 2.81, placing it within the inefficient control category. This reflects the complexity of estimating the workload in fluctuating thermal environments, especially when biomechanical assessments are not continuously available.

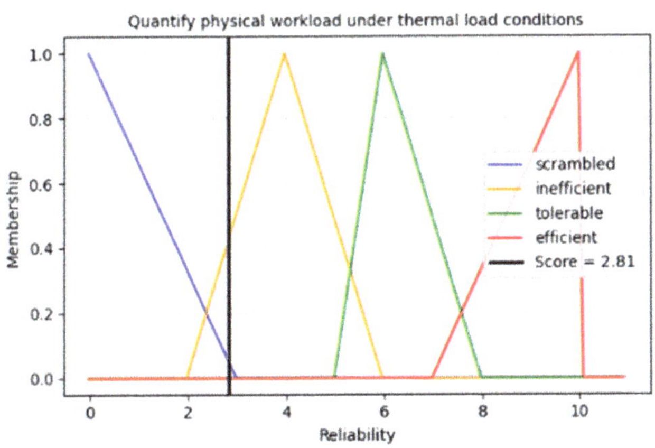

Figure 7. Quantify physical workload under thermal load conditions.

Workload is highly variable and depends not only on task type but also on posture, pace, hydration status, and recovery time—factors that are often underestimated during operational planning [4–6].

Conversely, the function in Figure 8 received a score of 4.76, like the thermal comfort assessment. Although the evaluation of postures may be partially supported by observational tools or ergonomic checklists, the context of mining—particularly in confined or irregular underground spaces—limits the consistency and repeatability of such assessments.

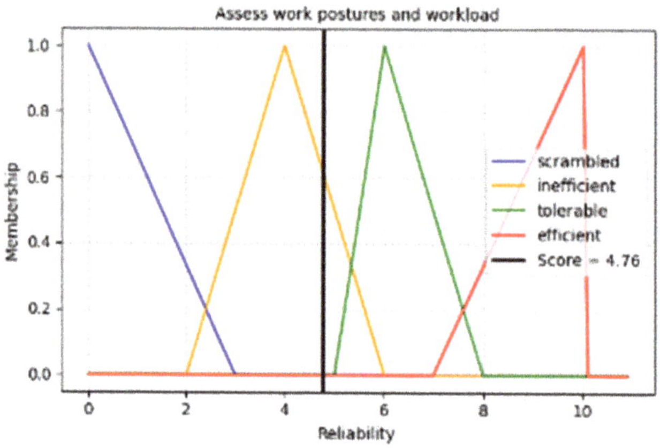

Figure 8. Assess work postures and workload.

As noted by [31], ergonomic functions that rely on visual inspection and self-reporting are particularly susceptible to contextual noise and evaluator bias.

Figure 9 shows a reliability score of 1.36, again within the scrambled category. This low score may be attributed to the lag between data collection and procedural implementation, as well as the need to adapt general recommendations to specific site characteristics and workforce variability.

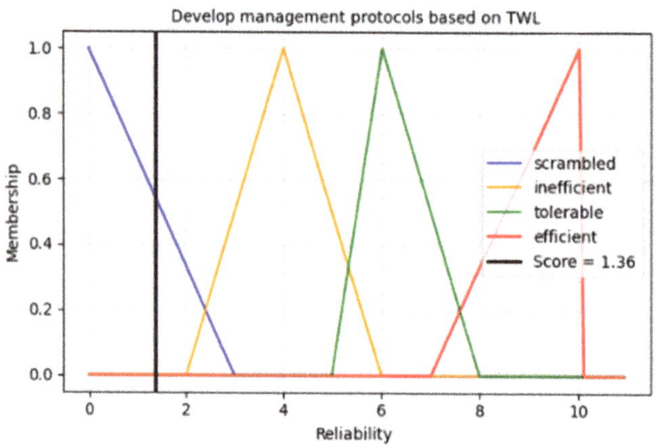

Figure 9. Develop management protocols based on the TWL.

TWL-based management demands a synthesis of meteorological data, metabolic load, and organizational capacity, making it highly sensitive to planning gaps and contextual instability.

In contrast, the function in Figure 10 classifies it as tolerable. This suggests that once procedures are defined (e.g., scheduled rest breaks, provision of fluids, shaded areas), they can be executed with moderate consistency.

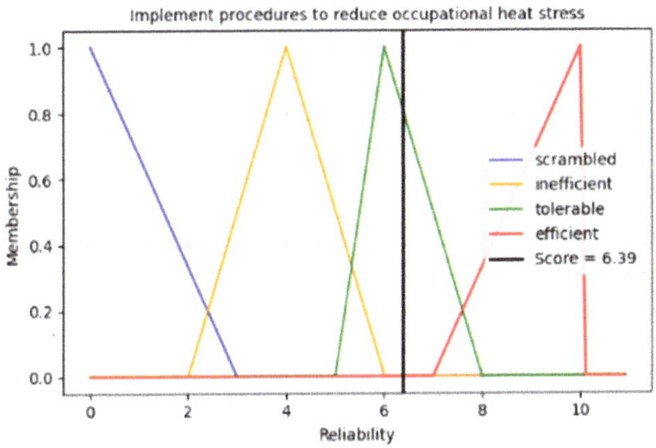

Figure 10. Implement procedures to reduce occupational heat stress.

However, their effectiveness still depends on adherence, supervision, and reinforcement mechanisms. This aligns with the view of [31], who emphasize the importance of enabling factors (e.g., leadership support, worker autonomy) in determining the actual reliability of procedural interventions.

Finally, the function "Establish control measures based on heat stress data" (Figure 11) matched the highest reliability score at 8.64, also in the efficient range.

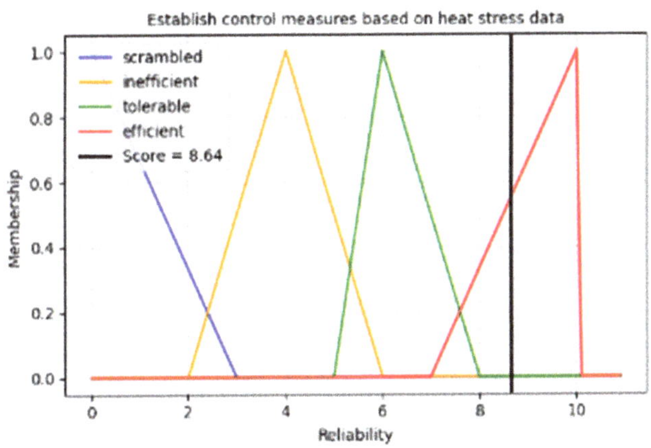

Figure 11. Establish control measures based on heat stress data.

This confirms that once empirical thresholds (e.g., WBGT limits) are reached, triggering predefined responses (e.g., work stoppage, alerts) tends to occur with high regularity. These mechanisms are often integrated into safety management systems and benefit from formal institutionalization, which explains their robust control.

3.3. Prioritization Using Gaussian AHP

Building upon the functional modeling performed using the FRAM and the human reliability scores obtained through the Fuzzy CREAM, the Gaussian Analytic Hierarchy

Process (AHP) was applied to derive a final ranking of critical functions. This integration offers a structured and quantitative approach to prioritization, accounting for both system connectivity and human performance under thermal stress conditions.

Three decision criteria guided the multicriteria evaluation as follows:

- C1—Thermal impact: the degree to which each function is sensitive to heat exposure and contributes to thermal risk propagation.
- C2—Human reliability: the fuzzy score derived from the CREAM assessment, indicating the cognitive robustness of each function.
- C3—FRAM connectivity: the level of interdependence and influence each function exerts in the functional network.

Each function was evaluated across the three criteria using Gaussian-distributed pairwise comparisons, which model judgment uncertainty more realistically than conventional crisp AHP. The normalized weights for each criterion were aggregated to obtain the final prioritization score for each function, as presented in Table 6.

Table 6. Final prioritization score.

Function	Thermal Impact (C1)	Human Reliability (C2)	FRAM Connectivity (C3)	Final Score
Evaluate heat stress conditions	0.92	0.91	0.90	0.91
Quantify physical workload	0.89	0.87	0.88	0.88
Identify psychophysiological stressors	0.85	0.86	0.84	0.85
Monitor and measure thermal conditions	0.80	0.82	0.83	0.82
Assess thermal comfort	0.78	0.79	0.81	0.79
Assess postures and workload	0.75	0.74	0.73	0.74
Establish control measures	0.70	0.72	0.71	0.71
Implement heat stress procedures	0.68	0.70	0.69	0.69
Develop protocols based on the TWL	0.65	0.68	0.67	0.67

Among the prioritized functions, "To Develop Management Protocols Based on TWL" received the lowest global score. This result reflects both methodological and contextual limitations. While the Thermal Work Limit (TWL) is widely used as a reference for managing heat exposure in industrial settings, it assumes relatively stable environmental and metabolic conditions [33]. In underground mining, however, thermal loads fluctuate dynamically due to ventilation changes, depth, workload variations, and unexpected operational demands. Static or pre-defined protocols, even when based on robust indices like the TWL, often fail to respond in real-time to such variability. Moreover, this function is dependent on the prior accuracy of environmental and physiological assessments—any deviation upstream (e.g., misclassification of heat stress or poor monitoring) compromises the relevance of the protocol generated. This fragility justifies its low reliability (C2), moderate thermal impact (C1), and limited systemic connectivity (C3), as shown in the model.

As the table shows, the function "Evaluate heat stress conditions in mining environments" was confirmed as the most critical across all three dimensions, emphasizing its centrality, high vulnerability, and strong thermal sensitivity. This reinforces the findings of the FRAM and fuzzy reliability analyses, validating the systemic risk associated with this function.

At the opposite end, functions like "Develop protocols based on TWL" and "Implement heat stress procedures" received the lowest aggregate scores. While these functions are crucial at the procedural level, their influence is limited in dynamic, real-time decision contexts, and they tend to operate under higher levels of structural support and predictability.

Overall, the Gaussian AHP results consolidate the insights from the previous phases and provide a decision-support basis for targeting the most fragile and influential functions in heat-stress-exposed mining systems.

3.4. Practical Implications for the Mining Industry

To enhance the practical relevance of this study, we propose a multifaceted implementation framework tailored to mining operations. First, the integration of real-time decision-support systems that fuse data from environmental sensors, wearable physiological monitors, and predictive analytics is crucial. These systems can proactively detect early markers of heat-induced cognitive decline, such as diminished attention and increased reaction time, which have been shown to impair worker safety and efficiency under thermal stress [34].

Second, organizational variables—notably shift scheduling, workload distribution, and leadership involvement—must be dynamically adapted to account for thermal load variability and its impact on executive function, vigilance, and memory [35].

Third, targeted technical training is recommended to bolster miners' ability to recognize and respond to symptoms of heat stress. This is particularly critical for roles involving decision-intensive tasks, which have been identified as most vulnerable in functional resonance analysis model (FRAM) simulations. Cognitive impairments under heat exposure include elevated omission errors and slowed response times, especially during extended shifts [36].

Finally, emergency protocols should be revised to include tiered interventions based on function-specific risk stratification. This ensures that critical operations receive priority support during thermal events, thereby enhancing operational resilience.

These strategies not only align with the priority functions highlighted in our hybrid modeling approach but also provide a structured roadmap for converting analytical insights into actionable, system-level resilience measures within the mining sector.

4. Limitations

This study offers a novel hybrid approach to model human reliability under thermal stress in mining environments. However, several limitations must be acknowledged to contextualize the scope and applicability of the findings.

First, the FRAM modeling was based on a restricted corpus of four scientific studies selected through a systematic review. Although these articles provided high-relevance empirical data on heat exposure and human performance in mining contexts, the small sample size may limit the generalizability of the functional model. Future studies should expand the literature base and consider broader operational settings—including surface mining, artisanal mining, and different climatic zones—to strengthen the external validity of the functional structure.

Second, while the functions modeled reflect domain-specific realities in mining, the structure and logic of FRAM have been successfully applied in other high-risk industries such as offshore oil and gas, commercial aviation, and maritime operations. Comparing results across such domains could help identify transferable patterns of human–system variability and refine the current model.

Third, the fuzzy inference and pairwise comparison methods (used in the fuzzy CREAM and Gaussian AHP, respectively) inherently rely on expert judgment and predefined linguistic rules. These assumptions introduce subjectivity, particularly when estimating contextual reliability or prioritizing functions under uncertainty. Although these techniques help manage imprecision, their outcomes are sensitive to how input scales and rule bases are defined. Incorporating sensitivity analyses or expert consensus panels may enhance robustness in future applications.

5. Conclusions

The results demonstrated that functions involving complex judgment, subjective assessment, or indirect indicators, such as evaluating heat stress conditions and identifying psychophysiological stressors, are the most vulnerable under thermal exposure. These functions showed both low reliability scores in the fuzzy analysis and high systemic connectivity in the FRAM model, indicating that performance failures in these areas may propagate variability throughout the system. In contrast, functions supported by standardized procedures and defined control mechanisms, such as monitoring thermal conditions or implementing control measures, demonstrate higher contextual reliability and greater operational consistency.

The integration of the three methods proved effective in translating qualitative system behavior into structured analytical insights. The use of the Gaussian AHP model further supported the prioritization of critical functions, emphasizing that thermal exposure compromises not only individual performance but also the structural stability of the operational network.

These findings confirm that human reliability in mining environments exposed to heat is influenced by multiple interacting dimensions, including organizational capacity, environmental dynamics, and functional dependencies. Addressing thermal risk in such settings requires more than individual protection or environmental thresholds. It demands a comprehensive approach that considers the specific vulnerabilities of each function and strengthens support mechanisms where variability is most likely to emerge.

Future developments may include the integration of sensor-based information, predictive algorithms, and empirical field validation to enhance the applicability and precision of the model. By identifying areas of fragility and prioritizing improvement efforts, this framework offers a valuable contribution to safety strategies and operational resilience in high-temperature mining scenarios.

Supplementary Materials: The following supporting information can be downloaded at: https://www.mdpi.com/article/10.3390/mining5030050/s1; Table S1: A complete list of functions, their descriptions, and associated aspects.

Funding: This research received no external funding.

Data Availability Statement: Data supporting the findings of this study, including the FRAM functional model and AHP matrices, are available from the corresponding author upon reasonable request.

Acknowledgments: During the preparation of this manuscript, the author used ChatGPT-4 (OpenAI, 2025) for assistance in refining the structure, improving language clarity, and ensuring adherence to journal formatting guidelines. The author has reviewed and edited all AI-generated content and takes full responsibility for the final version of the publication.

Conflicts of Interest: The authors declare no conflicts of interest.

References

1. Lazaro, P.; Momayez, M. Heat Stress in Hot Underground Mines: A Brief Literature Review. *Min. Metall. Explor.* **2021**, *38*, 497–508. [CrossRef]
2. Maurya, T.; Karena, K.; Vardhan, H.; Aruna, M.; Raj, M.G. Effect of Heat on Underground Mine Workers. *Procedia Earth Planet. Sci.* **2015**, *11*, 491–498. [CrossRef]
3. Yoon, Y.S.; Ham, D.H.; Yoon, W.C. A new approach to analysing human-related accidents by combined use of HFACS and activity theory-based method. *Cogn. Technol. Work* **2017**, *19*, 759–783. [CrossRef]

4. Fadeev, G.; Goryaev, D.V.; Zaitseva, N.V.; Shur, P.Z.; Red'ko, S.V.; Fokin, V. Health Disorders in Workers Associated with Health Risks at Workplaces in Mining Industry in the Arctic (Analytical Review). *Health Risk Anal.* **2023**, *2023*, 173–182. [CrossRef]
5. Härmä, M. Workhours in relation to work stress, recovery and health. *Scand. J. Work Environ. Health* **2006**, *32*, 502–514. [CrossRef] [PubMed]
6. Molek-Winiarska, D.; Kawka, T. Reducing Work-Related Stress Through Soft-Skills Training Intervention in the Mining Industry. *Hum. Factors* **2022**, *66*, 1633–1649. [CrossRef]
7. Guoshan, W.; Heqing, L.; Zhirong, W.; You, B.; Jufeng, Z.; Yiming, H. A heat stress control method for miners based on internal heat storage. *Sci. Technol. Built Environ.* **2024**, *31*, 484–493. [CrossRef]
8. Lazaro, P.; Momayez, M. Validation of the Predicted Heat Strain Model in Hot Underground Mines. *Min. Metall. Explor.* **2019**, *36*, 1213–1219. [CrossRef]
9. Hollnagel, E. *FRAM: The Functional Resonance Analysis Method*; CRC Press: London, UK, 2012.
10. Patriarca, R.; Di Gravio, G.; Woltjer, R.; Costantino, F.; Praetorius, G.; Ferreira, P.; Hollnagel, E. Framing the FRAM: A literature review on the functional resonance analysis method. *Saf. Sci.* **2020**, *129*, 1–23. [CrossRef]
11. Konstandinidou, M.; Nivolianitou, Z.; Kiranoudis, C.; Markatos, N. A fuzzy modeling application of CREAM methodology for human reliability analysis. *Reliab. Eng. Syst. Saf.* **2006**, *91*, 706–716. [CrossRef]
12. Marseguerra, M.; Zio, E.; Librizzi, M. Human reliability analysis by fuzzy "CREAM". *Risk Anal.* **2007**, *27*, 137–154. [CrossRef] [PubMed]
13. Shi, H.; Wang, J.H.; Zhang, L.; Liu, H.C. New improved CREAM model for human reliability analysis using a linguistic D number-based hybrid decision making approach. *Eng. Appl. Artif. Intell.* **2023**, *120*, 105896. [CrossRef]
14. Lee, D.; Kim, H.; Koo, K.; Kwon, S. Human Reliability Analysis for Fishing Vessels in Korea Using Cognitive Reliability and Error Analysis Method (CREAM). *Sustainability* **2024**, *16*, 3780. [CrossRef]
15. Alencar Silva Almeida Dantas, G.; Russo, A.C.; De Tomi, G.F.C. Integrating FRAM and Fuzzy Logic for the Analysis of Critical Functions and Human Reliability in Loading Operations in Underground Mining. *J. Saf. Sustain.* **2025**. [CrossRef]
16. Marins, C.S.; Souza, D.D.O.; Barros, M.D.S. O USO DO MÉTODO DE ANÁLISE HIERÁRQUICA (AHP) NA TOMADA DE DECISÕES GERENCIAIS—UM ESTUDO DE CASO. In Proceedings of XLI SBPO, Porto Seguro, Brasil, 1–4 September 2009; Available online: https://www2.ic.uff.br/~emitacc/AMD/Artigo%204.pdf (accessed on 23 July 2025).
17. Russo, A.C.; Russo, E. Application of the AHP-Gaussian method to support the prioritization of workers' health actions in Brazil, based on data from DATASUS. *Gest. Prod.* **2024**, *31*, e10423. [CrossRef]
18. Mahdi Rezaie, F.; Fakoor Saghih, A.M.; Motahari Farimani, N. A novel hybrid approach based on CREAM and fuzzy ANP to evaluate human resource reliability in the urban railway. *J. Transp. Saf. Secur.* **2021**, *13*, 1326–1364. [CrossRef]
19. Patriarca, R.; Di Gravio, G.; Costantino, F. A Monte Carlo evolution of the Functional Resonance Analysis Method (FRAM) to assess performance variability in complex systems. *Saf. Sci.* **2017**, *91*, 49–60. [CrossRef]
20. Maia França, J.E.; Hollnagel, E. Human Factors Approach to Assess Risks and Reliability in Offshore Operations with FRAM (Functional Resonance Analysis Method). In Proceedings of the Offshore Technology Conference Brasil, OTCB, Rio de Janeiro, Brasil, 24–26 October 2023. [CrossRef]
21. Fargnoli, M.; Murgianu, L.; Tronci, M. The Functional Resonance Analysis Method (FRAM) Application in the Healthcare Sector: Lessons Learned from Two Case Studies on Medical Device Management. *Appl. Sci.* **2024**, *14*, 9495. [CrossRef]
22. Huang, Y.; Liu, Y.; Zhao, D.; Song, X. Fuzzy comprehensive performance evaluation method for radar control. In CICTP 2012: Multimodal Transportation Systems—Convenient, Safe, Cost-Effective, Efficient, Proceedings of the 12th COTA International Conference of Transportation Professionals, Beijing, China, 3–6 August 2012; American Society of Civil Engineers: Reston, VA, USA; pp. 1845–1856. [CrossRef]
23. Mamdani, E.H. Application of Fuzzy Algorithms for Control of Simple Dynamic Plant. *Proc. Inst. Electr. Eng.* **1974**, *121*, 1585–1588. [CrossRef]
24. Dos Santos, M.; de Araújo Costa, I.P.; Gomes, C.F.S. Multicriteria Decision-Making In The Selection Of Warships: A New Approach To The Ahp Method. *Int. J. Anal. Hierarchy Process* **2021**, *13*, 147–169. [CrossRef]
25. Wyndham, C.H. Research in the human sciences in the gold mining industry. *Am. Ind. Hyg. Assoc. J.* **1974**, *35*, 113–136. [CrossRef]
26. Miller, V.S.; Bates, G.P. The thermal work limit is a simple reliable heat index for the protection of workers in thermally stressful environments. *Ann. Occup. Hyg.* **2007**, *51*, 553–561. [CrossRef]
27. Sakinala, V.; Paul, P.S.; Chandrakar, S. Assessment of Work Postures and Physical Workload of Machine Operators in Underground Coal Mines. *J. Inst. Eng. (India) Ser. D* **2023**, *104*, 87–98. [CrossRef]
28. Dey, S.; Dey, N.C. Determining factors affecting thermal comfort in underground coal mine. *J. Mines Met. Fuels* **2019**, *67*, 489.
29. França, J.E.M.; Hollnagel, E. Analyzing human factors and complexities of mining and O&G process accidents using FRAM: Copiapó (Chile) and FPSO CSM (Brazil) cases. *Process Saf. Prog.* **2023**, *42*, S9–S18. [CrossRef]
30. Hirose, T.; Sawaragi, T. Development of FRAM Model Based on Structure of Complex Adaptive Systems to Visualize Safety of Socio-Technical Systems. *IFAC-PapersOnLine* **2019**, *52*, 13–18. [CrossRef]

31. Vieira, L.C.; Saurin, T.A. Environmental disaster analysis: Case study using the functional resonance analysis method. *Eng. Sanit. Ambient.* **2018**, *23*, 373–383. [CrossRef]
32. Dekker, S.; Hollnagel, E. Computers in the cockpit: Practical problems cloaked as progress. In *Coping with Computers in the Cockpit*; Routledge: London, UK, 2018; pp. 1–6. [CrossRef]
33. Brake, D.J.; Bates, G.P. Limiting metabolic rate (Thermal Work Limit) as an index of thermal stress. *Appl. Occup. Environ. Hyg.* **2002**, *17*, 176–186. [CrossRef]
34. Yeoman, K.; Weakley, A.; DuBose, W.; Honn, K.; McMurry, T.; Eiter, B.; Baker, B.; Poplin, G. Effects of heat strain on cognitive function among a sample of miners. *Appl. Ergon.* **2022**, *102*, 103743. [CrossRef]
35. Schmit, C.; Hausswirth, C.; Le Meur, Y.; Duffield, R. Cognitive Functioning and Heat Strain: Performance Responses and Protective Strategies. *Sports Med.* **2017**, *47*, 1289–1302. [CrossRef]
36. Rastegar, Z.; Ghotbi Ravandi, M.R.; Zare, S.; Khanjani, N.; Esmaeili, R. Evaluating the effect of heat stress on cognitive performance of petrochemical workers: A field study. *Heliyon* **2022**, *8*, e08698. [CrossRef]

mining

MDPI

Article

Study of Weak-Acid-Dissociable and Free Cyanide Oxidation by Ozone Injection into Gold Mine Pulp

Coraquetzali Magdaleno López [1], Saúl Ortiz Landeros [2], Héctor Herrera Hernández [3], Eugenia Aldeco Pérez [4], Carlos Estrada Arteaga [4], Antonia Sandoval González [4] and Jorge Morales Hernández [4,*]

1 Tecnológico Nacional de México/ITS Abasolo, Blvd. Cuitzeo de los Naranjos #401, Col. Cuitzeo de los Naranjos, Abasolo 36976, Guanajuato, Mexico; coraquetzali.ml@abasolo.tecnm.mx

2 Fresnillo Plc, calzada Saltillo 400 989, Residencial Campestre la Rosita, Torreón 27250, Coahuila, Mexico; saulortizlanderos@gmail.com

3 Universidad Autónoma del Estado de México, Blvd. Universitario s/n, Atizapán de Zaragoza 54500, Estado de México, Mexico; hherrerah@uaemex.mx

4 Centro de Investigación y Desarrollo Tecnológico en Electroquímica, Parque Tecnológico Querétaro s/n, Sanfandila, Pedro Escobedo 76703, Querétaro, Mexico; ealdeco@cideteq.mx (E.A.P.); cestrada@cideteq.mx (C.E.A.); asandoval@cideteq.mx (A.S.G.)

* Correspondence: jmorales@cideteq.mx

Abstract

The effects of key variables on weak-acid-dissociable (WAD) and free cyanide oxidation by ozone injection in gold mine pulp were studied at laboratory scale to find an alternative cyanide treatment. A fractional factorial analysis of five process variables (O_3/O_2 flow, reaction time, NH_4HSO_3 concentration, temperature, and pH) informed a 60-run experimental matrix, in a 1 L cylindrical reactor, with the process variables controlled during the ozone injection. The findings may inform future strategies for safer cyanide management in gold mining processes. Free cyanide is the most toxic form of cyanide. Its oxidation increases with higher O_3/O_2 concentrations, longer exposure time, and higher pH. Maintaining a pH above 7 is crucial. Lower pH values favor the dissociation of cyanide into its toxic, free form. WAD cyanide oxidation depends mainly on the O_3/O_2 concentration, exposure time, and NH_4HSO_3 concentration. Increasing O_3/O_2 and time enhanced both WAD and free cyanide oxidation, while NH_4HSO_3 concentration affected oxidation rates differently. The results show that free cyanide was significantly more oxidized (84.1413%) than WAD cyanide (67.2423%). Controlling the WAD cyanide process yields excellent free cyanide oxidation. This represents ongoing improvement at an industrial scale. This approach quantifies the extent to which process variables affect the WAD and free cyanide oxidation under controlled conditions, thereby greatly reducing environmental impact.

Keywords: cyanide oxidation; ozone injection; environmental impact; gold mine pulp

1. Introduction

1.1. Generalities of Cyanide

Since the 19th century, the mining industry has used cyanide to extract gold from low-grade ores due to its stability in forming soluble cyanide–gold complexes and its efficiency in selectively dissolving gold [1]. However, the process generates leachates and sludges; that is, highly toxic cyanide complexes are produced, including cadmium, copper, zinc, free cyanide, and weak-acid-dissociable (WAD) cyanide [2,3]. Unfortunately, when these complexes are not properly disposed of, they cause serious harm to humans and

 https://doi.org/10.3390/mining6010009

aquatic ecosystems [4,5]. The persistence and toxicity of cyanide complexes necessitate the development and evaluation of treatment technologies. Among the most studied conventional methods are air/SO_2 oxidation (INCO process), hydrogen peroxide oxidation, alkaline chlorination, ultrasound, electrochemical oxidation, and biological treatments [6]. Although technically viable, these methods generate high solids loads, require expensive reagents, have low efficiency, and produce chemically complex byproducts that must be transformed into non-hazardous waste for disposal in accordance with applicable regulations [3,7]. On the other hand, ozonation, an advanced oxidation process, is a promising alternative for oxidizing free cyanide and WAD complexes due to its high oxidation potential (E° = 2.07 V) [8].

Cyanide complexes can be oxidized to form cyanate (CNO^-), CO_2, and nitrogen, depending on the operating conditions (pH, O_3 dosage, retention time, etc.). These oxidation reactions have been extensively studied [9]. Despite evidence of ozone's effectiveness in oxidizing cyanides in mining slurries at the laboratory level, our understanding of oxidation mechanisms and kinetics under real-world plant conditions remains limited [10]. Therefore, this research proposes to address ozonation by directly injecting ozone into the slurry of an operating mine. Specifically, the objective of this research is to evaluate the efficiency and oxidation mechanisms of free cyanide and weak-acid-dissociable (WAD) species.

The anion CN^- is a strong binder of metal cations solvated in an aqueous solution, a property used in many commercial processes, such as gold mining [1].

Cyanide compounds are used in large quantities, amounting to 2.6 million tons annually. Sodium cyanide, used in precious metals refining, the chemical industry, and electroplating, among other processes, accounts for 14% of global cyanide compound production [1]. Some cyanide compounds are generated as solid waste, released into the atmosphere, or associated with wastewater, as in aluminum production, the steel industry, and gas production, to name a few; however, as pollutants, they are most commonly found in water and soil. Due to the high toxicity of free cyanide (HCN and CN^-) in freshwater systems, the maximum level is 22 µg/L, as aquatic organisms are significantly more sensitive to cyanide than humans [1].

Cyanide compounds dissociating with a weak acid (at pH 4.5) are known as weak-acid-dissociable cyanide (WAD cyanide) and are simple salts that can produce free cyanide. Cyanide compounds that dissociate under the action of a strong acid (pH less than 2) are more complex salts that are known as strong-acid-dissociable cyanide (SAD cyanide). Under normal conditions, SAD cyanide is more resistant than WAD cyanide, forming a stable complex in solutions of gold, iron, and cobalt.

Total cyanide refers to all cyanide compounds present in solution. Table 1 shows the different cyanide species in the liquid phase [1], where metal–cyanide complexes are divided into weak and strong complexes for industrial applications.

Table 1. Classification of metal–cyanide complexes in aqueous solution [1].

Phase	Classification	Compound
Liquid (water)	Free cyanide	HCN, CN^-
	Metal–cyanide complexes	Weak complexes $Ag(CN)^{2-}, CdCN^-, Zn(CN)_4^{2-}, Cd(CN)_3^-, Cd(CN)_4^{2-}$ Strong complexes $Fe(CN)_6^{4-}, Fe(CN)_6^{3-}, Co(CN)_6^{4-}, Au(CN)_2$
	Cyanate, thiocyanate	CNO^-, SCN^-
	Organocyanides	Nitriles, cyanohydrins

https://doi.org/10.3390/mining6010009

The free cyanide corresponds to the soluble hydrogen cyanide ($HCN(aq)$), which is a weak acid and can dissociate into the cyanide ion (CN^-), according to Reaction 1. The thermodynamic equilibrium of this reaction, like a function of pH and electrochemical potential, is shown in Figure 1; at pH < 9.3, HCN is the dominant free cyanide species, whereas at higher pH, the cyanide ion (CN^-) is the dominant species. A positive electrochemical potential is required to form cyanate (CNO^-) under both acidic and alkaline conditions.

$$HCN(aq) = H^+ + CN^- \tag{1}$$

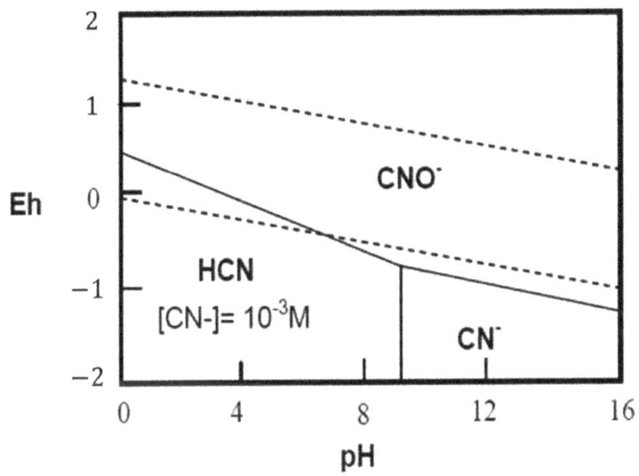

Figure 1. Electrochemical potential vs. pH diagram for the system CN-H_2O at 25 °C.

Hydrogen cyanide has a high vapor pressure; therefore, it volatilizes rapidly from liquid surfaces under ambient conditions, reducing the cyanide concentration in the solution. This means that cyanide leach systems are operated at pH values above 10 to minimize losses [3]. Another pathway is that hydrogen cyanide and free cyanide can be oxidized with little change in system potential, forming CN^-, CNO^-, and HCN (Figure 1).

The evolution of the species depends on the electrochemical potential and pH, creating opportunities for electrochemical oxidation, where the hydroxyl radical drives the development of special electrodes, such as doped diamond boron (DDB), which offer a wide potential window. However, this research establishes the basis for developing a new process that maximizes efficiency while minimizing costs.

The stability of metal–cyanide complexes varies, requiring pH conditions ranging from moderate to highly acidic for dissociation. As observed in reaction 2, the dissociation of metal–cyanide complexes releases free cyanide [4].

$$M(CN)_x^{n-} = M^+ + xCN^- \tag{2}$$

Weak metal–cyanide complexes are those in which cyanide ions are weakly bound to the metal cation, allowing them to dissociate under mildly acidic conditions (pH = 4 to 6) to release free cyanide. Because of their dissociative nature, these complexes are regulated in concert with free cyanide in water [11].

Strong metal–cyanide complexes include those with heavy transition metals, such as iron, cobalt, platinum, and gold. These complexes require highly acidic conditions (pH < 2) to dissociate and release free cyanide. Compared to weak complexes, strong metal–cyanide complexes are more stable in aqueous solution and are relatively less toxic [12].

https://doi.org/10.3390/mining6010009

As shown below (reactions 3 and 4), free cyanide can be oxidized to cyanate (CNO^-) or hydrogen cyanate ($HCNO$), depending on pH [5].

$$CN^- + H_2O = CNO^- + 2H^+ + 2e^- \quad (3)$$

$$HCN + H_2O = HCNO + 2H^+ + 2e^- \quad (4)$$

Strong oxidants, such as chlorine, hypochlorite, ozone, and hydrogen peroxide, can oxidize free cyanide. However, chlorine generates highly toxic byproducts. Under neutral to alkaline conditions, the final product is cyanate (CNO^-), a non-toxic compound [13].

Gurol M.D. et al. (1985) studied the ozonation of free cyanide and found that ozone (O_3) primarily reacts with the cyanide ion, while its reaction with HCN is minimal [5]. It was determined that the presence of free cyanide promotes the formation of free radicals (HO·, HOO·), where both radical-driven reactions and the direct reaction of free cyanide with ozone contribute to its oxidative destruction. Some of the responses identified by Gurol and Bremen for the ozonation of free cyanide are as follows [5].

$$O_3 + CN^- \rightarrow CNO^- + O_2 \quad (5)$$

$$HCN + \text{HO}\cdot \rightarrow HCNO + \text{HOO}\cdot \quad (6)$$

$$CN^- + \text{HO}\cdot \rightarrow CNO^- + \text{HOO}\cdot \quad (7)$$

$$CN^- + \text{HO}\cdot \rightarrow \cdot CN + OH^- \quad (8)$$

$$\cdot CN + \cdot CN \rightarrow (CN)_2 \quad (9)$$

$$(CN)_2 + 2OH^- \rightarrow CNO^- + CN^- + H_2O \quad (10)$$

The direct reaction of molecular ozone with the cyanide ion corresponds to Reaction 6. Other reactions of ozone with water, specifically with OH^-, produce the superoxide radical (O_2^-), which reacts strongly with ozone to form the hydroxyl free radical (HO) [14].

The ozonation reaction of free cyanide to cyanate is relatively slow; cyanate accumulates in the solution until all free cyanide is completely oxidized.

Weak metal–cyanide complexes are generally readily oxidized by oxidizing agents, such as ozone. Complexes with WAD-type bonds, such as those of nickel and silver, oxidize more slowly. Complexes with weaker bonds, including cadmium, copper, and zinc, decompose rapidly in ozone [14].

1.2. INCO Process

The INCO process (commercialized by INCO Ltd.) is widely used worldwide in the mining sector and is one of the two patented processes for cyanide destruction using sulfur dioxide. The INCO process is based on the oxidation of free cyanide and weakly complexed metal cyanides (WAD cyanides) into cyanate via a mixture of sulfur dioxide (SO_2) and air in the presence of a soluble copper catalyst at a controlled pH. Complex cyanides, such as iron cyanides, are reduced to ferrous and precipitate as insoluble copper–iron–cyanate complexes. Residual metals released from WAD cyanide complexes precipitate as their respective hydroxides [3].

Acid formation is possible during the reactions, so to maintain a pH between 8.0 and 9.0 in Reactions 11 and 12, lime is recommended for pH control.

$$CN^- + SO_2 + O_2 + H_2O \xrightarrow{Cu^{2+}\ Catalyst} CNO^- + SO_4^{2-} + 2H^+ \quad (11)$$

$$M(CN)_4^{2-} + 4SO_2 + 4O_2 + 4H_2O \xrightarrow[\rightarrow\rightarrow\rightarrow\rightarrow\rightarrow\rightarrow\rightarrow\rightarrow\rightarrow]{Cu^{2+}\ Catalyst} 4CNO^- + 4SO_4^{2-} + 8H^+ + M^{2+} \quad (12)$$

The theoretical SO_2 requirement is 2.46 g of SO_2 per gram of oxidized WAD cyanide, although practical reports indicate variations ranging from 3.0 to 5.0 g of SO_2 per gram of oxidized WAD cyanide [7].

The SO_2 required for the reaction can originate from liquid sulfur dioxide, sodium sulfite (Na_2SO_3), or sodium metabisulfite ($Na_2S_2O_5$). Ammonium bisulfite (NH_4HSO_3) solutions have shown promising results; however, one must be cautious about the potential impact of ammonia addition to wastewater. It can supply oxygen to optimize the reaction, or use high-pressure liquid oxygen; the choice of oxygen source significantly affects reaction time [4].

The soluble copper catalyst is added as a solution of copper sulfate pentahydrate ($CuSO_4{\cdot}5H_2O$) at a proportion of 10–20% relative to the WAD cyanide concentration.

This process is widely used to treat tailings with cyanide concentrations ranging from low to high. It aims to reduce cyanide levels to approximately 5 mg/L, thereby enabling the treated solutions to be safely discharged [8].

A typical INCO process consists of two stages. In the first stage, sulfur dioxide, lime, and copper sulfate are added, with continuous oxygen injection, to ensure complete oxidation of cyanide. In the second stage, metal precipitation occurs, requiring the addition of lime and other chemicals, such as ferric chloride [8].

The key process variables include retention time, air or oxygen feed rate, copper concentration, pH, and sulfur dioxide feed rate. The amount of copper sulfate or alternative reagents is adjusted based on the WAD cyanide concentration, determined through laboratory or pilot-plant evaluations.

Cyanide oxidation as a function of copper concentration and pH, showing greater effectiveness at copper concentrations exceeding 10 mg/L within a pH range of 8 to 10; at pH 11, Reactions 12 and 13 are reversible [8].

Table 2 presents the performance of the INCO process for cyanide destruction in various mining operations. The minimum concentration of CN_{TOT} was obtained with the lowest concentration of Cu^{2+} for which lime was necessary. The optimal balance between SO_2 and Cu^{2+} in the INCO process must be identified to improve cyanide oxidation efficiency, which requires a complete characterization of the ore, as its composition can vary greatly from one region to another.

Table 2. Performance of the INCO process for cyanide oxidation in the world [8].

Mining Unit in the World	CN_{TOT} (mg/L)		Reagent Usage (g/g) CH_{TOT}		
	Before	After	SO_2	Lime	Cu^{2+}
Colosseum	374	0.4	4.6	0.12	0.04
Ketza River	150	5.0	6.0	0	0.30
Equity	175	2.3	3.4	0	0.03
Casa Berardi	150	1.0	4.5	-	0.10
Westmin Premier	150	<0.2	5.8	-	0.12
Golden Bear	205	0.3	2.8	-	-

Iron plays a complex and multifaceted role in the INCO process, serving as both a reactant and a potential catalyst, and as a source of process complexity that warrants careful consideration. Unlike copper, which functions primarily as a homogeneous catalyst, iron in gold processing streams is predominantly present as strong metal–cyanide complexes,

https://doi.org/10.3390/mining6010009

particularly ferrocyanide ($Fe(CN)_6^{4-}$) and ferricyanide ($Fe(CN)_6^{3-}$) species [9,11]. These complexes are characterized by exceptionally high formation constants (log β values exceeding 30 for ferrocyanide), making them resistant to treatment by many conventional cyanide oxidation methods. The INCO process, however, demonstrates a distinctive capability to destabilize and remove these strong complexes, providing a significant advantage over alternative treatment technologies, such as alkaline chlorination, which is largely ineffective against iron–cyanide species [12,13].

The presence of iron–cyanide complexes in cyanide detoxification circuits presents both opportunities and challenges for INCO process operation. On the opportunity side, the ability to remove stable iron–cyanide complexes provides more complete detoxification than many alternative processes, resulting in lower total cyanide concentrations in discharged solutions [14].

This comprehensive cyanide removal is particularly valuable for operations subject to stringent discharge limits or those seeking to minimize environmental liability associated with residual cyanide species. However, the precipitation of copper–iron–cyanide complexes consumes the copper catalyst, potentially creating or exacerbating copper deficiency issues if not properly managed. Additionally, the stability of iron–cyanide precipitates under storage conditions must be considered, as these solids can decompose under ultraviolet light, releasing free cyanide and undermining the detoxification achieved during treatment [15].

2. Materials and Methods

Due to the high risks associated with SO_2 use, the mining industry has been using alternative chemical reductants, such as ammonium bisulfite (NH_4HSO_3), operating at pH 5–7. However, like the INCO process, this method also requires a high oxygen concentration, which currently results in high process costs (electric power, oxygen, copper bisulfite, and ammonium bisulfite) [16].

The NH_4HSO_3 concentration of 1 kg/ton was recommended as a best practice in the mining industry; it was added in the corresponding proportion to maintain a constant effluent volume of 1 L in a cylindrical reactor [17].

The experimental development was focused on exploring the use of ozone as an alternative oxidation source to improve the degradation efficiency of WAD and free cyanide, using a process based on the application of the INCO process with industrial gold mine pulp, ammonium bisulfite (NH_4HSO_3), and copper as a catalyst.

In this first stage, a factorial design (3 levels) was conducted using the Statgraphics 18 program. Laboratory tests demonstrated the feasibility of using ozone to convert CN^- to cyanate.

Figure 2 shows the experimental system conducted in a 1 L cylindrical reactor (Reacware, Albany, NY, USA) with a semispherical bottom to determine the transformation of CN^- upon ozone/oxygen injection. To start, 20 L drums were delivered to the mine and stored in a dark, protected area. Before each experiment, the drums were dynamically shaken for 1h. The system temperature was controlled by immersing the reactor in a thermal bath. A cooling system was implemented to maintain a controlled temperature in the alkaline trap, where the reaction gases were collected.

At the end of the ozone/oxygen injection exposure time, the pulp was filtered, and the filtrate was analyzed to determine the concentrations of free and WAD cyanide.

After obtaining all possible combinations of the six process variables in the factorial design, a fractional factorial design with 60 combinations was selected. The experimental design encompassed all possible combinations of levels and factors, enabling a statistical

https://doi.org/10.3390/mining6010009

evaluation of main effects and interactions. The experimental analysis was performed using a Pareto chart and analysis of variance (ANOVA).

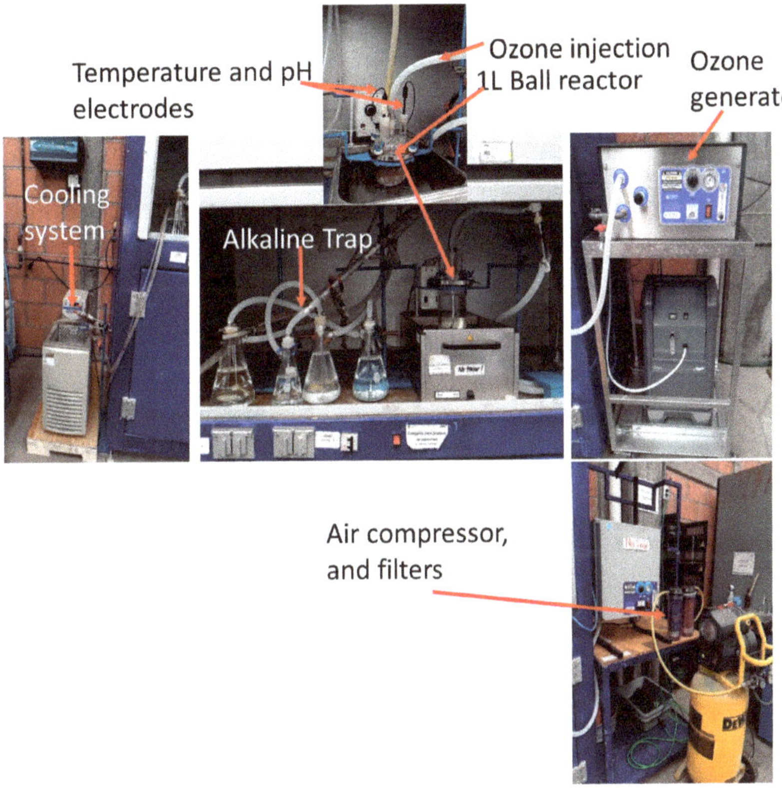

Figure 2. Experimental system for ozone injection in a 1 L cylindrical reactor with an alkaline trap as a safety measure.

The best results from the factorial design, based on the percentage of free cyanide degraded, were used to optimize free cyanide degradation.

3. Results

3.1. Volumetric Plots

Based on the main variables of the INCO process: ozone/oxygen mixture (flow rate), pH effect, temperature effect, ammonium bisulfite [NH_4HSO_3], and reaction time, an experimental matrix comprising 60 tests was established.

Table 3 shows the range within which each variable was evaluated across the 60 tests conducted in the fractional factorial analysis.

The O_3/O_2 ratio was expressed in standard cubic feet per minute (SCFM) and monitored in parallel with the ozone generation voltage, measured directly in the ozone generator within a constant volume system.

The pH range of 8–10 prevents the formation of HCN and CN^-. Preliminary results indicated that the temperature range (25 to 45 °C) and reaction time (8 to 24 min) are inversely related, yielding a more convenient reaction time for the oxidation of WAD cyanide and free cyanide.

https://doi.org/10.3390/mining6010009

Table 3. The evaluation range for each process variable implemented in the experimental matrix for the WAD and free cyanide oxidation with ozone injection.

Variable (Factors)	Units	Value to Be Evaluated (Levels)		
		Minimum	Medium	Maximum
O_3/O_2 flow	SCFM	3	6	9
pH	-	8	9	10
Temperature	°C	25	35	45
NH_4HSO_3 concentration	kg/ton	1	1.5	2
Reaction time	Minutes	8	16	24

SCFM—standard cubic feet per minute.

Table 4 reports the average oxidation percentage of WAD and free cyanide, based on 60 tests, grouped into sets of 10 samples. It can be observed that free cyanide oxidizes more rapidly, with higher oxidation percentages, than WAD cyanide.

Table 4. Average oxidation percentage of WAD and free cyanide, grouped into sets of 10 samples.

Group of 10 Samples	Average Oxidized WAD CN^- [%]	Average Oxidized Free CN^- [%]
1	65.0555 ± 5.092	89.5088 ± 3.149
2	70.7480 ± 6.401	92.7790 ± 3.535
3	68.4090 ± 5.556	88.4700 ± 5.876
4	61.2150 ± 8.510	64.9540 ± 7.576
5	65.8910 ± 8.092	81.0210 ± 6.734
6	72.1350 ± 10.101	88.1150 ± 4.891

Considering the average values reported in Table 4, free cyanide tended to oxidize 20% more under the optimal WAD cyanide oxidation conditions, reporting an average value of oxidation of 84.1413% for free cyanide and 67.2423% for WAD cyanide.

To identify the process variable that most strongly contributes to oxidation, volumetric plots were used to simulate the tendency of the variable towards cyanide oxidation.

The optimal values obtained for achieving 100% oxidation of WAD and free cyanide fall within a 90% confidence range, as shown in the volumetric plots in Figure 3. These plots display the interaction between O_3/O_2 flow, reaction time, and NH_4HSO_3 concentration for WAD cyanide oxidation, as well as the interaction between O_3/O_2 flow, reaction time, and pH for free cyanide oxidation.

According to the color coding in Figure 3, more than 60% of WAD cyanide can be oxidized with the maximum O_3/O_2 flow and reaction time. Combined with an NH_4HSO_3 concentration of over 1.8 kg/ton, it is possible to achieve more than 80% of WAD cyanide oxidation. Free cyanide oxidation between 80% and 100% requires an intermediate O_3/O_2 flow, an intermediate reaction time, and a pH in the range of 8 to 9.2, indicating that a high ammonium bisulfite concentration is not required for free cyanide oxidation.

These results suggest that it is necessary to maintain the maximum values in O_3/O_2 flow, reaction time, and NH_4HSO_3 concentration, with a pH of 8 to 9, to obtain the maximum free cyanide oxidation and a satisfactory WAD cyanide oxidation during the first step. It is possible to increase WAD cyanide oxidation by adding a second oxidation step before sending the residual material to the dam, ensuring the minimum cyanide concentration that could pose an ecological problem.

https://doi.org/10.3390/mining6010009

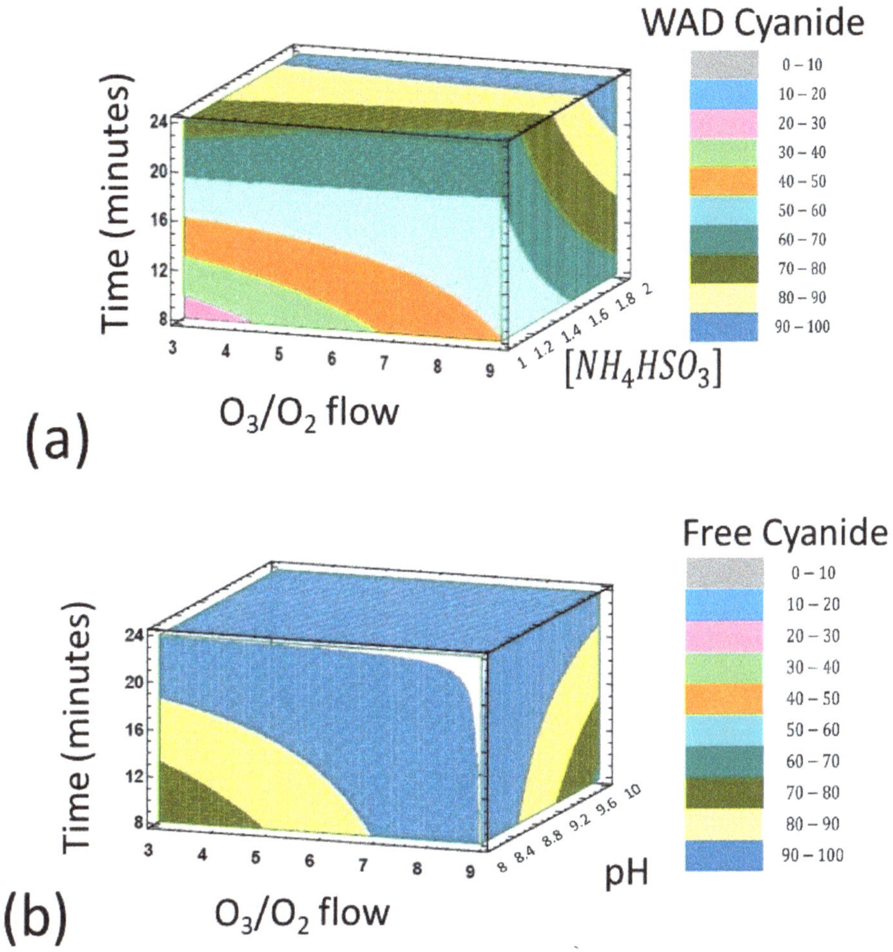

Figure 3. (**a**) Volume graphs with pH = 10 and temperature of 25 °C indicate the oxidation percentage of WAD cyanide, and (**b**) volume graphs with NH_4HSO_3 concentration of 1.0 and temperature of 25 °C for the oxidation of free cyanide.

Pareto diagrams for WAD and free cyanide oxidation show the order in which the variables impact the cyanide oxidation percentage. Being the more significant variables for the oxidation of WAD cyanide, the next order is as follows:

$$^{O_3}/_{O_2}\, flow > time > [NH_4HSO_3] > pH > temperature$$

For free cyanide oxidation, and according to prioritization in the Pareto diagram, the main variables follow the next sequence:

$$^{O_3}/_{O_2}\, flow > time > pH > temperature > [NH_4HSO_3]$$

3.2. Results in Surface Plots

The following series of graphs presents the oxidation response percentages of WAD cyanide and free cyanide with respect to two main variables: ozone concentration against O_3/O_2 flow rate. The graph in Figure 4a shows that low ozone concentrations and a low

https://doi.org/10.3390/mining6010009

O_3/O_2 flow rate have a slightly adverse effect on WAD cyanide oxidation compared to free cyanide oxidation (Figure 4b). However, at high ozone concentrations and high O_3/O_2 flow rates, the effect becomes slightly positive, with better oxidation conditions for WAD cyanide approaching 80% and for free cyanide reaching approximately 90%.

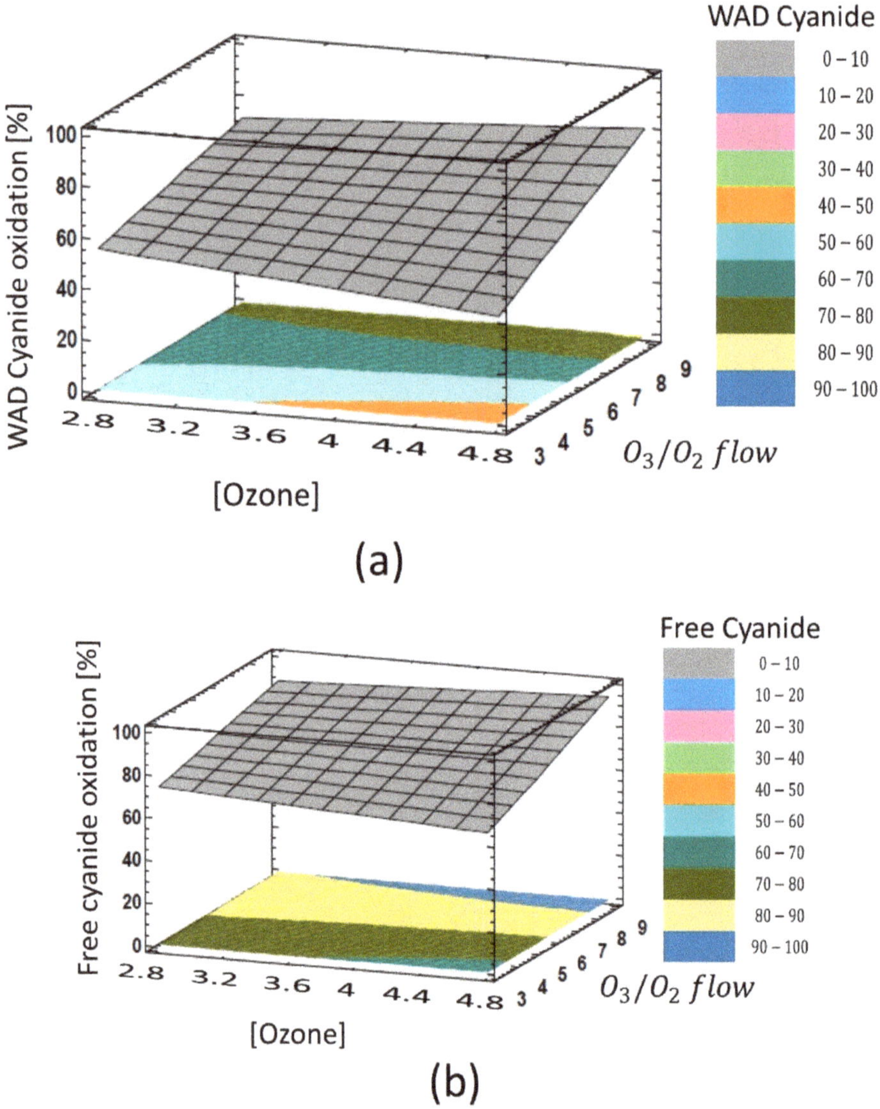

Figure 4. (**a**) Plot of the effect of ozone concentration and O_3/O_2 flow on (**a**) WAD cyanide and (**b**) free cyanide oxidation during 16 min of ozone injection, under a temperature of 35 °C for both cases.

The interaction effect between ozone concentration and pH for WAD cyanide oxidation is minimal, as shown in Figure 5a. There is no significant change in oxidation levels when ozone concentration is increased or decreased at higher or lower pH levels, with an oxidation range of 60–70%. WAD cyanide oxidation does not depend directly on pH variations; however, it is possible to increase the WAD degradation if the ammonium bisulfite and copper concentration increase.

https://doi.org/10.3390/mining6010009

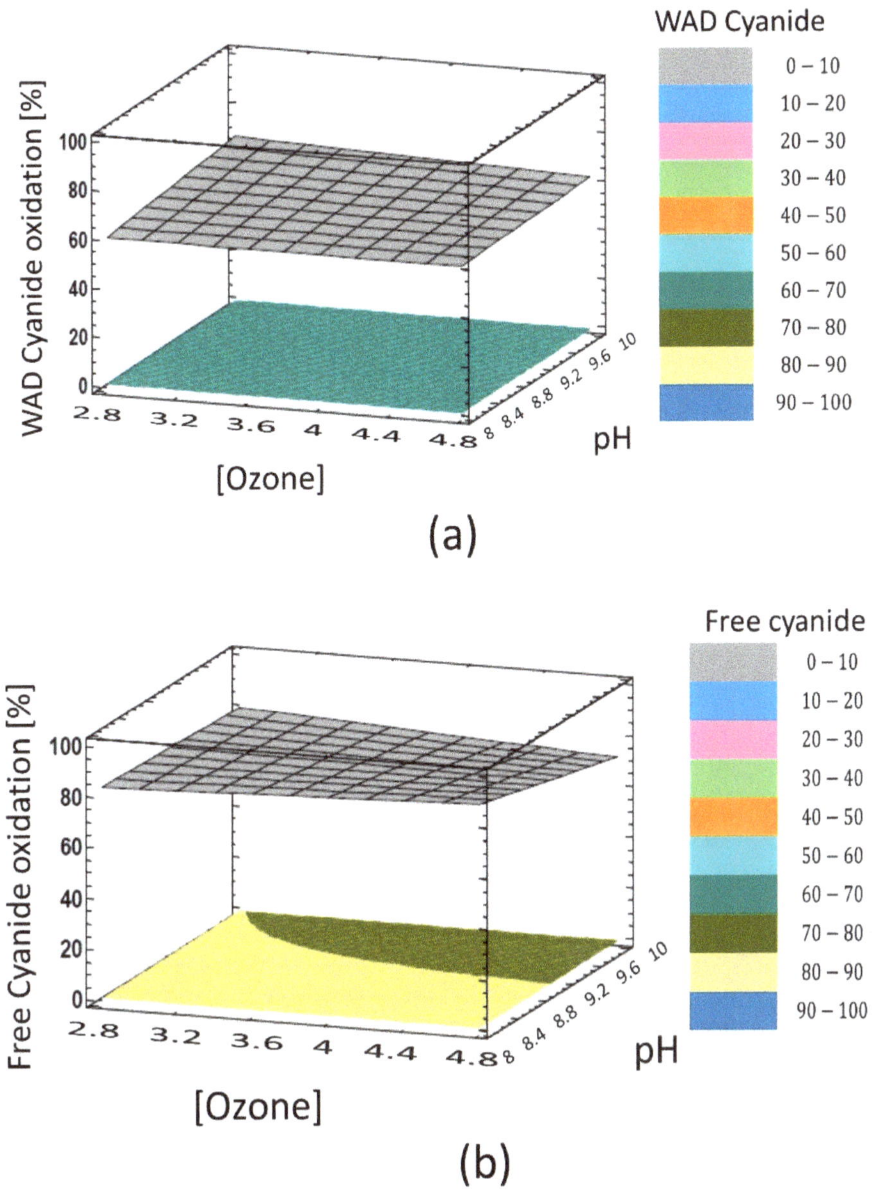

Figure 5. Effect of ozone concentration and pH on the oxidation of (**a**) WAD and (**b**) free cyanide during 16 min of ozone injections, under a temperature of 35° C for both cases.

In contrast, at low or high ozone concentrations and pH of 8 to 9.2, free cyanide oxidation ranged from 80 to 90% (Figure 5b), where pH control is essential to maintain thermodynamic equilibrium between HCN and CN^-.

The combined effect relationship between ozone concentration and temperature (Figure 6a,b) exhibited a similar behavior to that observed between ozone concentration and pH. The results showed no significant changes in WAD cyanide oxidation with the increases or decreases in ozone concentration and temperature, maintaining an oxidation

https://doi.org/10.3390/mining6010009

level of 60%. A slight rise in temperature between 41 and 45 °C reduces the oxidation percentage to 50%.

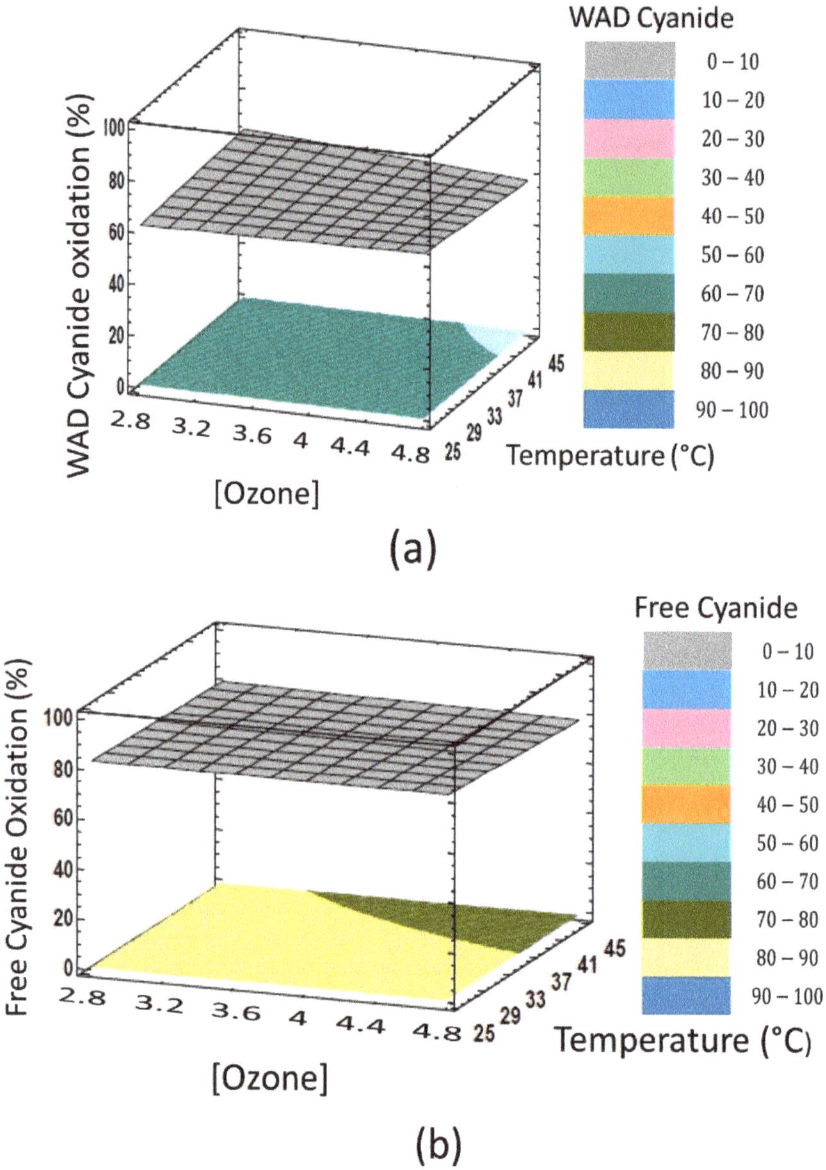

Figure 6. Effect of ozone concentration and temperature on the oxidation of (**a**) WAD and (**b**) free cyanide, with a pH of 9 and time of ozone injection of 16 min for both cases.

Conversely, a higher percentage of free cyanide oxidation (80–90%) was observed with increasing ozone concentrations at temperatures of 25–37 °C.

Increasing the temperature to 45 °C in both cases reduces the oxidation percentage for WAD and free cyanide by approximately 10%.

An increase in ammonium bisulfite consumption of around 2 kg/ton slightly increased the oxidation percentage of WAD cyanide to values around 70%, along with the high ozone

https://doi.org/10.3390/mining6010009

concentration (Figure 7a). In contrast, the rate of free cyanide oxidation increased with higher ozone and ammonium bisulfite concentrations (Figure 7b), reaching 80% and 90% oxidation, indicating that a maximum concentration of ammonium bisulfite is necessary to ensure WAD cyanide oxidation.

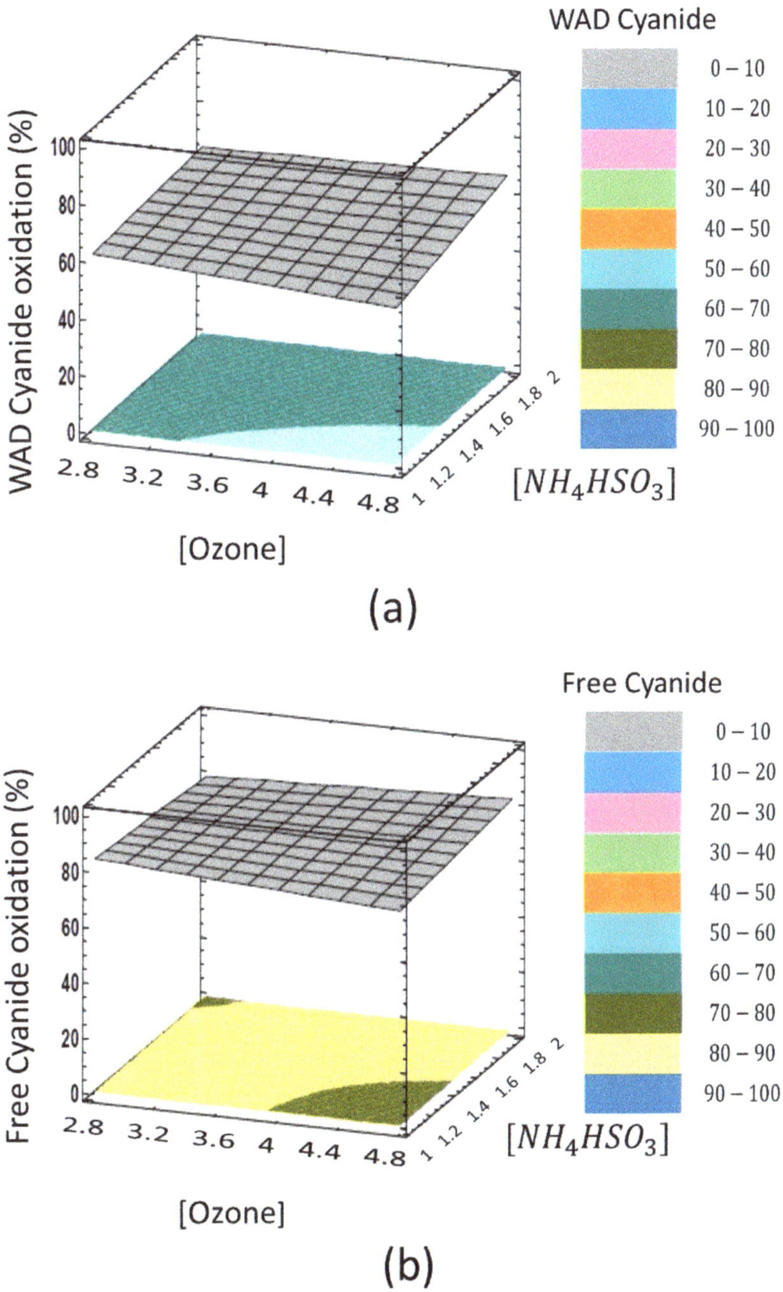

Figure 7. Effect of ozone concentration and NH_4HSO_3 consumption on the oxidation of (**a**) WAD and (**b**) free cyanide with a pH of 9, a temperature of 35 °C, and 16 min of ozone injection for both cases.

https://doi.org/10.3390/mining6010009

Increasing ozonation time results in a corresponding increase in the oxidation percentage of WAD cyanide, reaching approximately 70% at low ozone concentrations (Figure 8a). Free cyanide oxidation becomes more evident with an increasing reaction time, particularly at high ozone concentrations, with oxidation percentages reaching 90% (Figure 8b). The oxidation time determines the reaction kinetics, independent of ozone concentration. Greater reaction time results in higher oxidation percentages of free cyanide than of WAD cyanide.

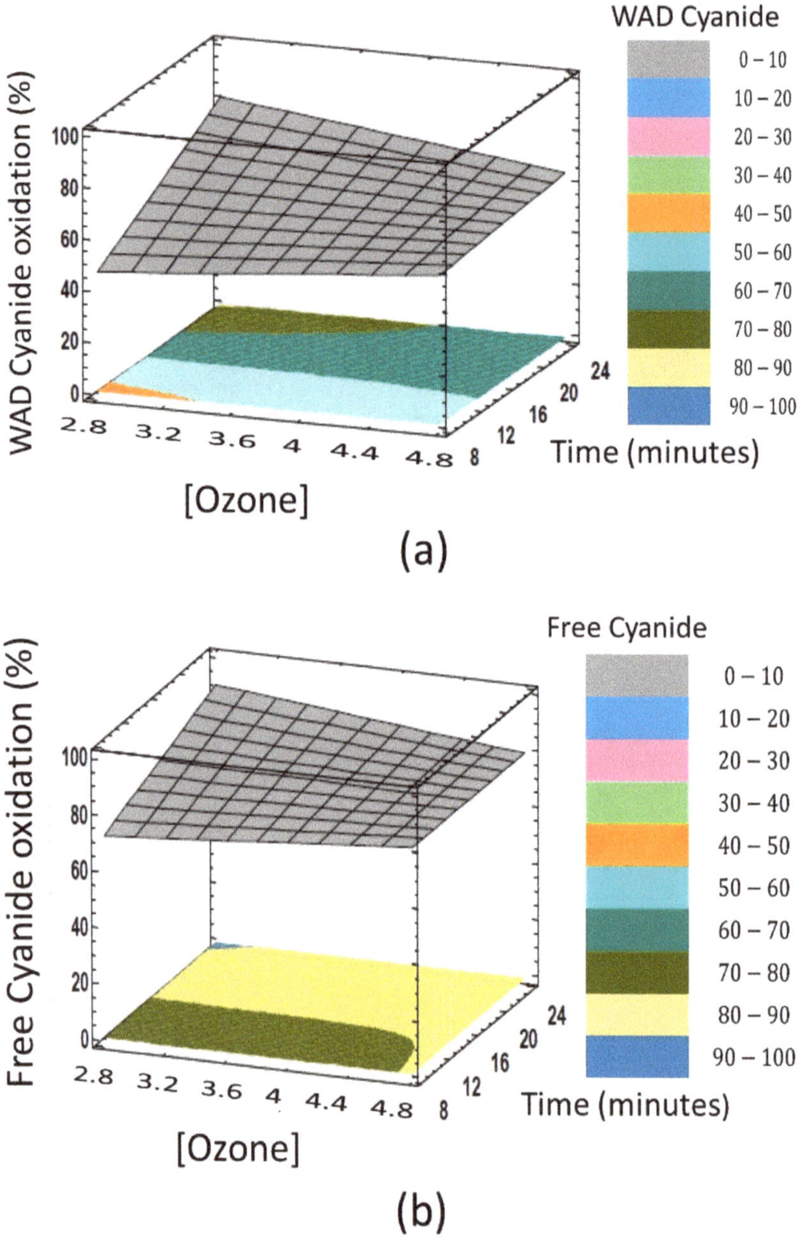

Figure 8. Effect of ozone concentration and ozone injection time on the oxidation of (**a**) WAD and (**b**) free cyanide with a pH of 9.0 and a temperature of 35 °C, for both cases.

https://doi.org/10.3390/mining6010009

In the following sequence of graphs, the O_3/O_2 flow was constant against ozone concentration, pH, temperature, NH_4HSO_3 concentration, and reaction time.

Based on the results of all combinations established in the experimental matrix, Table 5 presents the estimated optimal values for the oxidation of WAD and free cyanide. The optimal values for O_3/O_2 flow, temperature, and ozone injection time are the same for WAD and free cyanide. However, the optimal pH and NH_4HSO_3 concentration differ for WAD cyanide oxidation (10 and 2, respectively) compared with those for free cyanide oxidation (8 and 1, respectively).

Table 5. Estimated optimal values for WAD cyanide and free cyanide oxidation.

Factor	Low	High	Optimal Values for WAD Cyanide Oxidation	Optimal Values for Free Cyanide Oxidation
[Ozone] Volts	2.8	4.8	2.8	2.8
O_3/O_2 flow SCFM	3.0	9.0	9.0	9.0
pH	8.0	10.0	10.0	8.0
Temperature °C	25.0	45.0	25.0	25.0
[NH_4HSO_3] kg/ton	1.0	2.0	2.0	1.0
Time (minutes)	8.0	24.0	24.0	24.0

The volume graphs, shown in Figure 3, clearly represent the estimated optimal values for WAD and free cyanide oxidation. The volume corresponding to the maximum oxidation value of free cyanide is greater than that of WAD cyanide oxidation. However, in both cases, oxidation values above 70% are achieved, allowing for optimal WAD cyanide oxidation to achieve maximum oxidation for both cyanide types.

4. Discussion

The mining industry can contaminate water sources with cyanide, forming free cyanide, the most toxic form, at pH 9.3–9.5, where the anion (CN^-) and hydrogen cyanide (HCN) are in thermodynamic equilibrium. The free cyanide concentration that lethally affects the flora and fauna is 20–100 μg/L [18]. The World Health Organization recommends a maximum concentration of 0.5 mg/L to affect humans; however, the severity of effects depends on exposure duration and can include tremors and neurological effects [18].

The dissociation of WAD cyanide includes the free cyanide (HCN/CN^-) and the metal–cyanide complex at high- or low-acid conditions, depending on whether it corresponds to a strong or weak complex; however, the complex dissociation is accomplished by the release of free cyanide [19].

The WAD and free cyanide oxidation are promoted by the presence of free radicals (HO, HOO), when ozone is present, thereby contributing to the slow formation of cyanate (CNO^-), a non-toxic compound [20].

The average oxidation percentages of WAD and free cyanide were 67.2423% and 84.1413%, respectively.

The optimal values of the O_3/O_2 ratio, temperature, and reaction time could be maintained constant, thereby favoring the oxidation of both cyanide species.

A pH of 10 and an ammonium bisulfite concentration of 2 kg/ton corresponded with the optimal values for WAD cyanide oxidation.

As shown in Table 6, temperature and pH have a low impact on the oxidation of WAD cyanide; however, ammonium bisulfite, ozone flow rate, and reaction time significantly affect the reaction kinetics. The high concentration of ammonium bisulfite, maximum

https://doi.org/10.3390/mining6010009

O_3/O_2 flow, and reaction time offer the possibility to improve the WAD cyanide oxidation percentage, and, with the correct proportion of copper-like catalyst, obtain near 90% of degradation.

Table 6. The effect of the main variables on the oxidation of WAD and free cyanide.

Variable (Factors)	WAD Cyanide	Free Cyanide
O_3/O_2 (flow rate)	⇈	⇈
pH	⇊	⇈
Temperature	⇊	⇊
NH_4HSO_3 concentration	⇈	⇊
Reaction time	⇈	⇈

⇈—high impact; ⇊—low impact.

On the other hand, temperature and ammonium bisulfite concentration are not significant for the oxidation of free cyanide, indicating that ozone flow, pH, and reaction time are essential to prevent hydrogen cyanide formation and to achieve maximum free cyanide degradation (approximately 90%). Specifically, regarding temperature and ozone performance, the literature indicates that ozone solubility in aqueous solutions decreases considerably above 25 °C, which may be one reason for the observed reduced performance. In addition, pH is a significant factor, as hydronium ions can promote ozone decomposition, leading to less efficient oxidation at basic pH values [5,6].

pH is a relevant factor in maintaining thermodynamic equilibrium between the formation of HCN and CN^- and in improving free cyanide oxidation; however, it is not fundamental for WAD cyanide oxidation.

The statement regarding the two-stage oxidation of WAD cyanide is based on a mechanistic interpretation rather than on experiments conducted exclusively at alkaline pH.

In the present study, all experimental work was intentionally performed under alkaline conditions (pH 8–10), which are industrially relevant and necessary to prevent HCN volatilization. As shown in the volumetric and surface plots (Figures 3–8), the highest percentages of free cyanide oxidation were consistently observed at pH values near 8, confirming that free cyanide oxidation is kinetically favored under these conditions. However, WAD cyanide oxidation remained lower because, at alkaline pH, the dissociation of metal–cyanide complexes is limited, reducing the availability of free cyanide for subsequent oxidation.

Table 6 shows the correlations between the main factors and the high- or low-impact effects on WAD and free cyanide oxidation, showing that the temperature and pH have a low impact on WAD cyanide oxidation. On the other hand, temperature and ammonium bisulfite concentration demonstrated a low impact on free cyanide oxidation.

Ozone solubility in aqueous solutions decreases considerably above 25 °C, which could be one of the reasons for the reduced performance observed. In addition, pH is a significant factor, as hydronium ions can promote ozone decomposition, resulting in less efficient oxidation at basic pH values [9,10].

https://doi.org/10.3390/mining6010009

Mechanisms responsible for the observed trends are not studied in this work and could be further investigated in a future publication, given the many variables in this study.

5. Conclusions

Based on 60 experimental tests using a factorial design of the main variables in the ozone oxidation process, it was confirmed that free cyanide is oxidized to a greater extent, achieving an average oxidation of 84.1413% compared with WAD cyanide, which reported an average oxidation of 67.2423%. The oxidation of WAD cyanide requires two stages, considering that the dissociation of metal–cyanide complexes requires acidic pH conditions in the first stage to produce the free cyanide, which reduces the oxidation kinetics of WAD cyanide at pH values of 8, where the highest percentage of free cyanide oxidation was obtained.

To facilitate decision-making, the order in which process variables impact WAD cyanide and free cyanide oxidation is as follows:

For WAD cyanide oxidation,

$$Flow\ of\ ^{O_3}/_{O_2} > time > NH_4HSO_3 > pH > temperature$$

For free cyanide oxidation,

$$Flow\ of\ ^{O_3}/_{O_2} > time > pH > temperature > NH_4HSO_3$$

It is confirmed that pH ranks fourth in terms of impact on WAD cyanide oxidation at alkaline pH, whereas pH for free cyanide oxidation ranks third.

Ozone concentration, correlated with the O_3/O_2 flow rate and time, is the primary factor in the oxidation of both WAD cyanide and free cyanide.

The ozone molecule (O_3) primarily reacts with free cyanide to form cyanate (CNO^-), and its reaction with HCN is minimal. Therefore, it is recommended to reduce HCN levels by operating at high pH in the final stages of the process. The dissociation of the metal-cyanide complex to produce free cyanide allows for the formation of free radicals (HO·, HOO·), where both free cyanide and radicals contribute to the oxidative destruction of cyanide.

Although the ozonation reaction of free cyanide to cyanate is relatively slow, the six variables studied in the following order: ozone concentration, O_3/O_2 flow rate, pH, temperature, NH_4HSO_3 concentration, and time, positively affect the oxidation of free cyanide.

Working at the recommended optimal WAD cyanide oxidation values also enables free cyanide to reach 90%.

The reference to acidic pH is therefore conceptual and reflects well-established cyanide chemistry, in which the dissociation of WAD metal–cyanide complexes occurs more readily under mildly acidic conditions, generating free cyanide that can then be efficiently oxidized. Our experimental results indirectly support this interpretation by showing that, although free cyanide oxidation is maximized at pH $\approx$ 8, WAD cyanide oxidation is kinetically constrained under these conditions, thereby justifying the need for a two-stage approach.

Author Contributions: Conceptualization, S.O.L., C.E.A., and J.M.H.; methodology, E.A.P., C.E.A., and J.M.H.; software, H.H.H.; validation, A.S.G., C.M.L., and J.M.H.; formal analysis, S.O.L. and C.E.A.; investigation, C.E.A. and J.M.H.; resources, S.O.L.; data curation, two years; writing—original draft preparation, J.M.H.; writing—review and editing, J.M.H. and C.M.L.; visualization, A.S.G.; supervision, A.S.G.; project administration, C.E.A. and J.M.H.; funding acquisition, S.O.L. All authors have read and agreed to the published version of the manuscript.

https://doi.org/10.3390/mining6010009

Funding: This research was funded by the Fresnillo Group and Center for Research and Technological Development in Electrochemistry (CIDETEQ), grant number PCO-101. The APC was funded by the authors of this paper.

Data Availability Statement: The original contributions presented in this study are included in the article. Further inquiries can be directed to the corresponding author.

Acknowledgments: We would like to thank the Fresnillo Group for the economic support and the Center for Research and Technological Development in Electrochemistry (CIDETEQ) for their continuous support by providing their facilities, ensuring safety, and taking precautions during the execution of the PCO-101 Project. For any underlying research materials related to this paper (for example, data or models), please write to jmorales@cideteq.mx.

Conflicts of Interest: Author Saúl Ortiz Landeros was employed by the company Fresnillo Plc. The remaining authors declare that the research was conducted in the absence of any commercial or financial relationships that could be construed as a potential conflict of interest.

References

1. David, A.D.; Rajat, S.G.; George, M.W.C. *Cyanide in Water and Soil, Chemistry, Risk, and Management*; Taylor & Francis Group: Abingdon, UK, 2006; ISBN 978-1-56670-666-7.
2. Danielle, H.; Paul, B.; Coby, J. Cyanide detoxification of gold cyanidation tails and process streams. *Miner. Process. Extr. Metall.* **2012**, *121*, 228–236. [CrossRef]
3. Julio César, P.D.; Óscar Fabián, H.C. Comportamiento electroquímico del Cianuro. *Ing. Y Desarro.* **2008**, *24*, 63–76.
4. Edison, G.P.; Paola Andrea, C.M.; Claudia, C.J.; y Tomás Felipe, P.V. *Oxidación Fotocatalítica de Cianuro*; Universidad EAFIT: Medellín, Colombia, 2005; ISSN 1692-0694.
5. Gurol, M.D.; Bremen, W.M.; Holden, T.E. Oxidation of Cyanide in Industrial Wastewater by Ozone. *Environ. Prog.* **1985**, *4*, 46–51. [CrossRef]
6. Joseph, T.B.; Stephen, D.E.; David, A.D. Development of a Plant Uptake Model for Cyanide. *Int. J. Phytoremediat.* **2006**, *8*, 25–43. [CrossRef] [PubMed]
7. Ricardo, L.L. *Destrucción de Cianuro en Efluentes de la Industria Minera Mediante el Empleo de Reductores Químicos*; Catálisis Industrial S.A. de C.V.: Tizayuca, México, 2024; pp. 285–290. Available online: https://www.ocmal.org/wp-content/uploads/2017/03/destruccion_de_cianuro.pdf (accessed on 27 January 2026).
8. Terry, I.M.; Michael, M.B.; Adrian, S. Chemistry and Treatment of Cyanidation Wastes. In *Treatment and Recovery of Cyanide*; Chapter 6; Mining Journal Books Ltd.: London, UK, 2001; pp. 326–333.
9. Patricia, G.L. Degradación de Cianuros Mediante Oxidación Química en Efluentes Industriales. Master's Thesis, Departamento de Química Orgánica e Inorgánica, Universidad de Oviedo, Oviedo, Spain, 2012.
10. Roth, J.A.; Sullivan, D.E. Solubility of ozone in water. *Ing. Eng. Chem. Fundam.* **1981**, *20*, 137–140. [CrossRef]
11. Bin, A.K. Ozone Solubility in Liquids. *Ozone Sci. Eng. J. Int. Ozone Assoc.* **2006**, *28*, 67–75. [CrossRef]
12. Piervandi, Z. Pretreatment of refractory gold minerals by ozonation before the cyanidation process: A review. *J. Environ. Chem. Eng.* **2023**, *11*, 109013. [CrossRef]
13. Freitas, J.; Horta, D. Impacts of cyanide in gold mining and cyanide removal methodologies in liquid waste from gold processing. *Rev. De Metal.* **2023**, *59*, e247. [CrossRef]
14. Jiménez-Prieto, Y.J.; Esperanza-Pérez, G.; Ramírez-González, S.; Alomas-Vicente, I. Assessment of Technological alternatives for cyanides waste waters management in gold ores processing plant. *Rev. Cuba. De Química Habana Cuba* **2020**, *32*, 218–231.
15. Pandey, N.; Dutta, S.; Biswas, P. Cyanide mitigation at steel metallurgical process's effluent. *Water Air Soil Pollut.* **2023**, *234*, 682. [CrossRef]
16. Mamelkina, M.A.; Herraiz-Carbone, M.; Cotillas, S.; Lacasa, E.; Saez, C.; Tuunila, R.; Sillanpaa, M.; Hakkinen, A.; Rodrigo, M.A. Treatment of mining wastewater polluted with cyanide by coagulation processes: A mechanistic study. *Sep. Purif. Technol.* **2020**, *237*, 116345. [CrossRef]
17. Schneider, C.L.; Matiolo, E.; Neumann, R.; Gomes, O.F.M. *Beneficiamento de Minérios. Recursos Minerais no Brasil: Problemas e Desafios*; Academia Brasileira de Ciên-cias: Rio de Janeiro, Brazil, 2019; pp. 257–262. Available online: https://www.abc.org.br/IMG/pdf/doc-7006.pdf (accessed on 27 January 2026).
18. Zhang, H.; Zhang, Y.; Qiao, T.; Hu, S.; Liu, J.; Zhu, R.; Yang, K.; Li, S.; Zhang, L. Study on ultrasonic enhanced ozone oxidation of cyanide-containing wastewater. *Sep. Purif. Technol.* **2022**, *303*, 122258. [CrossRef]

https://doi.org/10.3390/mining6010009

19. Botz, M.M.; Mudder, T.I.; Akcil, A.U. Chapter 35—Cyanide treatment: Physical, chemical, and biological processes. In *Gold Ore Processing*, 2nd ed.; Elsevier: Amsterdam, The Netherlands, 2016; pp. 619–645.
20. Krylova, L.N.; Seliverstov, A.F.; Ryabtsev, D.A. Cleaning of industrial waters containing cyanides by using a combination of ozone and hydrogen peroxide. *Metallurgist* **2021**, *64*, 1331–1339. [CrossRef]

https://doi.org/10.3390/mining6010009

MDPI AG
Grosspeteranlage 5
4052 Basel
Switzerland
Tel.: +41 61 683 77 34

MDPI Books Editorial Office
E-mail: books@mdpi.com
www.mdpi.com/books

www.ingramcontent.com/pod-product-compliance
Lightning Source LLC
Chambersburg PA
CBHW061509210926
51850CB00008B/1861
* 9 7 8 3 7 2 5 8 8 0 4 9 2 *